MAGNETISM

Volume V

MAGNETISM

A Treatise on Modern Theory and Materials

EDITED BY **George T. Rado** **Harry Suhl**

Magnetism Branch
U.S. Naval Research Laboratory
Washington, D.C.

Department of Physics
University of California, San Diego
La Jolla, California

Volume I: Magnetic Ions in Insulators, Their Interactions, Resonances, and Optical Properties

Volume II: (in two parts): Statistical Models, Magnetic Symmetry, Hyperfine Interactions, and Metals

Volume III: Spin Arrangements and Crystal Structure, Domains, and Micromagnetics

Volume IV: Exchange Interactions among Itinerant Electrons
By Conyers Herring

Volume V: Magnetic Properties of Metallic Alloys

MAGNETISM

Volume V

Magnetic
Properties
of
Metallic
Alloys

EDITED BY **Harry Suhl**

Department of Physics
University of California, San Diego
La Jolla, California

 1973

ACADEMIC PRESS
New York and London

A Subsidiary of
Harcourt Brace Jovanovich,
Publishers

COPYRIGHT © 1973, BY ACADEMIC PRESS, INC.
ALL RIGHTS RESERVED.
NO PART OF THIS PUBLICATION MAY BE REPRODUCED OR
TRANSMITTED IN ANY FORM OR BY ANY MEANS, ELECTRONIC
OR MECHANICAL, INCLUDING PHOTOCOPY, RECORDING, OR ANY
INFORMATION STORAGE AND RETRIEVAL SYSTEM, WITHOUT
PERMISSION IN WRITING FROM THE PUBLISHER.

ACADEMIC PRESS, INC.
111 Fifth Avenue, New York, New York 10003

United Kingdom Edition published by
ACADEMIC PRESS, INC. (LONDON) LTD.
24/28 Oval Road, London NW1

Library of Congress Cataloging in Publication Data

Rado, George Tibor, DATE   ed.
    Magnetism.

    Includes bibliographies.
    CONTENTS: v. 1. Magnetic ions in insulators, their interactions, resonances, and optical properties. —v. 2A. Statiscal models, magnetic symmetry, hyperfine interactions, and metals.—v. 2B. Interactions and metals.—v. 3. Spin arrangements and crystal structure, domains, and micromagnetics. [etc.]
    1. Magnetism.   2. Nuclear magnetism.   3. Quantum theory.   I. Suhl, Harry, joint ed.
QC753.R3   538   63-16972
ISBN 0–12–575305–5 (v. 5)

PRINTED IN THE UNITED STATES OF AMERICA

Contributors to Volume V

Numbers in parentheses refer to the page on which the author's contributions begins.

P. W. **Anderson** (217), *Bell Telephone Laboratories, Inc., Murray Hill, New Jersey and Cambridge University, Cambridge, England*

M. T. **Béal-Monod** (89), *Laboratoire de Physique des Solides, Faculté des Sciences, Orsay, France*

A. **Blandin** (57), *Université de Paris-Sud, Centre d'Orsay, Laboratoire de Physique des Solides, Orsay, France*

W. **Brenig** (185), *Max Planck Institut für Festkörperforschung, Stuttgart, Germany, and Physik Department der TU München, Germany*

Bryan R. **Coles** (3), *Department of Physics, Imperial College of Science and Technology, London, United Kingdom*

Melvin D. **Daybell** (121), *Department of Physics, University of Southern California, Los Angeles, California*

Ø. **Fischer** (327), *Départment de Physique de la Matière Condensée, Université de Genève, Geneva, Switzerland*

D. R. **Hamann** (237), *Bell Telephone Laboratories, Inc., Murray Hill, New Jersey*

P. **Lederer** (89), *Laboratoire de Physique des Solides, Faculté des Sciences, Orsay, France*

M. Brian **Maple** (289), *Institute for Pure and Applied Physical Sciences, University of California, San Diego, La Jolla, California*

D. L. **Mills** (89), *Department of Physics, University of California, Irvine, California*

E. **Müller-Hartmann** (353), *Department of Physics, University of California, San Diego, La Jolla, California*

Albert **Narath** (149), *Sandia Laboratories, Albuquerque, New Mexico*

M. **Peter** (327), *Départment de Physique de la Matière Condensée, Université de Genève, Geneva, Switzerland*

J. R. **Schrieffer** (237), *Department of Physics, University of Pennsylvania, Philadelphia, Pennsylvania*

Dieter K. **Wohlleben** (3), *Department of Physics, University of California, San Diego, La Jolla, California*

Akio **Yoshimori** (253), *Institute for Solid State Physics, University of Tokyo, Roppongi, Tokyo, Japan*

Kei **Yosida** (253), *Institute for Solid State Physics, University of Tokyo, Roppongi, Tokyo, Japan*

G. **Yuval*** (217), *Joseph Henry Laboratories, Princeton University, Princeton, New Jersey*

J. **Zittartz** (185), *Institut für Theoretische Physik der Universität Köln, Germany*

* Present address: Racah Institute of Physics, The Hebrew University, Jerusalem, Israel.

Preface

Ferromagnetic metals have long presented a special challenge to solid state physics. This is especially true of those belonging to the transition series. Much of the difficulty in understanding these metals lies in the apparently dual character of their d-shell electrons: on the one hand (as in the free atom), these electrons are responsible for the observed magnetic moments which seem well localized on the lattice sites; on the other hand, they also participate to some degree in the conduction process. There is of course no fundamental contradiction here; after all, one has no difficulty accounting for an increased conduction electron charge density near a lattice site. The difficulty in the case of spin density appears to be twofold: First, there is the relatively minor difficulty that spin density is a vector quantity, while charge density is a scalar. Secondly, magnetic moment localization (in contrast with charge localization) owes its very existence to many-electron correlation effects that are hard to discuss simultaneously with conductivity, for which a single-particle (or quasi-particle) picture is both successful and deeply rooted in solid state physics.

For these reasons one may hope that some insight is to be gained by considering certain strongly paramagnetic situations. Thus in certain dilute magnetic alloys, a study of the immediate vicinity of the impurity might be expected to show how the electrons in traversing this vicinity manage to produce an effective local spin without relinquishing their mobile aspects.

Pending the arrival on the scene of radically new ideas we are obliged to deal with the complexities of this situation by using existing theoretical methods, and performing measurements of quantities related to traditional concepts. In this volume a status report is given by specialists. The intensive activity that followed Kondo's discovery of a serious divergence in the perturbative calculation of certain physical properties of magnetic alloys is described in detail in the theoretical chapters by Hamann and Schrieffer, Anderson and Yuval, Béal-Monod, Lederer, and Mills, Brenig and Zittartz, and Yosida and Yoshimori. The limited measure in which these theories jibe with experiment, and the considerably larger extent to which they fail to do so becomes apparent in the chapters by Coles and Wohlleben, Daybell, and Narath surveying the experimental situation.

The parallel problems encountered when the matrix is superconducting are discussed from the theoretical viewpoint by Müller-Hartmann. The measurements, especially those of the reduction in superconducting transition temperature as a function of parameters characterizing the impurity, are discussed in the chapter by Maple. A subject of long-standing concern, the question of coexistence of superconductivity and magnetism is described by Fischer and Peter.

The complexities alluded to above should not be allowed to obscure the fact that the systematics of magnetic moment formation can often be established, in outline at least, by the time-honored Hartree–Fock method. This method, discussed by Blandin, reduces the problem to a self-consistent single-particle picture, and thus regains all the simplifying features attendant upon such a picture. Perhaps the day is not too far off when it becomes clear precisely what minimal departure from the single-particle picture we must make in order to gain a detailed view of the magnetic situation around an impurity.

H. SUHL

Contents

Contributors to Volume V v

Preface vii

Contents of Previous Volumes xiii

Part I. CONDITIONS FAVORING LOCALIZATION OF EFFECTIVE MOMENTS

1. Formation of Local Magnetic Moments in Metals: Experimental Results and Phenomenology

Dieter K. Wohlleben and Bryan R. Coles

I.	Introduction.	3
II.	Experimental Observables in the Local Moment Problem	7
III.	General Features Revealed by Experiment	17
IV.	Solid Solutions of $3d$ Elements in Simple Hosts	31
V.	Transition Metals Containing $3d$ Solutes	34
VI.	Local Moment Formation in Rare Earth Metals	37
	References	51

2. Formation of Local Magnetic Moments: Hartree–Fock Theory

A. Blandin

I.	Introduction.	58
II.	Friedel's Approach: Resonance Scattering and Virtual Bound States	59
III.	Anderson's Approach	64
IV.	Discussion of the Hartree–Fock Solutions	70
V.	The Antiferromagnetic Coupling between Localized and Conduction Electrons	73
VI.	Transition Impurities in Transition Metals	78
VII.	Conclusions	86
	References	87

3. Spin Fluctuations around Impurities: Magnetic and Nonmagnetic Cases

D. L. Mills, M. T. Béal-Monod, and P. Lederer

    I. Introduction 89
    II. Theory of Local Spin Fluctuations Associated with Paramagnetic Impurities in Metals: Mean Field Description 92
    III. Renormalized Theories of Local Spin Fluctuations: Nonmagnetic and Magnetic Cases 100
    IV. The Effect of Local Spin Fluctuations on the Properties of Dilute Alloys 109
    References 116

Part II. THE s–d MODEL

4. The s–d Model and the Kondo Effect: Thermal and Transport Properties

Melvin D. Daybell

    I. Introduction 121
    II. Thermal and Transport Properties 125
    III. Interactions 140
    IV. Review Articles 144
    References 144

5. The s–d Model and the Kondo Effect: Magnetic Hyperfine-Interaction Studies

Albert Narath

    I. Introduction 149
    II. Theory of Magnetic Hyperfine Interactions in Dilute Alloys 150
    III. Impurity Magnetization Studies 157
    IV. Dynamic Response Studies 174
    V. Concluding Remarks 179
    References 180

6. Perturbative, Scattering, and Green's Function Theories of the s–d Model

W. Brenig and J. Zittartz

    I. Introduction 185
    II. Hamiltonian and Green's Functions 186

	III. Dispersion Theory	189
	IV. Equation of Motion Method	192
	V. Properties of the Solution	198
	VI. Thermal Properties	200
	VII. Electrical Conductivity	205
	VIII. Magnetic Field Effects	205
	Appendix A. Diagrammatic Methods	210
	Appendix B. Electronic Susceptibility	213
	References	214

7. Asymptotically Exact Methods in the Kondo Problem

P. W. Anderson and G. Yuval

	I. Introduction	217
	II. The Kondo Problem: A Discrete Path-Integral Approach	219
	III. Eliminating the Fermi Gas	221
	IV. The Method of Schotte and Schotte	224
	V. Summing Up over the Paths	225
	VI. Finite Temperatures	227
	VII. Numerical Results	228
	VIII. The Scaling Method	230
	IX. The Kondo Temperature	232
	X. Physical Implications	233
	XI. "Renormalization Group" Methods in the Kondo Problem	233
	References	235

8. Functional Integral Methods in the Magnetic Impurity Problem   237

D. R. Hamann and J. R. Schrieffer

	References	252

9. The Ground State of the s–d Model

Kei Yosida and Akio Yoshimori

	I. Introduction	253
	II. Perturbaion Theoretic Approach for the Singlet Ground State	258
	III. Bound State for the Anisotropic Exchange Interaction	264
	IV. Charge, Spin Polarization, and Spin Correlation Densities	270
	V. Bound State in the Presence of Magnetic Field	274
	VI. Local Electron Distributions and Magnetoresistance	278
	VII. Concluding Remarks	284
	References	285

Part III. MAGNETIC MOMENT EFFECTS IN SUPERCONDUCTORS

10. Paramagnetic Impurities in Superconductors
M. Brian Maple

I. Introduction	289
II. Long-Lived Local Moments in Superconductors	291
III. The Effect of Nonmagnetic Resonant States on Superconductivity	308
IV. The Effect of Localized Spin Fluctuations on Superconductivity	313
V. Magnetic–Nonmagnetic Transitions of Impurities in Superconductors	318
References	323

11. Recent Work on Ferromagnetic Superconductors
Ø. Fischer and M. Peter

I. Introduction	327
II. Magnetic Ordering in a Superconductor. The Low-Concentration Limit	328
III. Coexistence in More Concentrated Systems. The Case of $Ce_{1-x}Gd_xRu_2$	336
IV. Superconductor in a Molecular Field. Compensation of the Exchange Field by an External Field	343
V. Final Remarks	350
References	351

12. Recent Theoretical Work on Magnetic Impurities in Superconductors
E. Müller-Hartmann

I. Introduction	353
II. Model and Green's Functions	355
III. Nagoaka Approximation	358
IV. Solution of Hammann's Equation	361
V. Results at Low Impurity Concentration	369
VI. Treatment of Finite Impurity Concentrations	375
VII. Conclusion	380
References	381

Author Index	383
Subject Index	395

Contents of Volume I

Spin Hamiltonians
    K. W. H. Stevens

Exchange in Insulators
    P. W. Anderson

Weak Ferromagnetism
    Toru Moriya

Anisotropy and Magnetostriction of Ferromagnetic and Antiferromagnetic Materials
    Jûnjiro Kanamori

Magnetic Annealing
    John C. Slonczewski

Optical Spectra in Magnetically Ordered Materials
    Satoru Sugano and Yukito Tanabe

Optical and Infrared Properties of Magnetic Materials
    Kenneth A. Wickersheim

Spin Waves and Other Magnetic Modes
    L. R. Walker

Antiferromagnetic and Ferrimagnetic Resonance
    Simon Foner

Ferromagnetic Relaxation and Resonance Line Widths
    C. Warren Haas and Herbert B. Callen

Ferromagnetic Resonance at High Power
    Richard W. Damon

Microwave Devices
    Kenneth J. Button and Thomas S. Hartwick

Author Index—Subject Index

Contents of Volume II

PART A

Statistical Mechanics of Ferromagnetism
    R. Brout

Statistical Mechanics of Critical Behavior in Magnetic Systems
    C. Domb

Hyperfine Fields in Metals
    A. J. Freeman and R. E. Watson

Nuclear Resonance in Antiferromagnetics
    V. Jaccarino

Magnetic Symmetry
    W. Opechowski

Nuclear Resonance in Ferromagnetic Materials
    A. M. Portis

PART B

Antiferromagnetism in Metals and Alloys
    A. Arrott

Direct Exchange between Well-Separated Atoms
    C. Herring

Magnetism and Superconductivity
    M. A. Jensen and H. Suhl

On s–d and s–f Interactions
    T. Kasuya

Author Index—Subject Index

Contents of Volume III

Magnetism and Crystal Structure in Nonmetals
  JOHN B. GOODENOUGH

Evaluation of Exchange Interactions from Experimental Data
  J. SAMUEL SMART

Theory of Neutron Scattering by Magnetic Crystals
  P. G. DE GENNES

Spin Configuration of Ionic Structures: Theory and Practice
  E. F. BERTAUT

Spin Arrangements in Metals
  R. NATHANS and S. J. PICKART

Fine Particles, Thin Films, and Exchange Anisotropy (Effects of Finite Dimensions and Interfaces on the Basic Properties of Ferromagnets)
  I. S. JACOBS and C. P. BEAN

Permanent Magnet Materials
  E. P. WOHLFARTH

Micromagnets
  S. SHTRIKMAN and D. TREVES

Domains and Domain Walls
  J. F. DILLON, JR.

The Structure and Switching of Permalloy Films
  DONALD O. SMITH

Magnetization Reversal in Nonmetallic Ferromagnets
  E. M. GYORGY

Preparation and Crystal Synthesis of Magnetic Oxides
  C. J. KRIESSMAN and N. GOLDBERG

Author Index—Subject Index

Contents of Volume IV

Exchange Interactions among Itinerant Electrons
by Conyers Herring

I. Introduction
II. Ferromagnetic versus Nonmagnetic States of a Fermi Gas
III. Spin Susceptibility of a Free-Electron Gas
IV. The "Crystallized" Low-Density Electron Gas
V. Spin-Density Waves in a Fermi Gas
VI. Itinerant versus Localized-Spin Models for Ferromagnetic Metals
VII. Exchange among Bloch Electrons in the Hartree–Fock Approximation
VIII. Properties of Special Correlated Models Unlike Real Metals
IX. Extent of Polar Fluctuations in d-Band Metals
X. Formal Treatment of Strong Correlations
XI. Miscellaneous Topics in the Theory of Uniform Ferromagnetics
XII. Uniform Spin Susceptibility of Electrons in Metals
XIII. States of Metals with Oscillatory Polarization
XIV. The Bloch-Wall Stiffness and Ferromagnons

References

Author Index—Subject Index

Part I
CONDITIONS FAVORING LOCALIZATION OF EFFECTIVE MOMENTS

1. Formation of Local Magnetic Moments in Metals: Experimental Results and Phenomenology

Dieter K. Wohlleben

Department of Physics
University of California
San Diego, La Jolla, California

Bryan R. Coles

Department of Physics
Imperial College of Science and Technology
London, United Kingdom

I. Introduction	3
II. Experimental Observables in the Local Moment Problem	7
1. Operational Definition of a Local Moment	7
2. Experimental Techniques and the Information They Yield	10
III. General Features Revealed by Experiment	17
1. Quantitative Search for Single-Impurity Effects on Standard Systems	17
2. Qualitative Aspects of Local Moment Formation	22
IV. Solid Solutions of $3d$ Elements in Simple Hosts	31
V. Transition Metals Containing $3d$ Solutes	34
VI. Local Moment Formation in Rare Earth Metals	37
1. Abnormal Susceptibility of Intermetallic Compounds	39
2. Identification of the Intermediate Valence Phase	43
3. Resistivity in Rare Earth Systems with Unstable Valence	44
4. Temporal versus Spatial Mixture in the Intermediate Phase	45
5. Qualitative Consequences of Temporal Mixture	46
6. Implications of Nonmagnetic Rare Earth Shells on the Theory of Local Moment Formation	47
7. The Hirst Model	49
References	51

I. Introduction

Atoms with partially filled shells, when dilutely dissolved in nonmetallic hosts or when forming nonmetallic compounds or molecules, frequently assume a definite ionic valence state and show the corresponding Hund's rule magnetic moment, although the crystalline field may modify the magnetic behavior in fairly straightforward ways.

In the case of metallic hosts, however, the presence of the broad conduction band is clearly responsible for a plethora of more complex phenomena of which the most important is an apparent loss of the magnetic moment. There are no impurities with magnetic moments on incompletely filled s or p shells, and if transition metal or actinide impurities show signs of a magnetic moment at elevated temperatures, they often appear to lose it when the temperature is lowered, or occasionally when the composition of the host metal is altered or when pressure is applied. Only rare earth impurities were thought, until recently, to be immune against this metallic demagnetization.

The phenomena associated with local moments in metals may be roughly classified into local effects, studied by various kinds of magnetic measurements, and into modifications of the properties of the conduction-electron gas far from the local site. Information on the latter effects comes from transport properties, particularly resistivity, which provides a very sensitive, in fact, often embarrassingly detailed source of information. It is in this area that current theories, based on the s–d exchange Hamiltonian, on the s–d mixing approach, or on simple spin fluctuation ideas, have met with some success in interpreting the experiments. We mention the qualitative explanation of the resistance minimum, of the maximum in the thermal power and in the specific heat, the *prediction* of the saturation of the resistance at temperatures very much below the minimum, and last but not least, the almost completely successful interpretation of the influence of magnetic and nearly magnetic impurities on superconductivity [1]. Unfortunately, apart from the latter area no satisfactory *quantitative* merging of theory and experiment has yet materialized; in view of the large variety of phenomena it is probably unrealistic to expect a unified picture ever to develop. One is reduced to describing trends. It is perhaps worth emphasizing that in a large number of alloy systems and for a large number of properties, a surprisingly good picture of the data is yielded by the assumption that as $T \to 0$ the electrical and magnetic properties are those implied by the unmagnetized virtual bound state, while above some characteristic temperature a strongly correlated magnetic state is the more useful starting point. Magnetic measurements can normally not distinguish between intrinsic fluctuations of the magnitude or of the z component of a local moment. It appears therefore unwise to make a strict physical distinction between magnetic and nonmagnetic impurities (nonmagnetic in the sense that the z components of the individual local spins fluctuate independently of each other) merely because the so-called nonmagnetic impurities have a characteristic temperature near or above the upper end of the experimental temperature range while the socalled magnetic ones have it near or below the lower end. This purely operational distinction should be discarded, and in the interest of a more unified picture, a more vigorous effort should be made and perhaps new methods should be developed to study impurities with high characteristic temperatures.

As to the theories which set out to solve the problem of moment *formation* from first principles (e.g., Friedel [2], Anderson [3], Wolff [4], Suhl [5], Coqblin and Blandin [6]) (in contradistinction to explaining transport phenomena with a given spin, interacting with the conduction electrons in the spirit of the s–d model) it is questionable whether they really provide appropriate models for existing systems. Convincing consistency checks are missing since the energy parameters thought to be critical are nearly inaccessible to measurements. Not surprisingly, therefore, there is a marked shortage of quantitative predictions concerning which impurity has what magnetic moment in which host above what temperature. It appears that this deficiency is not merely due to a limited knowledge of the basic parameters. Rather, the feeling is winning acceptance that the traditional models are producing some qualitatively wrong trends because they are strongly overemphasizing the itinerant aspects of the problem. It seems to be unnecessary, nay harmful for proper understanding, to strip the impurity of all its valence electrons (and the Hund's rule correlations) first and then to let its core potential interact with a free conduction-electron sea without further regard to intraionic correlations between different *local* electrons or to local crystal fields. If transition metal, rare earth, and actinide impurities are able to retain their magnetic character in metals, they seem to do so precisely because of strong correlations between *several* local electrons within these shells of high degeneracy, which seem to reduce the effective level width of the impurity state very much below crude Hartree–Fock (HF) estimates and produce a much stronger tendency to integral occupation numbers of such shells than traditionally assumed. In other words, for real magnetic impurities, the vast majority of which have many electron shells, ionic Hund's rule correlations appear more important than conduction-electron–local-electron correlations. (Just two shells are known to be magnetic in which only conduction-electron–local-electron correlations can be at work: Ce and Yb. Both, incidentally, have a full orbital moment $L = 3$.)

In short, with hindsight it appears unfortunate that at the outset the problem was defined as that of moment *formation*; it would have considerably sped up our understanding had one spoken of the problem of moment *survival* instead. With respect to this problem we suggest that more attention be paid to ionic models like that of Hirst [7]. This model leaves the original Hund's rule structure of the impurity untouched and predicts much more specifically than the Friedel–Anderson model, under what conditions the moment does not survive. These predictions are borne out by experiment for rare earth shells.

While ten years ago nearly all the data relevant to the local moment problem were confined to solid solutions in simple metals like copper, silver, gold, and aluminum, today the range of solvents in which such effects have been sought is almost coextensive with the periodic table, and includes concentrated solid solutions and intermetallic compounds, semimetals, and even semiconductors. A rich variety of striking new phenomena has emerged: resistance minima in intermetallic compounds, some above room temperature, concentration dependence of the characteristic temperature of the moment, giant moments which are temperature and concentration dependent, crystal-field effects in metals, loss of the magnetic moments of rare earth metals under pressure, sometimes accompanied by giant resistance anomalies, resolution of insulator-like hyperfine spectra on magnetic impurities in good metals by EPR and Mössbauer measurements, and so on. These phenomena are clearly overstraining the capacity of the existing theoretical models and are waiting to catch the imagination of theoreticians who wish to disentangle themselves from the less and less fruitful staggering in the jungle of traditional Kondoism.

This article makes no effort to provide an exhaustive survey of the experimental literature concerned with local moments, which is clearly unnecessary in view of the existence of a large number of excellent reviews (e.g., Daybell and Steyert [8], Heeger [9], Van Dam and Van den Berg [10]). Instead, we concentrate on the following points:

In the next section we attempt to discuss in a phenomenologically self-consistent manner what is (or might be) actually measurable about a local magnetic impurity by various experimental techniques. In the third section, we briefly indicate the experimental difficulties which so far have prevented our making experimental distinctions between the various theoretical models as applied to traditional systems. These are mainly connected with impurity–impurity interactions. We then go on to cite the evidence which shows that the itinerant aspect has been overemphasized in the currently prevalent theories of local moment formation. A broad outline of some new experimentally established trends of general interest is then given under three separate headings: simple solvents containing $3d$ elements, transition elements in transition element solvents, and metals containing incomplete $4f$ shells.

II. Experimental Observables in the Local Moment Problem

1. Operational Definition of a Local Moment

In view of the ever-increasing and quite confusing variety of experimental phenomena associated with local magnetic moments in metals it seems appropriate to remark on the problem of measurement. We shall attempt to keep in focus the observables only, in particular the effective interaction energy of the electrons on the local impurity sites with the conduction electrons, which can range between several electron volts and less than 10^{-6} eV, and is the quantity of central interest in the problem of the existence of a local magnetic moment in a metal. We shall not dwell on the effect of local moments on the transport properties of alloys since these have been reviewed extensively elsewhere.

For the time being, we adopt the rather restrictive point of view that a local magnetic moment exists if the expectation value of the z component of the moment operator over a volume of the order of the lattice cell has a finite value in the limit $H \to 0$, $T \to 0$. We consider the z component of the moment rather than its magnitude, which is the focus of interest of the theoretician, since every observation done with the aid of an externally applied field is an observation of the z component only. In such measurements fluctuations of the magnitude are indistinguishable from those of the z component.

A static local moment in the sense discussed above exists in metals only below a magnetic ordering temperature, that is, in practice only in systems with a sufficient concentration of impurities with partially filled d or f shells. In the dilute limit, on the other hand, such a moment is not expected to exist, even if an effective moment can be measured via a Curie–Weiss law at higher temperatures. Any possible Zeeman degeneracy of the system ought to be lifted at $T = 0$, i.e., μ_z cannot remain a good quantum number as $T \to 0$ simply because the ground state of the system composed of the local cell with central symmetry interacting at least residually with the conduction electrons with the translational symmetry of the lattice must be nondegenerate at $T = 0$. Apart from direct Coulomb and exchange interactions between local and conduction electrons, there are, for instance, residual crystal fields and also at least a residual orbital moment and therefore residual spin–orbit coupling on both the local cell and the host. There must exist a temperature T_f below which the motion of the axis of the moment is no longer dominated by thermal fluctuations (which results in Curie–Weiss like behavior, $\chi = N\mu_{\text{eff}}^2/3k(T + \theta)$) but by intrinsic fluctuations due to residual interactions with the conduction electrons. The axis of the moment will tumble in time, the susceptibility of the impurity in a finite field will be less than without such interactions, and the zero field susceptibility cannot diverge at $T = 0$ ($\chi \to \mu_{\text{eff}}^2/kT_{\text{f}}$ for $T \ll T_{\text{f}}$, $H \to 0$). This intrinsic fluctuation of μ_z may be very complicated; there may in general be a whole frequency spectrum, discrete or continuous, and the average frequency ω_{f} can range over many orders of magnitude. Nondiverging susceptibilities on isolated impurities have been observed on systems which also often do but sometimes do not show a resistance minimum. (Nonmagnetic states at $T = 0$ of course follow from various models: They are a natural consequence of the s–d exchange and a possible consequence of the virtual bound-state approach, but cannot in the latter lead to quasimagnetic behavior at finite temperatures, requiring the *ad hoc* superposition of local spin fluctuations.)

Clearly, then, an isolated local moment in the sense described above is not expected to exist in a metal. However, the definition remains useful since it points to the significance of the very general phenomenological intrinsic fluctuation frequency of μ_z. There is, in fact, no phenomenological distinction in the intrinsic fluctuation of μ_z between host and impurity cell. For individual conduction electrons S^2 and μ_{eff}^2 are, of course, constants of the motion up to 0.5 MeV. The effective moment is in principle measurable above the Fermi temperature T_{F}, where the static susceptibility obeys a Curie law. The z-component of the moment becomes "visible" as well (that is, it becomes a quasi constant of the motion) when the measurement is done on a sufficiently short time scale ($t < h/kT_{\text{F}}$) and over a sufficiently small volume, say smaller than the volume of the host cell. Thus a polarized neutron beam with a kinetic energy of about 20 eV sent through a metal with $E_{\text{F}} = kT_{\text{F}} \approx 10$ eV should be partially depolarized by spin flip with conduction electrons. In short, on the host cell μ_z may be viewed to fluctuate intrinsically with $\omega_{\text{f}} \approx E_{\text{F}}/h$. Phenomenologically the most important distinction between host and impurity cell is then the potentially very much smaller intrinsic frequency on the latter. Clearly the often enormous difference between the local susceptibilities of host and impurity cell, which gave rise to the notion of a local moment, is caused by a large difference in fluctuation frequencies, not in moments. The magnitude of the fluctuating moment can only vary by a factor of about ten. On the impurity cell the effective moment, related to the magnitude, is either that of a single-electron spin or that of a configuration of several electrons which correlate their motions on the cell according to Hund's rules or in some more complicated way. The largest Hund's rule moments are those of Dy or Ho ($10\mu_B$). If the energy of correlation E_c of an ionic configuration in the metal remains of the order of electron volts, as in the isolated ion, while $T_f \ll E_c/k$, the Hund's rule effective moment of the composite does not disintegrate at practical temperatures and the notion of a moment of magnitude other than that of a single-electron spin is useful (**Cu**Mn). However, it is conceivable that the intrinsic fluctuation energy becomes of the order of electron volts, that is, large enough to break up the Hund's rule composite (e.g., **Nb**Fe). Then another correlated composite might form or the individual spins on the impurity cell might be left to fluctuate completely independently. The latter presumably is Friedel's picture of the nonmagnetic state. The correlation energy may also become very low in certain parts of a local system, of the order of the temperature employed. That is one way of viewing the temperature-, concentration-, and field-dependent giant moments (**Pd**Fe, Gerstenberg [11], Clogston *et al.* [12], Maletta and Mössbauer [12a]) where the constituents of the polarization cloud are only very weakly coupled with each other and with the impurity site. Since in practice the resulting correlation seems to remain incomplete even at the lowest concentrations and temperatures, the notion of an effective or saturation moment of fixed magnitude becomes impractical here, and a description of $\mu_z(H, T, c)$ is the only sensible thing left. It also may happen that two or more ionic configurations are nearly degenerate, while their correlation energies remain large, so that a fluctuation (resonance) between two or more distinct effective moments takes place (SmB_6; see Section VI).

At this point it seems appropriate to distinguish between the operational definition of the moment via its z component discussed here, (which can be fairly directly linked with the type of local spin fluctuation concept, for example, used by Rivier and Zuckermann [13]) and the local spin fluctuation concept used for example by Lederer and Mills [14] and Kaiser and Doniach [15] where the only feature of the impurity is a local exchange enhancement in an otherwise unperturbed band. The latter spin fluctuation theories only seem to cover a subset of situations admissible in our operational definition, namely when there is no correlated many-electron composite fluctuating on the impurity cell. The first situation is **Al**Mn (Hund's rule correlations presumably still exist on Mn) while the second is possibly realized in **Pd**Ni.

During the measurement, μ_z undergoes intrinsic fluctuations, thermal fluctuations, and fluctuations due to the measuring process itself. In practice, the motion of μ_z can also be affected by impurity–impurity interactions. However, we assume that the impurity concentration is sufficiently low to keep the average interaction energy small compared with the intrinsic fluctuation energy. The measure of "sufficiently low" is that the observed deviation from the free moment behavior should be independent of concentration. Crystal-field effects on certain non-S-state ions, on the other hand, may be much more difficult to separate from intrinsic fluctuations due to conduction-electron–local-moment interactions. Actually the effect of weak crystal fields can be viewed as a coherent fluctuation of μ_z. We shall for the moment disregard systems where such effects are important (**LaCe**).

The highest of the fluctuation frequencies (thermal, intrinsic, or measuring frequency) dominates the motion. Since the intrinsic fluctuation spectrum contains the most interesting physical information, it should be explored at some stage of the experiment by sweeping the thermal and/or measuring frequency through the relevant frequency range. This is not always possible with a given measuring technique.

In the following we discuss the suitability of various techniques for the detection of a local magnetic moment and especially for the intrinsic fluctuation frequency spectrum of μ_z. Each technique has a frequency of maximum sensitivity and a frequency window whose width depends on practical factors, either on systematic errors or on counting statistics.

All experimental observations of a moment involve transitions between at least two states which are split in energy by $\hbar\omega_m$ (ω_m is the measuring frequency) and have a finite width δ. The splitting can be larger or smaller than the width. The intrinsic fluctuation modifies this spectrum with respect to the free moment behavior, that is, it adds to the thermal broadening, if it is incoherent or it shifts the splitting to a value $\hbar\omega_m'$, if it is coherent, or both. The inverse number of transitions necessary to detect this modification, N^{-1}, is a measure for the sensitivity of the technique for the intrinsic fluctuation. The number N is, at least in principle, not much larger than 1 if the modified width δ' or the change of splitting $\hbar(\omega_m - \omega_m')$ is larger than δ; that is, if $\hbar\omega_f \gg \delta$. On the other hand, if $\hbar\omega_f \ll \delta$, as is often the case, then ω_f is unobservable in a single transition. A minimum number of transitions must then be observed independently and under identical conditions, which must satisfy $N_{min}^{1/2}\hbar\omega_f/\delta > \frac{1}{2}$. Clearly this type of measurement is relatively insensitive and will eventually run into a limit given by counting statistics, if nothing else.

2. Experimental Techniques and the Information They Yield

a. *Magnetic susceptibility.* Local magnetic moments were first detected by static susceptibility measurements. This method is still the most important one, and is always employed before any of the other more refined techniques which are actually capable of local measurements. On the example of this technique it is also easiest to point out the ambiguities in the concept of a local moment which stem from the existence of the intrinsic fluctuation of μ_z.

In practice one speaks of a local moment if the intrinsic fluctuation energy is near the lower limit of the accessible temperature range, or below, that is, if one can observe Curie–Weiss behavior. Conceptually this is clearly an unsatisfactory state of affairs. If $T_f \approx 100°K$ (**ThU**, Maple et al. [16]), an experimenter whose susceptibility apparatus is only equipped with liquid-nitrogen cryogenics will find a local moment, whereas one who only works at helium temperatures will not, because the susceptibility would be only weakly temperature dependent sufficiently far below T_f, with an approximate value of $\chi \approx N\mu^2/kT_f$ at $T = 0$. Both would speak of a local moment if $T_f \approx 10^{-2}°K$ (**CuMn**, Hirschkoff et al. [17]), and none if $T_f \approx 600°K$ (**AlMn**, Caplin and Rizzuto [18]). Nevertheless, physically there would be no *qualitative* distinction between these systems if the effective intraatomic correlation energy E_c were of the order of electron volts, since in all cases $kT_f \ll E_c$, so that all three systems possess a temperature window $kT_f < kT < E_c$ where Curie–Weiss behavior would be observable in principle. Quite a few systems have been classified as nonmagnetic on the basis of low-temperature measurements ($T < 300°K$) but may actually be magnetic in the sense that 10^{-2} eV $< kT_f \ll E_c \approx 5$ eV.

In static susceptibility measurements the measuring frequency is given by $\omega_m = g_J \mu_J H/\hbar$. The linewidth is due exclusively to thermal motion and intrinsic fluctuations since the applied field exists over practically infinite time. The absolute sensitivity is best and nearly constant when $\mu H \gg kT, \hbar\omega_f$; that is, when the linewidth is smaller than the measuring energy. Otherwise it decreases rapidly proportional to $\mu H/kT$ or $\mu H/\hbar\omega_f$, whichever is smaller. The technique has the optimal sensitivity for intrinsic fluctuations when $kT \ll \mu H \approx \hbar\omega_f$, that is, in saturation, when the Zeeman energy is of the order of the fluctuation energy. If the field is very much larger ($\mu H \gg \hbar\omega_f > kT$), the relative deviation of the magnetization from free moment behavior becomes small. If it is very much smaller ($kT < \mu H \ll \hbar\omega_f$), again the deviation, nut now also the magnetization, that is, the absolute sensitivity, becomes small. In both cases the influence of systematic errors increases. In other words, for $\mu H \approx \hbar\omega_f$ the "contrast" of the result of a measurement with fluctuation against what is expected without fluctuation is largest. This is what was meant by "sweeping the measuring frequency through the fluctuation spectrum."

In static susceptibility measurements ω_m can be easily set anywhere below 10^{11} Hz ($\sim 10^{-4}$ eV) to obtain optimal sensitivity (contrast) for intrinsic fluctuations at $\omega_m \approx \omega_f$. This makes the technique very powerful to detect intrinsic fluctuation at low but not at high frequencies. The lower limit of detectability of ω_f depends on the lowest accessible temperature, the upper admissible concentration, and on the relative accuracy of the magnetization measurement $\Delta M/M$. It is roughly given by $\hbar \omega_{f_{min}} \approx k T_{min} \cdot \Delta M/M$. With present-day equipment (dilution refrigerator, superconducting flux detector) it is theoretically possible to detect intrinsic fluctuation frequencies of the order of 10^6 Hz (10^{-4}°K) in samples with dilution as high as 10^{-8}. (We have made use of a rule given by Star [19] that the concentration should be lower than T_f/T_F in order to have the fluctuations dominate over impurity interactions.) Unfortunately in this concentration range, there is a strong bottleneck for progress in sample preparation since the static susceptibility is dominated by the magnetic impurity which has the largest concentration, so that the concentration of all other magnetic impurities must be much less than that of the one under study.

b. Electron paramagnetic resonance. EPR is another measurement directly involving the interaction of the local moment with an applied static magnetic field. The technique uses the same measuring frequency as normal static susceptibility (10^{10}–10^{11} Hz) but suffers from much lower sensitivity than static susceptibility when the linewidth is large compared to the measuring frequency. Measurements are therefore restricted to helium temperatures where the sensitivities can become comparable. On the other hand, to our knowledge no measurements have been done below 1°K, so that practically the available fluctuation window is very narrow. Since only those transitions are observed which are coherent with the incoming radiation, a high degree of coherence of the motion of the observed moments is required, again contrary to static susceptibility. Such coherence is favored if the spin-lattice relaxation is slow compared to the Zeeman period and if the g factor of the impurity is close to that of the conduction electrons. Experiments have therefore been until recently almost entirely limited to S-state ions (Gd^{3+}, Mn^{2+}, Eu^{2+}). The technique, where it can be successfully applied, is very sensitive to details of the interaction between the moment and the conduction electrons. The interpretation of the spectra is, however, often ambiguous and therefore controversial. Among the few clear contributions of EPR measurements to the local moment problem per se are some recent experiments which indicate that rare earth impurities dissolved in good metals can behave very much as they do in insulators; that is, they show crystal-field splittings [20, 21] and even resolved hyperfine spectra [22]. This is very interesting because it demonstrates that the Zeeman levels can be very much sharper than the hypothetical resonances in the single-particle spectrum which constitute the virtual bound state. AuYb, which shows these effects, is a particularly surprising case since the appearance of a resistance minimum as Ag is substituted for Au [130] makes it probable that a real mixing takes place here.

c. Neutron scattering. Neutron diffraction is the only available tool to measure the microscopic magnetization distribution everywhere in the lattice cell (rather than only the uniform macroscopic magnetization or the magnetization on nuclear sites). This information comes from elastic scattering. In the case of dilute impurities and in the absence of an external field one should obtain the radial dependence of the magnetization squared averaged over all possible orientations of the axis, that is, this is a measurement of $\mu^2(\mathbf{r})$ rather than μ_z. The measurement is obviously insensitive to intrinsic fluctuations, when done in this way. However, with an applied field the scattering becomes appreciably anisotropic if $\mu H \gtrsim kT, \hbar\omega_f$. Thus the technique could be employed to detect low intrinsic fluctuation frequencies (below 10^{11} Hz) at low temperatures by observing the switchover from isotropic to anisotropic scattering when μH goes through $\hbar\omega_f$ while kT is small compared to both. This, however, is more easily done by static susceptibility, since the sensitivity of neutron scattering is severely limited by counting statistics, especially at the necessary low concentrations. Secondly, one might expect inelastic neutron scattering on the fluctuations, for instance an energy broadening of the beam if $\hbar\omega_f \gg \delta$. The energy width of the incoming beam is usually not much smaller than a few degrees Kelvin. Therefore fluctuation frequencies below 10^{11} Hz should be very hard to detect by inelastic neutron scattering. On the other hand, this technique seems to be the best available to detect *high* fluctuation frequencies. The measuring energy is given by the kinetic energy, E_k, which is of the order of a few hundred degrees Kelvin. Thus the greatest relative sensitivity is for fluctuations in this range. The absolute sensitivity is nearly constant for $\hbar\omega_f \ll E_k$ (assuming $kT < \hbar\omega_f$), it is largest near $E_k \approx \hbar\omega_f$ and decreases like $E_k/\hbar\omega_f$ for $E_k \ll \hbar\omega_f$. Therefore, with E_k properly chosen slightly above kT_f it is possible to detect both the moment and the intrinsic fluctuation frequency in a region where static susceptibility and other measurements are less conclusive. This is a rather important point in view of the present uncertainty about the existence of moments with fluctuation frequencies higher than a few hundred degrees Kelvin. This type of measurement could answer the following central question of the local moment problem: In the case of a relatively high fluctuation frequency, for instance for Mn in Al, does the Mn impurity, commonly called nonmagnetic, preserve its internal ionic structure, for instance the Hund's rule ground state of Mn^{2+} with $S = \frac{5}{2}$, while the direction of the axis fluctuates with a frequency of a few times 10^{12} Hz? Or is the Hund's rule structure completely lost, as suggested by the virtual bound-state model, where the width of the state is expected to be of the order of several electron volts, sufficient to excite many other ionic configurations and thus wiping out the Hund's rule correlations? Neutron scattering studies on **Al**Mn have been carried out by Kroo [23] but their interpretation is still open to discussion.†

The greatest contribution of neutron scattering so far has been to cases where the system is ferromagnetic, or nearly so, and magnetic diffuse scattering reveals differences of the spatial distribution of the magnetization around the impurity and the host in the elegant experiments of Low [24]. Another contribution is the detection of a d-shell-like form factor on Fe in dilute **Cu**Fe alloys in the neighborhood of T_f (no compensating conduction-electron polarization cloud was found by Stassis and Shull [25]).

d. *Measurements via the hyperfine interaction.* The techniques employing the interaction of the electronic shell of the impurity with its nucleus provide the most localized probes conceivable (in fact, somewhat more localized than desirable). They have their maximum sensitivity at a lower frequency window than the techniques which look directly at the electronic shall. Mössbauer and γ–γ angular correlation measurements (but not NMR) can be done at extremely low concentrations so that impurity–impurity interactions may in general be ignored, and the measurement is highly selective, that is, is on one type of nucleus only, so that the samples need not be very pure with respect to other types of impurities. A major drawback of these techniques is the restricted number of suitable nuclei in the periodic table.

(1) *Mössbauer effect.* Indirect evidence of the ionic state of the impurity is sometimes provided by the isomer shift [26], but the most useful and direct information on the local magnetic moment is through the Zeeman split levels of the nucleus in the hyperfine field provided (partly by core polarization) by the incomplete shell. The measuring frequency $\omega_m = \mu_n H_s/\hbar \sim 10^9$ Hz, where H_s is the saturation hyperfine

† Recently neutron diffraction data have become available for dilute Mn in Al [23a, b]. The data are claimed to be consistent with the Hund's rule magnetic groundstate of the Mn^{2+} configuration ($S = 5/2$), with considerable conduction electron polarization extending to about 5 Å, i.e., far outside of the Mn cell. The fluctuation frequency is of order 10^{12} Hz and appears to be temperature dependent.

field. The linewidth, can be very small ($\delta/\hbar \gtrsim 10^6$ Hz). Thus, the electronic fluctuation spectrum can be sampled easily down to $\omega_f > \delta/\hbar$, that is, to $\omega_f \approx 10^6$ Hz. In fact, if the intrinsic and thermal fluctuation frequencies of the impurity are sufficiently small ($\delta, \hbar\omega_f \lesssim \mu_n H_s$), a full hyperfine spectrum can be measured by Mössbauer spectroscopy *without* application of an external field. Such a hyperfine spectrum gives rather unambiguous information about the magnetic configuration even if crystal fields are present. In fact, the existence of crystal fields is helpful in identifying the configuration and the information can be more detailed than that extracted from the effective moment obtained from the Curie–Weiss behavior of the susceptibility. One might, of course, feel that a well-defined Hund's rule magnetic moment in a crystal field with very small intrinsic fluctuation frequencies presents nothing new in view of the large body of knowledge about magnetic ions in insulators. However, the fact that nearly complete decoupling of the ionic shell from the conduction electron spins of a good metal like Au or Zr is observable at all [27] is in itself a challenge to the theory of magnetic impurities in metals. It certainly is hard to reconcile with the notion of a virtual bound state.

If the main fluctuation frequency of the electronic moment is large ($\hbar\omega_f > \mu_n H_s$), the hyperfine spectrum without magnetic field is wiped out. However, a residual hyperfine field can be reestablished by application of an external field, which causes an imbalance in the distribution over the Zeeman levels of the impurities and thus an effective field due to the electronic shell at the nucleus. The condition for observability of this spectrum is

$$\mu_n H_{\text{eff}} = \mu_n H_s \cdot \mu_e H / \hbar\omega_f = \text{const} \cdot H \cdot \chi > \delta.$$

The splitting is then directly proportional to the local susceptibility of the impurity: that is, if the thermal fluctuation is large compared to the intrinsic fluctuation and the electronic Zeeman energy $\mu_e H$, the splitting follows a Curie–Weiss law (and is related to the temperature-dependent part of the Knight shift in NMR); and if the electronic Zeeman energy is larger than both intrinsic and thermal fluctuation energies (saturation), the hyperfine spectrum is again fully developed. Although at high fluctuation frequencies the Mössbauer technique is inferior to static susceptibility with respect to *absolute* sensitivity, it may become superior with respect to *relative* sensitivity because of the selectiveness of the measurement, since the static susceptibility can be evaluated only if all contributions from other sources (host and other impurities) are controllable to sufficient relative accuracy. (This is rarely the case when the impurity contribution is less than 1% of the total susceptibility.) Actually, we feel that much more experimental effort should be spent in impurity Mössbauer spectroscopy at higher temperatures because as long as Curie–Weiss behavior is observable, there is a good chance that the impurity should be classified as magnetic with a high intrinsic fluctuation frequency rather than as nonmagnetic.

(2) *γ–γ angular correlation.* Fewer results are available for this technique presumably because of experimental complexity (e.g., Campbell *et al.* [28], Rao *et al.* [29], Flouquet [30]). It has much in common with Mössbauer spectroscopy since it also exploits the hyperfine interaction on long-lived metastable excited nuclear states. It has the potential of even better frequency resolution than Mössbauer spectroscopy. A very interesting case in question is the study of Rh impurities in Pd by Rao *et al.* [29]. These authors were able to detect a deviation of the fluctuation spectrum of the Rh impurity from that of the Pd host cells. They could measure relative shifts of the nuclear levels with changing effective electronic hyperfine field which were *smaller* than the linewidth δ of the metastable nuclear state by employing a technique which is reminiscent of phase-sensitive detection in electronic engineering. They reached a frequency resolution of 10^5 Hz with a linewidth of 4×10^6 Hz. This measurement points up clearly the advantage of exploiting the hyperfine interaction at high temperatures for the study of local moments with high intrinsic fluctuation frequency where static susceptibility measurements may fail.

(3) *Nuclear magnetic resonance.* Nuclear magnetic resonance also exploits the hyperfine interaction, but, contrary to Mössbauer and γ–γ techniques, in the absence of intrinsic or thermal fluctuations the natural linewidth is zero for the same reason as in static susceptibility. It has therefore potentially an unlimited sensitivity to low fluctuation frequencies. However, unfortunately the absolute sensitivity for low concentrations is poor. The technique is therefore more often applied to the study of the host than of the impurity nuclei, in which case information on the state of the impurity is rather indirect. The sensitivity to higher fluctuation frequencies of the impurity is then very low, because the low-frequency NMR measurement averages over fast events in the conduction-electron gas and the small residual changes of the NMR linewidth due to conduction-electron impurity effects are hard to track down, and even harder to interpret unambiguously. A more useful observable is the host Knight shift, which can be an observable function of the distance of the host nucleus from the impurity. This phenomenon has been successfully exploited in the study of conduction-

electron polarization clouds around giant moment impurities. However, it gives little information about the fluctuation spectrum of the impurity itself. One might think that a study of the temperature dependence of the line shift of the impurity itself should give similarly valuable information as Mössbauer spectroscopy with a magnetic field. However, near saturation of an impurity with *low* intrinsic fluctuation the line shift will become very large, so that the resonance is far beyond the upper frequency limit of normal NMR equipment. On the other hand, NMR is in principle a potential tool to study impurities with high intrinsic fluctuation frequencies and has been used by Narath [31] in **Au**V and **Mo**Co and by Alloul and Launois [32] in **Al**Mn. For such impurities practically no field dependence and a tolerable temperature dependence of the line shift are expected, and the limited sensitivity of the technique is no severe obstacle since relatively large impurity concentrations are admissible without violating the condition that the impurity–impurity interaction energy be smaller than the intrinsic fluctuation energy. Thus NMR should be suitable to study "questionable" local moment cases.

III. General Features Revealed by Experiment

During the last few years the experimental activity has taken two main directions, namely to obtain quantitative results on a few standard local moment systems (mainly 3d impurities in simple metals) by a large variety of measuring techniques on alloys prepared with great care and to explore qualitatively the behavior of new systems by a few standard techniques (mainly resistivity and susceptibility).

1. Quantitative Search for Single-Impurity Effects on Standard Systems

The attempts at quantification were stimulated by the large outpour of theoretical literature following Kondo's solution [33] of the old puzzle in the correlation between resistance minimum and Curie–Weiss behavior of the susceptibility. Theoreticians began to make interesting predictions on new untested aspects of the local moment problem, mainly on the saturation of the resistivity and the loss of magnetic moment sufficiently far below the characteristic temperature T_f. The search for the characteristic temperature of a single impurity and unambiguous quantitative measurements of its effects turned out to be much more difficult than finding a resistance minimum or a Curie–Weiss law of the susceptibility. The position of the resistance minimum is very insensitive to concentration of the impurity and is normally not far from 5 to 15°K simply because the small magnetic impurity resistance does not become clearly observable until the phonon resistivity has almost completely vanished. The characteristic temperature, however, can be many decades of temperature away from this region, in both directions. For reasons of experimental convenience, most systems studied in detail initially had characteristic temperatures *below* that of the minimum. It has become clear now (unfortunately too late for the period of great enthusiasm of the mid-sixties), that in the case of low characteristic temperatures of 3d impurities in simple metals extremely pure matrices with very small impurity concentrations are of paramount importance to obtain results which are genuinely single-impurity effects, that is, characteristic of interactions between the *matrix* and the impurity rather than of interactions between impurities.

a. Interference of the RKKY interaction with quantitative measurements of single-impurity effects

(1) *Susceptibility*. In a well-randomized (unclustered) solution of impurities at a dilution sufficiently high to make occurrence of a nearest neighbor impurity pair unlikely, the paramagnetic Curie–Weiss temperature drops with concentration much faster than linearly [17, 34, 35] in a way as yet unexplored quantitatively by theory. The mean interaction energy kT_i due to the RKKY interaction can then no longer be found reliably from the Curie–Weiss temperature θ_p by susceptibility measurements above the interaction temperature T_i; it becomes noticeable only when the temperature is actually lowered through T_i, by a maximum or a saturation of $\chi(T)$. Then T_i is found to be much higher than θ_p and is indeed linear in concentration. The deceptive behavior of the Curie–Weiss temperature misled many workers to confuse interaction effects in the susceptibility with single-impurity effects. At $T < T_i$ the bulk susceptibility may carry all the earmarks of an apparent loss of magnetic moment rather than what is naively expected from impurity–impurity interaction effects. In fact, in a highly dilute random alloy the freezing out of the Zeeman degeneracy by impurity–impurity interactions results in a magnetic structure where the moments point permanently in random directions from site to site. Such an ordered state cannot be called ferro- or antiferromagnetic without confusing the issue (although it bears more resemblance to the latter). We prefer the term "magnetic glass." It opposes moment alignment in the direction of the applied field with a characteristic energy kT_i for $T \lesssim T_i$ but not for $T \gg T_i$, so that one obtains

$\chi \approx N\mu_{\text{eff}}^2/kT_i$ (for $\mu H \ll kT_i$) just as expected for a Kondo condensation ($T_i \stackrel{?}{=} T_f$). When $\mu H > kT_i$, the susceptibility becomes field dependent in qualitatively the same manner as it would if the field destroyed the "Kondo polarization cloud." In all experiments which rely on an applied external field (not only static susceptibility but also Mössbauer or γ–γ angular correlation measurements; e.g., Kitchens et al. [36], Campbell et al. [28]) it is difficult to distinguish these *spatial* fluctuations of μ_z from the expected *temporal* fluctuations or Kondo "condensation" predicted by theory. The ordering of a magnetic glass is also accompanied by a broad anomaly in the specific heat and thermopower. The only safeguard against mistaking impurity–impurity interactions for impurity–matrix interactions is a thorough check for concentration independence over *decades* of concentration. This is of course very laborious, and often impossible because of a lack of sensitivity or of pure metals and was rarely done in the early experiments after 1964. Even if the results are definitely independent of concentration, they may often simply reflect a crystal-field effect rather than a loss of moment due to interactions with conduction electrons.

In the case of **Cu**Mn the concentration dependence of the interaction energy is known to be $dT_i/dc \approx 20°$K at. % Mn from specific heat and magnetic measurements [17, 37, 38] while the characteristic single-impurity temperature is $T_f \approx 10^{-2}$°K (Fig. 1). Results meaningful to test quantitative theoretical predictions of the susceptibility of single impurities near and below T_f can therefore only be expected in this system at concentrations less than 5 ppm ($c\,dT_i/dc < T_f$). They have not yet been done. One of the recent experiments which exhibits the difficulties clearly is reproduced in Fig. 1.

(2) *Resistivity.* The difficulties in obtaining quantitative single-impurity data at finite impurity concentrations are also serious for resistance measurements. For instance, early work by Pearson and Kjekshus [39] showed a well-defined resistance minimum on **Cu**Co only for Co concentrations above 0.5%, while Souletie [40] suggests that $d\Delta\rho/dT > 0$ for isolated Co atoms. The work of Star and Nieuwenhuys [41] has clearly shown that even if $c\,dT_i/dc \ll T_f$, the resistivity is still sensitive to impurity–impurity interactions. It is reassuring, however, that if proper care is taken in keeping the impurity concentrations low, the low-temperature dependence of the resistivity is very simple [$\rho \sim (1 - \alpha(T/T_f)^2$], independent of the magnitude of the characteristic temperature (**Cu**Fe, $T_f \sim 10°$K, **Al**Mn, $T_f \sim 600°$K). The data for the **Cu**Fe system are reproduced in Fig. 2.

One might think that a possible way out of the difficulties with

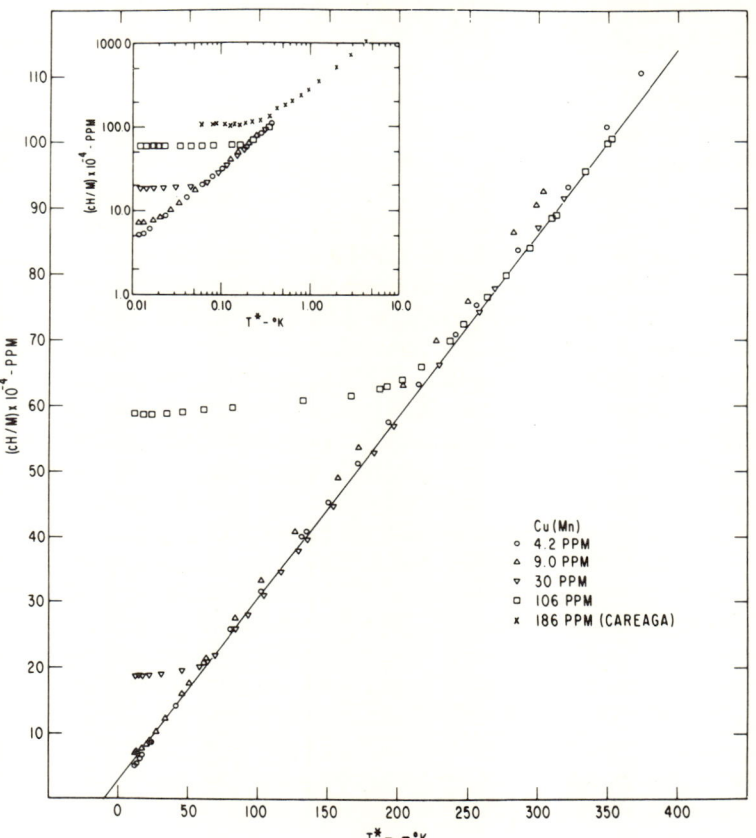

FIG. 1. The inverse low-field susceptibility of very dilute Mn in Cu at very low temperature [17]. The impurity interaction becomes noticeable at low temperature by saturation ($T_i/c \approx 20°$/at. % Mn), but the high-temperature susceptibility does not reflect this interaction by any appreciable paramagnetic Curie–Weiss temperature. The intrinsic characteristic temperature is $T_f \sim 8 \times 10^{-3}°$K. Thus, single-impurity effects can only be expected for concentrations less than a few parts per million.

impurity interactions is to study systems with higher characteristic temperatures. The condition $c\, dT_i/dc < T_f$ (or the more stringent $c < T_f/T_F$) then allows for higher concentration of impurities without interaction effects, and the requirements on the purity of the alloy are in general relaxed. Indeed, while the transition temperature T_c versus concentration curves in superconductors with *magnetic* impurities ($T_f \ll T_c$) are often strongly affected by impurity interactions, these

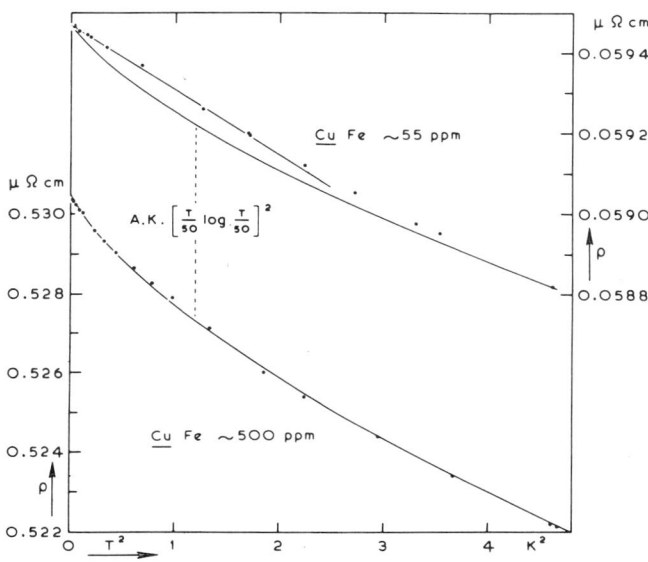

FIG. 2. The resistivity of dilute Fe in Cu [41]. Interaction effects cause drastic deviations from the low-temperature T^2 dependence already at a concentration of 0.05 at. % Fe. Underlining in the figure corresponds to boldface letters in the text.

curves taken with so-called *nonmagnetic* impurities ($T_f \gg T_c$) never are [1]. However, both susceptibility and resistivity measurements become much less sensitive for impurities with high T_f, the first because the susceptibility is inversely proportional to T_f and the second because of the large phonon scattering. Nevertheless, this approach has become more popular recently. We refer the reader to the resistivity studies of **AlMn** by Caplin and Rizuto [18], the susceptibility studies by Hedgcock and Li [42], and the NMR investigations by Alloul and Launois [32] and Narath and Gossard [43]. As pointed out in the last section, quantitative studies of systems with high T_f are important from a fundamental point of view since they ought to shed light on the question of where actually the borderline between magnetism and nonmagnetism is lying; that is, whether the existence of a Curie–Weiss moment is merely a question of temperature range or whether really more profound modifications of the electronic structure of the impurity may occur.

b. Modification of the characteristic temperature by the environment. The by far most concentrated and persistent effort to obtain quantitative magnetic information on single-impurity effects was made by the Grenoble group around Tournier and Souletie, who collected a vast amount of static magnetization and specific heat data for dilute 3d impurities in simple metals [44-48]. In no case could the data be interpreted on the basis of single-impurity effects alone. However, as a most interesting and unexpected by-product of their search they found strong evidence for drastic modifications of the local characteristic temperatures when two or more impurities are nearest neighbors; that is, that the impurity interaction not only leads to an ordered state but may also determine the local characteristic fluctuation temperature. For instance, in Au for single Co impurities $T_f \approx 225°K$, for pairs $T_f \approx 25°K$, and for triplets $T_f < 1°K$. This phenomenon was observed in **AuCo**, **CuCo**, **CuNi**, **CuFe**, and **AuV**. While in most systems the impurity interaction *decreases* the local characteristic temperature, T_f is *increased* in **AuV**, from about $T_f \approx 225°K$ for truly isolated impurities to about $T_f \approx 1120°K$ for nearest neighbors, presumably by nearest neighbor d–d overlap [49]. More distant interactions via RKKY lower T_f from 225°K [50].

Similar abrupt modifications of the characteristic temperature by a *nonmagnetic* environment are known in ternary alloys. Jaccarino and Walker [51] showed rather early from a statistical analysis that in NbMo alloys with Fe and in PdRh alloys with Co the appearance of the moment on the magnetic impurity might be discontinuous (in our language, T_f might drop discontinuously through the temperature range of the experiment). In the first case Fe is only magnetic when it has less than seven Nb nearest neighbors and in the second Co is magnetic with more than two Pd nearest neighbors. This behavior was verified by Brog and Jones [52] with NMR measurements on NbMo and TiMo hosts with Co impurities.

2. Qualitative Aspects of Local Moment Formation

Our interpretation of the observations with respect to the formation of moments on dilute magnetic impurities has to begin with strongly oversimplified models. A good measure of the theoretical difficulties is the large number of existing treatments which start from various radically different basic assumptions and proceed, within a given model, along widely diverging routes. (For a recent review of these theories, see Fischer [53].) In spite of the activity in the field, contact between theory and experiment on a fundamental level has so far remained superficial at best. This may be attributed to the mathematical difficulties involved in making quantitative predictions from first principle models, and on the experimental difficulties in measuring the energy parameters which are thought to be critical for moment formation. However, there is also the possibility that the currently prevailing theories start out from insufficiently realistic assumptions. We feel that the influence of itinerant single-electron parameters $[V_{kl}, \rho(E_F)]$ on the magnitude of the moment and on the critical conditions of its appearance is strongly overemphasized in the traditional models. The experimental evidence points increasingly to the importance of such local quantities as valence, Hund's rule correlations, crystal fields, and spin–orbit effects, which must be described in terms of many electron configurational energies.[†]

The traditional preference of the theories for a single-particle description derives perhaps from the original interest and success in explaining resistance phenomena in dilute alloys. However, while a single-particle picture is undoubtedly very appropriate for conduction electrons, it is much more difficult to retain when describing the local shell, especially with respect to magnetic phenomena. This was recognized by Anderson, who introduced the intraionic Coulomb repulsion as a correlation energy and treated an orbitally nondegenerate local shell from this point of view in the HF approximation. The result can be fairly easily translated back into the single-particle picture, with $E_l - E_F$, $\Delta = V_{kl}^2 \rho(E_F)$, and U_{eff} determining the critical conditions for magnetism in this case. The difficulties in going beyond the HF approximation are formidable, and attempts in this direction again have been largely confined to the case of the nondegenerate local orbital, which, for real impurities with highly degenerate local shells, presumably retains some validity for shells with only one electron or hole (d^1, d^9, $4f^1$, $4f^{13}$). However, most real magnetic impurities have more than one electron or hole in the local shell, and their description in the Anderson model would require the fully correlated treatment of the degenerate orbital case. This problem has not been solved. If it were, it is by no means obvious that the three simple parameters above would remain the relevant ones. In fact, the demagnetization of rare earth shells can be quite successfully described qualitatively by an entirely different set of energy parameters (Section VI).

Phenomenologically many electron shells obey Hund's rules when they are magnetic. The energy levels associated with Hund's rules are many-electron levels and have nothing to do with the band structure of

[†] A similar trend away from the purely itinerant to a more local point of view is discernible in the treatment of concentrated magnetic systems, exemplified by the recent shift from the rigid band to the minimum polarity model for CuNi alloys [54, 55].

the host. There is experimental evidence which suggests that Hund's rule correlations may be preserved in some sense even on impurities which are commonly called nonmagnetic (see below and Section VI). If a sufficiently strong case could be built on this assumption, a reordering of priorities would seem to be required. A theory should then start from the ionic Hund's rule-correlated many-electron state and merely explain why the susceptibility of such ions often becomes temperature independent below some characteristic temperature due to interaction with the conduction electrons.

In the following we discuss some basic difficulties with the Friedel–Anderson model and the experimental evidence which points to a more ionic model.

a. *The critical energy parameters for moment formation.* The lack of contact between existing first principles theories and experiment is serious with respect to the critical conditions for magnetism. Assuming the Friedel parameters Δ and J or the Anderson parameters Δ, U_{eff} and $E_l - E_F$ to be realistic critical parameters for local moment formation on $3d$ and $4f$ impurities, we are faced with the fact that no such *complete* set has ever been measured directly, for instance, by optical spectroscopy, on any system. Measurements of magnetic, thermal, or transport properties leave too many adjustable parameters and can therefore only give a bootstrap kind of consistency with theory. Among these even low-temperature specific heat data, which come closest to measuring the energy parameters, are insufficient for independent determination of the position $E_l - E_F$ and width Δ of the local state, since they do not explore the crucial regions of the energy spectrum far away from the Fermi surface. No prescription is known for direct measurements of J and U_{eff}.

Optical absorption measurements of some of the Friedel–Anderson parameters have become available for reasonably dilute Mn and Pd in Cu and Ag [56]. While a broad virtual bound state was found for Pd, which is nonmagnetic in these matrices, surprisingly no positive evidence for such a level could be detected for Mn, which is magnetic.

In view of only a few scattered optical measurements with such puzzling results the basic relevance of the traditional first principles theories to real systems can by no means be regarded as confirmed.

b. *The s–d model for 3d solutes in simple hosts.* The situation is somewhat improved for the *s–d* exchange interaction approach, which, however, renounces prediction of the structure of the moment from first principles (i.e., from a single-particle picture) by starting with a given local spin (not local moment, incidentally) and only tries to predict the characteristic temperature. In this theory T_f depends exponentially on the product of the effective exchange integral J and the density of states at the Fermi level $\rho(E_F)$ $[T_f = T_F \exp\{-(\rho(E_F)J)^{-1}\}]$. One important feature of this Kondo–Suhl expression is the correct prediction of an extremely drastic dependence of T_f on the energy parameters. However, in practice the prediction of T_f from this expression becomes an order of magnitude guess at best because J and $\rho(E_F)$ are hard to obtain from independent measurements with the high required accuracy. This problem is similar to predicting the superconducting transition temperature from normal state data, but alas no comparable effort has gone into its solution.

As shown by Daybell and Steyert [8] the Kondo–Suhl expression for T_f seems to be roughly obeyed for 3d impurities in Au, Ag, and Cu if one chooses $J = S \cdot J_d$ with J_d and $\rho(E_l)$ as constant. Then S varies by about a factor of 5 between Ti and Mn, reflecting Hund's first rule rather well. The v-shaped curve of ln T_F versus atomic number is reminiscent of the depression of the superconducting transition temperature due to rare earth impurities in superconductors via spin–spin interactions [57, 58].

It must be emphasized that there is no guidance from the Anderson model (via the Schrieffer–Wolff transformation, Schrieffer and Wolff [59]) for why S seems to follow Hund's first rule in these systems. Although it is gratifying that the experimental evidence points to the relatively simple phenomenon of conservation of these rules, the need to invoke them *ad hoc* in the theoretical treatment [60] exhibits a central weakness of first principles models which ignore intraionic correlations between *several* local electrons from the start in favor of local-electron–conduction-electron correlations only.

c. Moment formation of Fe in 4d metals. The effect of intraionic many-electron correlations is also apparent in the classical series of binary 4d alloys with dilute Fe impurities, which was the starting point of the Anderson and Wolff models.

First, the moment on iron is rather independent of the matrix. It seems to have about the same value all the way from NbMo to PdAg alloys, even in Rh, where the moment seemingly vanishes somewhere below 20°K, and even under the disguise of the giant moment near Pd, where Mössbauer measurements [36] and neutron diffraction [61] find the same local magnetic structure of the Fe cell as, for instance, in Cu. Moreover, the Fe moment in 4d metals is the same as on Fe impurities in simple metals, and it is close to the moment in the ferromagnetic element itself. In other words, there seems to be no continuous variation of the microscopic Fe moment with the strongly varying parameters $E_l - E_F$, Δ, and U_{eff}.

Secondly, it is now established that the appearance of the moment on Fe in the critical regions depends on the relative distribution of the two kinds of nonmagnetic nearest neighbor host atoms [52, 62], as suggested by Jaccarino and Walker [51]. There is no continuous variation of the *microscopic* moment; a sudden change of the characteristic temperature of a fixed moment is the more appropriate picture, and this change, moreover, is not triggered by a smooth change of occupation of translationally invariant conduction-electron states, but by the symmetry of the local environment.

These facts point into the same direction as the trends found by Daybell and Steyert [8] for the characteristic temperature in simple hosts: The iron effective moment may be thought to exist always with a relatively fixed magnitude, while the characteristic fluctuation temperature varies widely, depending on the host.

d. *The effective width of the magnetic state.* While a Curie–Weiss law is the hallmark of a local moment for the experimenter (who is usually, hard put to measure the zero-temperature saturation moment, for reasons of experimental convenience or because of the interference of a finite intrinsic fluctuation temperature), it is very difficult to arrive at Curie–Weiss behavior with the traditional first principles theories even for magnetic impurities! Only a few such calculations exist which start either from a structureless potential interacting with a free electron gas [63] or from the Anderson model, but without orbital degeneracy [64]. These studies have established that a Curie–Weiss-like susceptibility can be obtained at *high* temperatures, while below a characteristic temperature the susceptibility is no longer expected to depend much on temperature. Unfortunately these treatments do not attempt to produce realistic many-electron effective moments (including orbital contributions) for d or f impurities in real matrices nor realistic characteristic temperatures. We consider the latter point very important because in the absence of unambiguous predictions of the characteristic temperature it is impossible to ascertain merely from the temperature independence of the susceptibility of an impurity in the accessible temperature range that it really is nonmagnetic in the sense of the prevailing models.

Phenomenologically the susceptibility of a magnetic impurity is expected to become temperature independent if the thermal energy drops below the intrinsic effective width of the Zeeman levels. Experimentally the width of the Zeeman levels of impurities in metals can be exceedingly small, many orders of magnitude narrower than the width of the virtual bound state. This is not easy to understand because it happens in spite of the fact that the impurity is strongly interacting with the conduction-electron spins. Impurities with incomplete $3d$ and $4f$ shells have direct exchange integrals of 1.0 and 0.1 eV, respectively, in metallic matrices, as evidenced by magnetic impurity–impurity interactions and by the suppression of the transition temperature of superconductors with magnetic impurities. This interaction causes the well-established long-range oscillatory RKKY spin polarization in the conduction-electron gas. One might naively expect that the ionic impurity would experience a strong spin randomizing molecular field due to the indiscriminate impact of conduction electrons with equal occupation of spin up and down (independent of whether $J \geq 0$), so that the impurity susceptibility should saturate at $\chi_{\text{imp}} \approx \mu^2/3 J^2\rho(E)$, that is, only about an order of magnitude above the Pauli susceptibility, and EPR lines should be as wide as $J^2\rho(E)$, that is, utterly unobservable. The fact that EPR is observable in a pure metal was defended against an analogous argument by recognizing [65] that the Zeeman energy of different conduction-electron spins is exactly the same, so that a highly coherent motion (spin precession) in the field becomes possible in which the exchange effects on the spins are canceled out in time average to a very high precision, which increases with J. Since narrow EPR lines due to impurities in a metal are observable, a similar argument should hold there. However, the g factor of impurities, even if it is identical from one impurity to the next, is not *a priori* identical to that of the conduction electrons. It was then argued that if all J's are large enough compared to μH and if spin–lattice relaxation is small, a sharp resonance occurs at a frequency which is a weighted average of the impurity and conduction-electron resonances [66]. Thus the spin interaction (scattering) between conduction electrons and impurity spins helps to narrow the line *if it is coherent*. The only way to broaden a line is by incoherent scattering, which occurs on imperfections of the lattice. A coherent motion of the impurity moment in the field *together* with the long-range RKKY conduction-electron magnetization is necessary in order to observe sharp EPR lines. In particular, such motion is not impossible if both impurity and conduction electrons have orbital contributions to their moment. Recently, resonances of non-S-state magnetic impurities have been observed in several good metals [67]. Crystal-field splittings of the impurity state were resolved [20], and even hyperfine splittings [68]. The corresponding EPR linewidth is often less than 10^{-6} eV. Similarly narrow effective widths of magnetic impurity states can be derived from the Curie–Weiss law of very dilute impurities at very low temperatures (e.g., **Cu**Mn, Hirschkoff et al. [17]). This coherent precession of the impurity moment together with the conduction-electron polarization is perhaps the magnetic analog of the zero resistance of metals at $T = 0$.

One should of course not confuse the width of the virtual bound state with the effective width of the Zeeman levels. The former is made up of uncorrelated single-electron states, while the latter are sets of many-electron levels.

The majority of the first principles theories insists on the single-particle picture as the starting point,[†] but these treatments are apparently unable to obtain quantitatively the above-mentioned highly correlated low-temperature state, the only state which is really accessible to experiments. While the single-particle levels can only become "visible" in rare high-energy measurements like optical absorption (i.e., measurements which proceed at such a fast time scale that they *destroy* the correlations), the usual susceptibility and resistance measurements are low-energy measurements, and their observables are dominated by the correlations (if they exist). There is a perhaps crucial lack of contact between theory and experiment because phenomenologically the overwhelming majority of the magnetic and so-called nonmagnetic impurities fit a picture according to which the impurity is in a Hund's rule correlated state with the z component of the corresponding magnetic moment fluctuating at an intrinsic frequency somewhere below 2×20^{13} Hz ($\approx 1000°$K, above this temperature a possible Curie–Weiss behavior of dilute "nonmagnetic" impurities is normally not measurable). It is important to realize that 2×10^{13} Hz is still about two orders of magnitude below the energy range in which the demagnetization should occur in the single-particle models for $3d$ impurities (where the width of the virtual bound state is presumed to be of the order of a few electron volts).

The correlated low-temperature state is extraordinarily difficult to calculate starting from single-particle states, but such a calculation may actually be unnecessary since we know the approximate result, namely the ionic multiplet levels, very well from atomic spectroscopy. If one only accepts the idea that the strong Hund's rule correlations are not broken up by the conduction electrons (as the s–d model does, but we are concentrating here on the question of moment formation, not transport phenomena), the task of theory is reduced to finding the

---

[†] To call the single-particle free electron states of the conduction electrons a first principles description is somewhat inappropriate in itself since these states are the consequence of a highly correlated motion which cancels the Coulomb repulsion.

modifications of the Hund's rule state by the metal (quenching of the orbital moment, conduction-electron spin polarization, characteristic temperature, etc.) by perturbation theory.

How then should one visualize nonmagnetic impurities in such an ionic scheme? Phenomenologically the experimental evidence does not necessarily call for a high-energy demagnetization as proposed by the Friedel–Anderson model; a moderate broadening of the Hund's rule Zeeman levels to kT_f would be sufficient. Relatively small such effects could be obtained by residual spin–orbit coupling and crystal-field effects, as discussed in Section II, 1. A more drastic effect is mixing. It should be clear that in a strictly ionic model the impurity has a definite integral occupation number of the local shell, and that the only "fractional moment" effects can come from conduction-electron polarization, quenching of orbital moment and excitations within a multiplet. Mixing with conduction electrons, on the other hand, would be strictly inadmissible in zeroeth approximation. If *weak* mixing is allowed for, (that is, one electron at a time hopping from the shell to the conduction sea or back at a low frequency $\omega_f \ll E_{cn}/h$, where E_{cn} is the correlation energy of the shell with n electrons), the initial and final states of the system will in first approximation involve the conduction band and most likely no more than two well defined many-electron configurations of the shell, which differ in occupation number by only one electron, in the spirit of the minimum polarity model [54]. Then one simple way of arriving at a nonmagnetic impurity would be the uncertainty principle. The shell lives in any one of the two configurations with a given set of Zeeman levels for a limited time τ only, the *effective* width of the Zeeman levels becomes of order h/τ, and the susceptibility flattens out below $T_f \approx h/k\tau$. Apparently, then, in this picture *only* the nonmagnetic state will have broad Zeeman levels, and a nonmagnetic–magnetic transition should be accompanied by a drastic reduction in h/τ. This provides a very attractive explanation for the observed narrow Zeeman levels on magnetic impurities and for the frequent appearance of pseudomagnetic behavior on so-called nonmagnetic impurities at elevated temperatures, which may be viewed as a crossing of kT and h/τ. While the effective width h/τ has nothing to do with the width of the virtual bound state, it has some albeit distant relationship with the width due to mixing in the Anderson model, and may be identified with the spin fluctuation energy postulated *ad hoc* by Rivier and Zuckerman [13], for which it provides a simple and attractive physical reason. This model is strongly supported by the magnetic behavior of rare earth compounds with unstable valence (Section VI).

e. Impurities with one electron or hole in the local shell. Since the theoretical treatments of local moment formation usually ignore intraionic many-electron correlation effects or include them only in a highly formal way without going to the trouble of producing some (maybe modified) Hund's rules, one might try to test the existing nondegenerate orbital calculations experimentally by restricting oneself to impurities with only one electron (or hole) in the magnetic shell. Such shells, although still afflicted by orbital degeneracy, at least do not suffer from local many-electron correlations. In the absence of magnetic impurities with incomplete s or p shells, one is left with impurities like Sc, Y, La, and Lu at the beginning and Ni, Pd, and Pt at the end of the transition series, and with Ce and Yb among the rare earths. All other magnetic impurities apparently have more than one electron (or hole) in the shell, since in no such case is the effective moment consistent with $J = S = \frac{1}{2}$ (it is always larger) or with a d^1, d^9, f^1, or f^{13} configuration. Unfortunately, none of the transition metals mentioned has so far been found magnetic in high dilution (a very high T_f cannot be excluded in **BeNi**; see Klein and Heeger [69]) but it is highly interesting in this context that with respect to their magnetic behavior in the metallic element or in compounds or concentrated alloys, the d^1 and d^9 elements clearly belong to a common class which is qualitatively distinct from the rest of the transition metals. There is, for instance, the old differentiation between the local (Heisenberg) ferromagnet Fe and the itinerant ferromagnet Ni, which has d^9 character. The absence of superconductivity [70] and the tendency to form large polarization clouds around magnetic impurities in d^9 and d^1 metals [11, 12, 35] gave rise to the notion of paramagnons [71, 72] and one kind of localized spin fluctuation [73]. The missing intraionic many-electron effects seem to be at the root of the phenomenological distinction of d^1 and d^9 metals from the other d metals, and an orbitally nondegenerate model may have some justification here. However, so far d^1 and d^9 shells were found to be truly magnetic only at high concentrations; that is, they magnetize only cooperatively (NiPd, Ni$_3$Al, Sc$_3$In). Therefore as *impurity* they cannot provide an experimental testing ground for magnetic–nonmagnetic transitions. The only impurities in the one-electron (hole) class which show magnetic moments in dilute alloys are Ce and Yb. They, moreover, are sometimes observed to lose it when pressure is applied [74] or upon small changes of composition of an alloy host [75]. But when they are magnetic, they show very good Hund's rule moments (f^1 or f^{13}) including full orbital contributions, and also often exhibit crystal-field effects and resistance minima. The loss of moment seems to occur simply because of a change of valence of the impurity ion, that is, because of temporary or permanent switchover of an electron from a localized f to a conduction state, accompanied by a drastic change of the local potential. Thus even these simple one-electron shells, when they are magnetic, do not fit very well into the framework of theories which ignore orbital moment, Hund's rules, and local effects like crystal fields. Moreover, the change of valence in the magnetic–nonmagnetic transition is a much more drastic change of state than implied in the simple Friedel–Anderson demagnetization which does not take into account the possibility of a change (within the sum rule) of the relative amounts of screening in the various phase shifts.

In short, there seems to be not a single type of *magnetic* impurity to which application of the models of local moment formation in their present oversimplified (and theoretically manageable) form is justified without ignoring qualitatively important features. These theories can be tested on real systems only in the nonmagnetic limit.

IV. Solid Solutions of 3d Elements in Simple Hosts

In spite of the strictures discussed above the general rules governing the observation or nonobservation of moments in these systems are still those enunciated on the basis of the virtual bound-state approach by Blandin and Friedel [76] many years ago. As the valence electron concentration (and hence the Fermi energy and density of states at the Fermi surface) increases along the series Cu → Zn → Al the probability of finding a moment on a $3d$ solute decreases, although moments persist longest for the solutes with nearly half-filled d shells. For a given host metal with various solutes or a given solute in various hosts the tendency to be magnetic can be regarded as inversely proportional to some characteristic temperature T_f derived by naive fitting of the experimental data for susceptibility or resistivity to some theoretical expression. Perhaps the least objectionable of such fits is that of susceptibility data at $T \gg T_f$ to $\Delta\chi = C/(T + T_f)$ where $\Delta\chi$ is the susceptibility increment for very dilute alloys, but great significance should never be attached to absolute values. Table I gives approximate values for Cu, Ag, Au, Zn, Cd, and Al as hosts; some of these are drawn from resistivity data. Zinc-based alloys with $3d$-element additions have been the subject of extensive analysis by the Genoa group [77]. They conclude that characteristic temperatures of about 3 and 1°K are appropriate for Cr and Mn solutes, while a value about 10^2 times higher is required for Fe, and not less than 10^3 times larger for V and Co. Recent experimental work in such systems has largely been, on the one side, a matter of

TABLE I

ORDER OF MAGNITUDE "KONDO TEMPERATURES" FOR 3d ELEMENTS IN SIMPLE METAL HOSTS[a]

Solvent \ Solute	V	Cr	Mn	Fe	Co	Ni
Cu	—	1.0	0.01	25	2000	>5000
Ag	—	~0.02	0.040	3(?)	—	—
Au	300	~0.01	<0.01	0.3	200	—
Zn	—	3	1.0	90	—	—
Cd	—	—	~0.02	—	—	—
Al	—	1200	530	>5000	—	—

[a] Values in degrees Kelvin [139].

seeking "Kondo temperatures" at rather low temperatures in materials previously regarded as good moment systems (**Cu**Mn, Hirschkoff et al. [17]; **Au**Fe, Loram et al. [78]; **Zn**Mn, Ford et al. [77]), or, on the other, a matter of clarifying the behavior of those systems which, while not having obvious low-temperature moments, show feeble memories of moments in their properties. In this latter group come **Al**–Mn, where the elegant resistivity work of Caplin and Rizzuto [18] first raised the concept of the local spin fluctuation in such a system, and **Au**–V, where the earlier work of Kume [79] has been followed by extensive studies aimed at testing simple power laws for various physical properties [e.g., $\chi = \chi_0(1 - (T/T_f)^2)$]. In both these systems significant contributions have been made by NMR and other microscopic measurements, the general indications of which have been shown by Narath [31] to give no support to a compensation cloud picture, favoring rather a spin fluctuation view.

An interesting application of NMR techniques is the correlation by the Budapest group of quadrupole effect "wipe-out" numbers with resistivity data for the same system. Thus Tompa [80] finds good agreement for **Cu**–Mn and **Cu**–Fe alloys between experimental "wipe-out" numbers and those calculated from a resonant scattering theory using phase shifts implied by resistivity data. Furthermore Gruner and Janossy [81] show that the temperature dependence of such effects in **Al**–3d alloys implies a resonant scattering as a function of 3d-element atomic number that is single-peaked (like the residual resistivity) at low temperatures but double-peaked (as if the virtual bound state on Mn in Al were spin-split) at high temperatures.

A very interesting correlation has been obtained by a number of workers between the temperature dependences of the susceptibility and resistivity increments of the impurity, following the method of Daniel [82]. The apparent temperature-dependent effective spin $S^2(T) \sim \mu_{\text{eff}}^2(T) = 3kT\Delta\chi(T)$ causes a temperature-dependent difference in the phase shift for spin-up and spin-down conduction electrons which appears in the resistivity via $\Delta\rho = A - B\cos[2S(T)\pi/5]$. Souletie [40] has treated the noble metal hosts. He shows that a *resistance minimum* follows for most values of the $l = 2$ phase shift, and that a good correlation can be obtained with experimental data for **Au**–V, **Cu**–Cr, **Cu**–Fe, **Au**–Fe, **Cu**–Mn, and **Au**–Co.

It should be noted that, as first pointed out by Loram et al. [83], when the phase shift required by the sum rule is close to zero or to π, the resistance should *increase* on passing (with temperature) from the nonmagnetic to the magnetic regime. In noble-metals–$3d$ systems this transition takes place over too wide a range of temperature to be detectable experimentally, although Souletie points out signs of this effect for very dilute **Cu**–Co alloys. (See also Section V in this context.)

Very recent work of particular interest is the demonstration by Van Dam and Gubbens [50] that a T^2 dependence of the form above is shown by the susceptibility of dilute ($<0.5\%$) **Au**–V alloys. The simple power law approach to the very low temperature, very low concentration regime has been emphasized strongly by the Leiden group (especially Star [19], see also Van den Berg [84]), and has been shown to be helpful even for **Cu**–Fe where a T^2 dependence has been found for 55 ppm of solute although already modified at concentrations ten times larger [41] (see Fig. 2).

Alloys of this type in the "feeble moment" category have naturally been the principal actors in demonstrations of the enormous importance of solute–solute interactions. Tournier and the Grenoble group have produced overwhelming evidence of the role of such interactions (and hence of the local environment) in **Cu**–Co, **Au**–Co, and **Cu**–Fe systems [45]. While the assignment of different "Kondo temperatures" to different types of neighborhoods should probably not be taken too literally, it is clear that in more concentrated alloys the range of temperatures over which a loss of moment occurs is very wide.

One must bear in mind the possibility of interactions of different types. We have already referred to the stabilization of moments by RKKY interactions between solute atoms, and Van Dam and Gubbens [50] have recently shown that this effect takes place in **Au**–V alloys as the concentration increases from 0.2 to 2.0%. At higher concentrations, however, they confirm the earlier conclusions of Creveling and Luo [49] and Narath and Gossard [43] that near-neighbor V atoms effectively demagnetize one another. Presumably this type of interaction is associated with the greater radial extent of $3d$ wave functions on V atoms and the onset of d–d overlap effects which so efficiently demagnetize V atoms in pure V.

V. Transition Metals Containing 3d Solutes

The choice of such complicated systems for experimental investigation must seem to some like a simple act of revenge for the experimentalist's inability to understand the higher mysteries of the theoretical developments. In fact, they have significance for three important reasons.

1. When resonances in the band structure are expected to have widths close to that of the host band the specifically exchange character of impurity site effects should manifest itself more clearly.

2. If the dilute alloy problem is, as often suggested, to yield an approach to bulk magnetism in transition metals, we cannot tie ourselves always to simple band structures.

3. Such systems played a vital historical role. It was Matthias's concern with the very different effects of iron on the superconductivity of Nb and Mo [85] that drew Anderson's attention to the dilute alloy problem [3], while the transport properties of Pd–Ni alloys [86] led theoreticians to a consideration of the local spin fluctuation as an important scattering mechanism.

On the basis of a naive application of the Friedel virtual bound-state viewpoint one should not of course expect to find moments on $3d$ impurities in $4d$ metals, for such states would have widths equal to the host bandwidth and little hope of magnetization. From this viewpoint the loss of moment on an Fe atom in vanadium is no more surprising than the loss of the atomic moment on V in vanadium (V_{kd} is of the same order as the d–d overlap integrals of a tight-binding calculation for the metal) and that in turn is little different from the loss of the atomic moment of a lithium atom in the Li_2 molecule. Evidence that in **V**–Fe and **Nb**–Fe the impurity site correlations of a Hund's rule type have become negligible is provided not only by the complete absence of anomalies in susceptibility, Mössbauer data, and resistivity but also by the failure of the superconducting transition temperature to fall at a rate greater than seems completely understandable in terms of band structure effects like those operating in **Nb**–Mo. On similar grounds

[87] we may judge the body-centered cubic **Ti**–Fe and **Ti**–Mn alloys to be truly nonmagnetic. The local moment character of hcp **Ti**–Mn is thus the more surprising [88], and is the first indication that local state symmetries may be important.

The most striking sign of the failure of a naive approach to d–d alloys is provided by the **Mo**–Fe alloys. In resistance [89], susceptibility [12], interaction effects, and so on, these are practically indistinguishable from **Au**–Fe alloys, and Mössbauer measurements [36] give a very similar $T_f \sim 0.25°\text{K}$. A simple argument in terms of a low density of states at E_F for Mo cannot hold; although in **Al**–Fe the value of $n(E_F)$ is smaller and the band certainly has less d-like character, no sign of a good local moment is visible on the iron. It seems likely that consideration will have to be given to the detailed symmetry of the Fe d-electron states in comparison with those of the dominant sub-band or bands at the Fermi surface in molybdenum.

The possibility of fairly narrow resonant levels at the Fermi level in Ni, Pd, and Pt has been the basis of discussions of solid solutions of $3d$ elements in them by Friedel [2] and Moriya [90]. In the spirit of the Friedel–Anderson approach we should then expect fairly good moments on Fe and Mn solutions in Pd as in Au, and this agrees well with experiment.

Although both **Pd**–Ni and **Pd**–Fe are good long-range ferromagnets at higher concentrations, a good local moment does not seem to exist on Ni in the extreme dilute limit, analogous again with Au–Ni. The ferromagnetism of **Pd**–Fe is a long-range coupling of overlapping polarization clouds centered on Fe atoms and involving many Pd atoms, but at very low iron concentrations there is evidence [91] for the usual magnetic glass phase of good moment systems like **Cu**–Mn and **Au**–Fe. **Pd**–Mn for concentrations up to about 3% Mn resembles **Pd**–Fe (but with Curie temperatures almost a factor of 10 smaller); at concentrations above 5%, however, the ferromagnetism disappears and an enfeebled version of the **Cu**–Mn magnetic glass appears. This striking difference between Fe and Mn must be due to a difference in sign of the nearest neighbor Fe–Fe and Mn–Mn interactions which becomes important as the concentration increases.

The analogies (**Pd**–X with **Au**–X) seem to hold for Co as solute as well, since **Pd**–Co resistivity studies [92] suggest strongly that in the dilute limit local spin fluctuations of the **Pd**–Ni type provide a better description than a "good" local moment, although these are easily stabilized as their polarization clouds overlap to give ferromagnetism. (This type of behavior seems to be shown by Fe in Pt [83] as well as by Co and Ni, so that as far as solute moment character is concerned Pd and Au are more similar than Pt and Au. It is clear therefore that the presence of a significantly exchange-enhanced d band in the solvent plays no part in governing the "quality" of the local moments, however great its effects on the interactions between them.)

Just as solutions in Ni change their character on going back to Cr and V (the famous deviations from the Slater–Pauling curve) so also do those of Cr and V in Pd. Good local moments are no longer found (in fact, enhancement effects in the host are strongly reduced) and the resistivity minima can be understood in terms of nonmagnetic virtual bound states narrowed by local spin fluctuations of much weaker character, and resembling the resistance minimum of **Al**–Mn. (It is worth noting that resistivities decreasing with temperature in Pd alloys can be understood in simple one-electron terms applied many years ago to $s \rightarrow d$ scattering in Pd–Ag, and that recently clear-cut evidence of minima in Pd alloys not involving local moments has been provided for **Pd**–Rh alloys by Rowlands *et al.* [93].)

For solutes at the end of the $3d$ transition series, however, it seems more natural to emphasize local exchange enhancement effects and to play down resonances in the band structure for isoelectronic solutes Thus **Pd**–Ni and **Pt**–Ni seem obvious candidates for the type of treatment originated by Lederer and Mills [14] where the solute atom is seen as the site of local spin fluctuations (LSF), the buildup of which and eventual stabilization provide a natural description of the transition to itinerant ferromagnetism. The correlations among the T^2 contribution to the electrical resistivity, the susceptibility, and the specific heat in the **Pd**–Ni and **Pt**–Ni systems have been examined in detail by Schindler [94]. (It is interesting to note that the critical composition for the transition to ferromagnetism in **Pd**–Ni is much more readily located in the electrical resistivity [95] than by susceptibility measurements.)

The most detailed consideration of the LSF with an attempt to examine their consequences at finite temperature in terms of the spectral density of their excitation spectrum is that of Kaiser and Doniach [15] who calculate a universal curve (scaled by a characteristic temperature) for the resistivity which is proportional to T^2 at low temperatures but with an extensive linear regime at higher temperatures. These results are invoked to explain the rather surprising behavior of **Rh**–Fe [96] and **Ir**–Fe [97] resistivities, involving rather low characteristic LSF temperatures. The approach yields a rather simple relation between the mean square effective moment of susceptibility data $\Delta\chi T$ and the resistivity increment, a relationship that had been noted experimentally and justified heuristically by Knapp [98]. At higher temperatures the situation is less satisfactory, for one might expect (at $T \gg T_{\text{LSF}}$) that the effective moment would become constant with $\Delta\rho$ yielding the temperature-independent scattering characteristic of a fully developed magnetic moment.

An approach, at first sight diametrically opposed to this, has been suggested (see also Section IV) by Loram et al. [83], who in the spirit of Daniel [82] (and following Blandin and Friedel) see the scattering in the nonmagnetic (low-temperature) and magnetic (high-temperature) limits as expressible simply in terms of the phase shifts required for up- and down-spin electrons if the Friedel sum rule is satisfied. Unlike the situations considered by Souletie [40] they find Pt-based alloys, where the change from nonmagnetic to magnetic regime takes place at fairly low temperatures but the phase shifts at low temperatures are sufficiently small (or large), to yield a positive $d\Delta\rho/dT$ as observed for **Pt**–Fe and **Pt**–Co. A similar approach [99] would predict correctly the **Rh**–$3d$ resistivity behavior [100], where positive $d\Delta\rho/dT$ is found for Mn, Fe, and Co but a resistance minimum for **Rh**–Cr. It thus seems that once the phenomenological leap of expressing an effective moment in terms of spin-up and spin-down phase shifts has been taken one can understand the resistivity behavior in a wide variety of systems. The approach seems more satisfactory than the earlier one of Fischer [101] and Kondo [102] of introducing a large potential scattering into the Kondo model, since as pointed out by Nagasawa, $\Delta\rho$ for $T \to 0$ is rather small for **Rh**–Fe. On almost any grounds, however, a T^2 dependence of ρ is to be expected at low temperatures for **Rh**–Fe, and this seems at last to be appearing in experimental data of Laborde and Radhakrishna [103] and of Rusby and Coles [104], if the concentration is low. A complete theoretical treatment that subsumes the Kaiser and Doniach [15] aspects and the phase shift aspects is clearly desirable.

None of the theoretical work on this problem, however, seems to provide any guidance to the temperature range required to take one from the effectively nonmagnetic to effectively magnetic regime. Why T_f in **Mo**–Fe and **Pd**–Fe should be comparable with that of **Au**–Fe, and that of **Rh**–Fe with **Cu**–Fe is likely to be a very difficult question for some years. What we can hope to understand is the variation from solvent to solvent of the behavior of a solute *either* at $T \gg T_f$ *or* $T \ll T_f$.

VI. Local Moment Formation in Rare Earth Metals

When rare earth atoms are dissolved in metals, they normally carry the same ionic Hund's rule magnetic moments as in insulators. The interaction between the local shell and the conduction electrons is considerable

($J \sim 0.1$ eV), but it results merely in the RKKY polarization of the conduction electrons and does not lead to any appreciable modification of the local moment itself. Therefore rare earth shells have been traditionally regarded as uninteresting from the point of view of the local moment formation.

Recent developments, however, suggest that just the opposite is true. Rare earth shells apparently demagnetize in metals whenever the ions are about to change their valence. This type of demagnetization is clearly more in the spirit of the Anderson transition than of the singlet-ground-state loss of moment suggested by the *s–d* model, since the total spin S cannot be well defined or "hard" during a valence change. The effect provides a welcome alternative opportunity of studying the basic mechanism of local moment formation, an endeavor which proved so difficult in transition metals. There are some great advantages of demagnetizing 4*f* shells over similar transition metal systems.

1. The structure of the moment in the magnetic state is well defined.
2. The demagnetization occurs on a lower energy scale and can sometimes be studied in a continuous manner by varying the pressure, a very clean. parameter.
3. The local demagnetization can profitably be studied even in concentrated systems, since overlap between neighboring 4*f* states is negligible.

Some indications for "soft" moments on rare earth ions are fairly old. In cerium metal the large discontinuous volume contraction under pressure (17%, Lawson and Tang [105]) and a discontinuous drop of the susceptibility around 200°K [106] were recognized as manifestations of an electronic phase transition involving the 4*f* electron. More recently the discovery of an abnormally high depression of the superconducting transition temperature T_c in dilute **La**Ce alloys [107], a strong pressure dependence of this effect [108], and a resistance minimum [109] hinted at an incipient electronic transition for *dilute* Ce ions as well, which was eventually revealed by a pronounced maximum in the depression rate of T_c at higher pressure [74]. Meanwhile, great experimental and some theoretical interest has evolved for the effects of Ce ions in dilute and concentrated alloys and intermetallic compounds. Drastic anomalies in susceptibility, resistivity, lattice constant, and superconductivity are found to be the rule rather than the exception with such metals. The effects in superconducting Ce alloys are discussed by Maple [1] and some work on concentrated Ce alloys and intermetallic compounds was reviewed by Van Daal and Buschow [110].

While at present the bulk of our knowledge pertinent to soft rare earth moments comes from cerium systems, the effect appears to be much more general. Indications for soft moments exist now for Ce, Pr, Sm, Eu, Tm, and Yb ions, mostly in compounds, but also in dilute Ce and Yb alloys.

Soft rare earth moments occur near valence instabilities of the ions. In metals $4f$ ions may appear at unusual valences (other than 3) at the beginning (Ce^{4+}), in the middle (Sm^{2+}, Eu^{2+}), and at the end (Tm^{2+}, Yb^{2+}) of the lanthanide series. Apparently in some such cases, the state with valence 2 or 4 is energetically very close to the trivalent state and small changes of the free energy induced by pressure or by slight shifts of the composition of the system may trigger a transition. This transition has the very important feature that it seems never to be from one integral valence state to the other. At least in concentrated systems *a thermodynamically stable intermediate valence phase exists* over a finite and often considerable area of the phase diagram. Thus in general such systems have at least two, and sometimes more phase boundaries with one or more nonintegral valence states between two normal, integral valence states.

The best known intermediate valence phase is αCe, which is bordered at low pressure and high temperature by the trivalent magnetic γ and β phases [111] and at high pressure by the quadruvalent superconducting α' phase [112]. (Fig. 6) In other systems only a limited region of the intermediate phases has been studied. One observes a sharp transition under pressure from integral to nonintegral valence which is sometimes discontinuous, as in SmS [113, 114], and sometimes continuous, as in SmSe, SmTe, TmTe, EuTe, and YbTe [115, 116, 117]. The high-pressure end of these intermediate phases has so far remained inaccessible to magnetic or resistive investigation.[†] Occasionally one finds compounds in the intermediate valence phase at atmospheric pressure, for example, SmB_6 [118], $YbAl_2$, YbC_2 [119], $CePd_3$ [120], TmSe [121], $CeSn_3$ [122], Ce below 100°K, and so on.

The following general experimental features pertain to concentrated lanthanide systems with unstable valence.

1. Abnormal Susceptibility of Intermetallic Compounds

The susceptibility, instead of showing Curie–Weiss behavior with ordering at lower temperature, is abnormal. At high temperature the susceptibility may be temperature dependent but lies between that of

---

[†] The "high pressure" end of the intermediate phase of Ce has recently been studied by x-ray analysis and subceptibility in the system $Ce(Pd_xRh_{1-x})_3$ [117a].

the two energetically degenerate valence states; that is, it is lower than that of the lower valence state for elements on the left and higher for elements on the right of Gd. It gradually flattens out at some characteristic temperature T_f, without any signs of magnetic ordering as $T \to 0$. In analogy with those transition metals which have a high but temperature-independent low-temperature susceptibility without showing magnetic order (e.g., Pd, Sc, V) such rare earth metals may be considered as nonmagnetic. The nonmagnetic 4f character is clearest in the case of Ce and Yb compounds (Fig. 3a, b). The susceptibility is nearly temperature independent already at room temperature (sometimes with indications of Curie–Weiss behavior above). It exhibits often a very weak broad maximum near 100–200°K, and a moderate rise at lower temperature, but does *not* diverge as $T \to 0$. (Proper account must be taken of the susceptibility of other magnetic rare earth impurities [123]. Figure 4c shows the similar abnormal temperature dependence of the susceptibility of SmB_6 [118] and SmS [114]. Figure 4 shows the pressure dependence of the susceptibility of TmTe at helium temperatures, which is continuous and reversible in the intermediate phases B and C above 20 kbar and temperature independent below about 4°K [116].

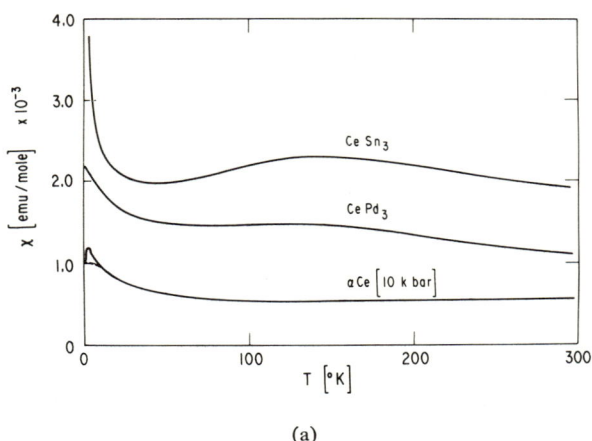

(a)

FIG. 3. The susceptibility of nonmagnetic rare earth systems with intermediate valence. Note the absence of Curie–Weiss behavior or magnetic ordering. In spite of a moderate rise the susceptibility definitely does not diverge at $T \to 0$ in (a) αCe [123], $CePd_3$ [120], (b) $YbAl_2$, YbC_2 [119], and (c) SmS [114].

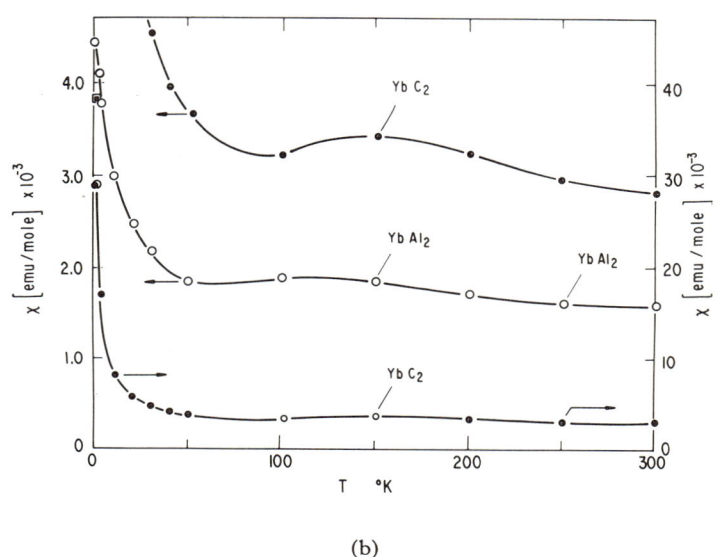

(b)

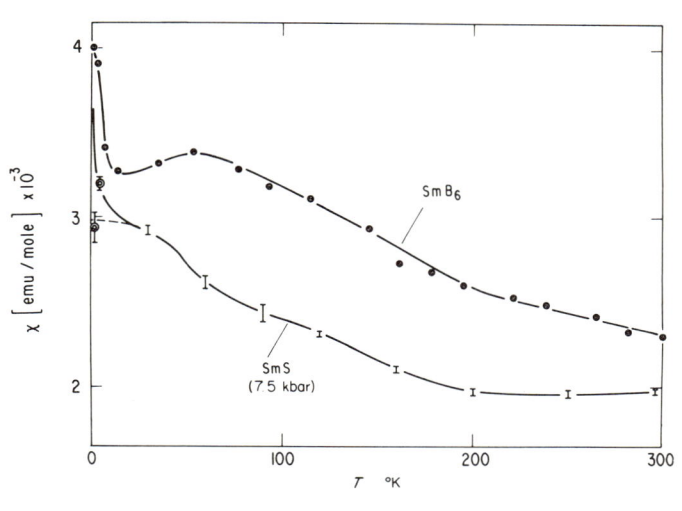

(c)

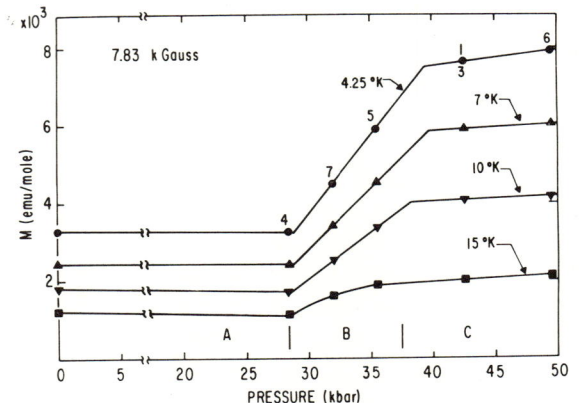

FIG. 4. The pressure dependence of the low-temperature susceptibility of TmTe. Pressure was applied in the indicated sequence 1-7. Note the continuous and reversible variation in the intermediate phases B and C [116].

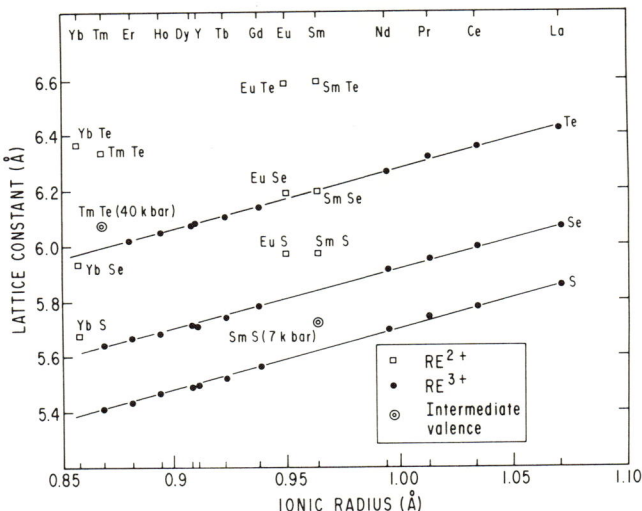

FIG. 5. Lattice constant as function of ionic radius in rare earth monochalcogenides [124]. Divalent and trivalent rare earth ions are clearly distinguishable. Intermediate lattice constants are found for SmS and TmTe under pressure. (Reprinted, by permission, from A. Iandelli, in "Rare Earth Research" (E. V. Kleber, ed.). MacMillan, New York, 1961. Copyright 1961, by the MacMillan Company.)

2. Identification of the Intermediate Valence Phase

Systems with such abnormal magnetic behavior invariably show abnormal lattice constants. The anomaly of the lattice constant is identified by comparison with related rare earth compounds. Figure 5 (mostly due to Iandelli and co-workers [124]) shows the lattice constants of the rare earth monochalcogenides. These plots are exemplary for the dependence of the lattice constants on the so-called ionic metallic radius. All compounds have the NaCl structure. Note that one single monotonous (but not linear) sequence of radii as function of atomic number produces the same linear dependence of the lattice constant for all three series of compounds. The majority of the ions (on the full lines) are trivalent according to the magnetic susceptibility. Compounds with

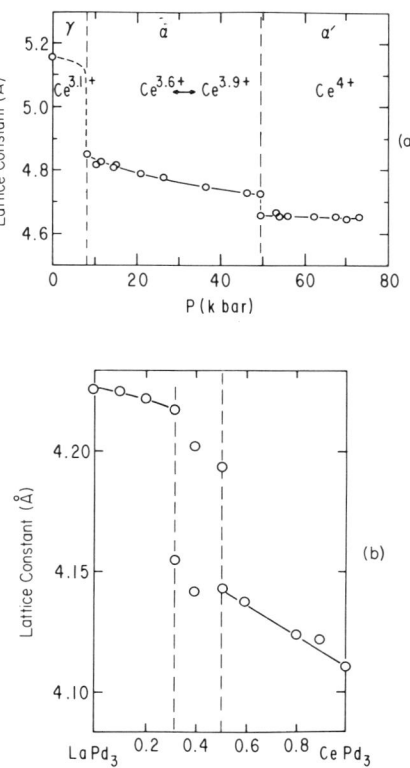

FIG. 6. Lattice constant anomalies as function of pressure {Ce metal, Franceschi and Olcese [125]} and concentration {$(La_{1-x}Ce_x)Pd_3$, Rao et al. [126]}. αCe and $CePd_3$ are in intermediate valence states. The transition near $x = 0.5$ in (b) corresponds to the (hysteretic) γ–α transition in (a).

divalent magnetic behavior (Eu, Sm, Yb, and sometimes Tm) have a much larger lattice constant. The distinction between divalency and trivalency is such that *intermediate* valences stand out clearly (TmTe, SmS). The fractional valence ε is measurable to $\pm 5\%$ from such plots. The same lattice constant anomaly can also show up directly as a function of pressure or concentration. Figure 6 shows the anomaly in Ce [125] as function of pressure and in $(La_xCe_{1-x})Pd_3$ [126] as function of concentration. Note that the fractional valence ε seems to change continuously with pressure in αCe. This continuity also appears in the susceptibility of αCe [123] and in the resistivity and susceptibility of SmSe, SmTe, and TmTe [115, 116, 127] (Fig. 4). In the case of SmB_6 the intermediate valence is corroborated by Mössbauer isomer shift [26] and soft x-ray absorption [128].

3. Resistivity in Rare Earth Systems with Unstable Valence

A soft moment on a rare earth is accompanied by resistance anomalies, which are often much more dramatic than those caused by dilute transition metal impurities in simple hosts but are on the whole not well understood.

In dilute systems on the magnetic side ($T \gg T_f$), one observes the ρ vs. ln T character familiar from $3d$ impurities (**La**Ce, Sugawara and Eguchi [109]; **AuAg** Yb, Boes *et al.* [129]; **Au**Yb, Murani [130]) and on the nonmagnetic side ($T \ll T_f$) the $\rho \sim [1 - (T/T_f)^2]$ behavior usually attributed to spin fluctuations (**Y**Ce, Maple and Wittig [131]).

The resistance anomalies are very much larger in concentrated magnetic systems and become more complicated, especially in intermetallic compounds where they show strong qualitative and quantitative variations from one system to the next. Some of these anomalies were interpreted with "Kondo side-band scattering" on several crystal-field split impurity levels [132].

In the intermediate valence phases, resistance anomalies are observed as well, but less frequently. Thus αCe has no resistance minimum [133] while nonmagnetic $CePd_3$ has a resistance minimum above room temperature [134]. Nonmagnetic $CeMe_2$ compounds (Me = Co, Rh, Fe, Ir, Ru) show deviations from normal metallic behavior, but no minima [110]. In nonmagnetic SmB_6 the resistivity rises by more than two orders of magnitude between room and helium temperature. This was interpreted as an indication of a very small semiconducting gap ($\Delta = 25°K$, Nickerson *et al.* [118]) but could also be due to anomalous magnetic scattering as in $CePd_3$. Behavior similar to that of SmB_6 has been observed in all known intermediate valence cases of lanthanide monochalcogenide compounds (SmS, SmSe, TmSe, TmTe, [135]). As a general rule one finds that giant resistance anomalies occur in the nonmagnetic intermediate valence state when the number of conduction electrons per rare earth ion is less than one while there is no minimum, but somewhat anomalous behavior if this number is larger than one.

4. Temporal versus Spatial Mixture in the Intermediate Phase

The phenomena associated with soft rare earth moments are complex and the experimental evidence is by no means complete. However, some simple common features are standing out clearly enough to justify a qualitative analysis of the underlying mechanism. These features are: (a) intermediate valence phases with continuous reversible variation of the fraction with pressure or concentration; (b) intermediate susceptibility at high temperature; (c) constant susceptibility as $T \to 0$; (d) absence of magnetic order. The phenomena in question occur when the rare earth ions change their valence from one ionic state with definite Hund's rule character to another. It therefore seems reasonable to assume that the basic ingredients of the intermediate valence state are these two ionic states and the conduction band *only*. The first question to ask is whether the intermediate state is spatially homogeneous, that is, a time average identical on each ion (temporal mixture), or whether it is a static distribution of ions with different integral valences at different sites (spatial mixture). The latter state goes against physical intuition, especially in a metal. It is a state of low spatial symmetry, and a continuous variation of the fraction as function of pressure in a spatial mixture is especially hard to visualize in very dilute alloys and in intermetallic compounds. Indeed, the maximum of the depression rate of superconductivity in **La**Ce [74] is inconsistent with a simple decrease of the concentration of moment-bearing Ce atoms,[†] and the absence of magnetic order as seen by susceptibility, or occasional Mössbauer ($CeSn_3$, Shenoy *et al.* [136]; SmB_6, Cohen *et al.* [26]), or neutron diffraction experiments (TmSe, Cox *et al.* [137]) are difficult to reconcile with any spatial mixture approach. (The magnetic order in a spatial mixture should be very similar to that in concentrated alloys.) No two-phase regions are discernible in x-ray patterns of carefully prepared compounds in the intermediate valence state.

[†] Incidentally, if the notion of a change of valence has any merit at all, one must abandon the idea that this maximum is due to a crossing of the superconducting transition temperature and the Kondo temperature since the spin of the ion is changing and is at this point as "soft" as it can possibly be. A slow spin fluctuation frequency going through resonance with the gap frequency seems to be a more natural explanation for this maximum. It is consistent with the idea of a changing valence.

5. Qualitative Consequences of Temporal Mixture

We shall accept the temporal mixture. It implies a periodic (or nearly periodic) switchover from one integral valence state with n local electrons on the $4f$ shell to the other with $n-1$ electrons on the shell and with the extra electron either in another ionic shell or in the conduction band. The basic period τ of the temporal mixture defines a characteristic fluctuation energy $E_f \approx h/\tau$ (which presumably equals the effective width $\varDelta$ of the local state) and a fluctuation temperature $T_f = h/k\tau$. From the uncertainty principle it can be inferred that if a measurement is fast compared to τ, it might "see" the individual valence states. Evidence that the individual ionic states are indeed well preserved comes from the intermediacy of the susceptibility at high temperature, when the thermal fluctuations are faster than the intrinsic fluctuations. This is especially clear when both ionic states carry moments $[\chi \approx \varepsilon \chi_1(T) + (1-\varepsilon)\chi_2(T)$ for $T > T_f$ in SmS, Sm_2Se_3, SmB_6, TmTe, TmSe]. Further evidence comes from a soft x-ray absorption measurement on SmB_6 by Vainshtein et al. [128] which is very fast and clearly distinguishes the two valence states (Sm^{2+} and Sm^{3+}) with the appropriate relative weight. The mixture is apparently independent of temperature [128]. This experiment is very instructive since it shows that during one period the system spends a long time in the two ionic configurations and a relatively short fraction of the period in the transition. Obviously the fractional valence ε may be changed continuously in the temporal mixture by relative adjustment of these times $[\tau_1 = \varepsilon\tau, \tau_2 = (1-\varepsilon)\tau]$, and there will be *two* independent fluctuation temperatures in the system, both of which may cause observable effects. One might ask why the intermediate phase is stable in concentrated systems. Since an appreciable fraction of an electron per lattice cell is transferred into the broad conduction band from the narrow $4f$ states, a great deal of kinetic energy must be supplied by whatever mechanism drives the transition, and the system must strike a balance between these competing processes. A similar argument can of course not be made for dilute nonmagnetic $4f$ shells. In such systems the existence of an intermediate phase with sharp boundaries is uncertain.

It is clear that the fluctuation impedes magnetic ordering. Thus if T_f is larger than the interaction temperature T_i of the integral valence states, the fluctuation dominates the motion and no ordering is possible. On the other hand, cooperative emission and absorption of the extra electron by the $4f$ shells may induce a phase-coherent motion of the entire electronic system in the intermediate valence state of intermetallic compounds, leading to new and very complex magnetic phenomena at low temperature. Such coherent motion is much less likely in concentrated alloys because of the reduced spatial symmetry which favors incoherent hopping. Even in intermetallic compounds coherent motion seems unlikely if the conduction bands are broad. Then the low-temperature susceptibility will be dominated by T_f via $\chi(T) \to \mu_{\text{eff}}^2/kT_f = $ const with $\mu_{\text{eff}}^2 = \mu_B^2(\varepsilon\mu_1^2 + (1 - \varepsilon)\mu_2^2)$ (μ_i designates the Hund's rule effective moments on the two ionic states). The susceptibility will therefore switch gradually from intermediate behavior to a constant value when the temperature is lowered through T_f. In αCe and YbAl$_2$, T_f is larger than 300°K. In SmB$_6$ and SmS (Fig. 4c) it is of the order of 50°K, and in TmTe and TmSe of the order of 10°K. Complex magnetic behavior is expected if $T_f \approx T_i$, and this is indeed observed in TmSe and TmTe.

In the absence of any theoretical guidance it is much more difficult to foresee even qualitatively the behavior of the resistivity in the intermediate valence cases. There are, however, two classes of systems for which $\rho(T)$ should differ drastically:

a. Systems which are semiconductors or insulators in the lower integral valence state, that is, systems where the conduction in the intermediate valence is dominated by the very electrons which are periodically emitted and reabsorbed by the unstable 4f shells (SmS, SmB$_6$, TmTe, TmSe).

b. Systems which are already metallic in the lower integral valence state, that is, systems where the conduction is dominated by s- or p-type electrons which only scatter off the fluctuating shells but are not involved in the mixing process (αCe, CeMe$_2$, Me = Ru, Rh, Ir, etc.; YbAl$_2$, YbC$_2$, etc.).

Phenomenologically the resistance anomalies are much more drastic in the first than in the second class.

6. Implications of Nonmagnetic Rare Earth Shells on the Theory of Local Moment Formation

What does the intermediate valence phase teach us with respect to moment formation in general? The emission of an electron from the local shell into the conduction band followed by reabsorption of another electron at a later time is synonymous with hybridization or mixing between 4f and conduction-electron states and is in the spirit of the model of Anderson [3] for local moment formation. The degree of mixing can to some extent be read off directly from the lattice constant. Then the abrupt onset of the intermediate phase as function of pressure, after a clear integral valence state with magnetic ionic Hund's rule behavior, indicates that *mixing does not occur in the integral valence state*. Apparently, in the magnetic state one has a sharp distinction between local- and conduction-electron states, and only direct potential and exchange scattering without appreciable mixing, compared to the nonmagnetic state. The near absence of mixing in the magnetic state implies that *the effective width of the magnetic state is very much less than that of the nonmagnetic state*. This important fact has never been explicitly discussed in the context of the Friedel–Anderson model (see also Section III, 2).

Translating the temporal mixture of the intermediate valence state into the static (or very low frequency) average implies that two local $4f$ configurational levels of width approximately kT_f are intersected by the Fermi level such that the fractional occupancy of one is ε and of the other $1 - \varepsilon$.[†] Obviously in the very low frequency limit ($T \to 0$) this nonmagnetic state is practically indistinguishable from the nonmagnetic state visualized in the Friedel–Anderson model as far as susceptibility measurements are concerned. However, the degeneracy of the local levels derives from that of the two Hund's rule configurations, and is not $(2l + 1)$ as in the traditional independent electron models [6]. The underlying Hund's rule structure becomes apparent in the high-temperature susceptibility ($T > T_f$). Thus, at least nonmagnetic rare earths and possibly also transition metal impurities have well-preserved Hund's rule correlations *even in the nonmagnetic state*, that is, the resonant levels have a quite specific structure which is not expected in the Friedel–Anderson model. We remind the reader that similar ideas are very old. They were incorporated in the "forgotten model of ferromagnetism" [54] and have recently been revived in the minimum polarity model for CuNi alloys [55].

Finally, one might ask what happens if a system is driven from an integral through an intermediate to the next integral valence state. So far this has only been achieved in cerium, where unfortunately the second integral valence state has an empty $4f$ shell. A glance at Fig. 5 shows, however, that such an event is to be expected for all divalent or intermediate valent systems at sufficiently high pressure. It is difficult to doubt that a compound with the trivalent lattice constant (corrected, of course, for compressibility) will show the usual trivalent magnetic behavior. A sequence of magnetic phases interrupted by nonmagnetic ones has never been predicted on the basis of the Friedel–Anderson

---

† A very large density of states at the Fermi level (62 states per eV) was recently found by specific haet measurement in the intermediate valence phase of SmS [138].

model, where the possibility of various ionic valences is not explicitly considered.

7. The Hirst Model

A model for local moments in metals which produces in a natural fashion most of the features discussed above was recently proposed by Hirst [7]. According to this model the ionic many-electron states are zero-order ingredients, which can normally only be perturbed and can only rarely be wiped out by interactions with conduction electrons. The integral occupation of the magnetic shell is a central feature and is assumed to be a consequence of Hund's rule correlations which can only be broken up at large cost of energy. The relative positions of the ionic energy levels are given by the values of a parabola $E(n)$ (n is the occupation number of the shell) at the integral values n_i (Fig. 7). The parabola $E(n)$ follows from the sum of the nuclear potential energy V_0 and the electron–electron term F_0 on the shell:

$$E(n) = \tfrac{1}{2}[n(n-1)F_0] + V_0 n + \text{const}$$
$$= \tfrac{1}{2}[(n - n_{\min})^2 F_0] + \text{const}.$$

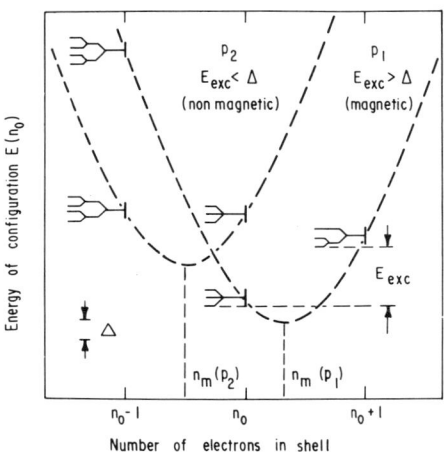

FIG. 7. Energy levels in the ionic model as function of the occupation number of the shell [7]. The occupation numbers are normally integral, in which case the impurity is magnetic. They are nonintegral only when the excitation energy E_{exc} between two neighboring configurational levels is of the order of or smaller than $\varDelta = V_{kl}^2 \rho(E)$, where V_{kl} is an effective local–conduction-electron interaction. Only then is mixing appreciable and the impurity becomes nonmagnetic.

The value $n_{\min}$ is a continuous variable and depends on the local environment and the band structure (pressure). The conduction electrons interact with the local electrons through a matrix element V_{kl}, resulting *potentially* in an effective level width $\Delta = V_{kl}^2 \rho(E_l)$ due to mixing. However, mixing, that is, electron transfer from the local shell to the conduction band and back, can only occur if V_{kl} excites the ion from the ground state $E(n_i)$ to the next higher state, either $E(n_i + 1)$ or $E(n_i - 1)$. Clearly there exists an effective gap for mixing:

$$E_{\text{exc}} = E(n_i \pm 1) - E(n_i),$$

which depends of course on $n_{\min}$. It is near zero whenever $n_{\min}$ is a half-integer. In the effective width due to mixing E_{exc}^{-1} then replaces the usually assumed $\rho(E_l)$, that is, dramatic mixing effects can only occur if $\Delta \gtrsim E_{\text{exc}}$ (Fig. 7). The dependence of E_{exc} on $n_{\min}$ is sketched in Fig. 8. Also shown are order-of-magnitude values of Δ for rare earth and transition metals. Presumably whenever $\Delta \gtrsim E_{\text{exc}}$ the local shell becomes nonmagnetic. The criterion $\Delta \gtrsim E_{\text{exc}}$ therefore replaces the Stoner-type criterion for magnetism used by Friedel and the Anderson criterion involving Δ, U_{eff}, and $E_l - E_F$. The nonmagnetic windows are very much narrower for rare earths than for transition metals.

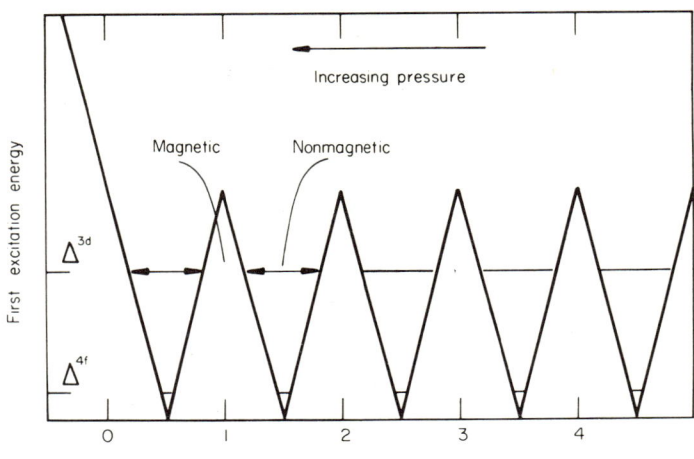

FIG. 8. Schematic dependence of the excitation energy E_{exc} on the occupation number $n_{\min}$ for minimum configurational energy [7]. Magnetic impurity states alternate with nonmagnetic ones. The width of the nonmagnetic regions is determined by Δ, which is larger for $3d$ than for $4f$ impurities.

The exact structure of the resonant nonmagnetic state and also the precise mechanism for the demagnetization remain open in the Hirst theory. However, the success of the Hirst model at predicting qualitative features borne out experimentally by rare earth shells with soft moments is impressive:

1. The theory predicts correctly that only two well-defined many-electron configurations are resonating in the nonmagnetic case. The experimental evidence, moreover, seems to point to a nonmagnetic state which is simpler than originally feared [7]. The two resonant states actually retain their individual Hund's rule structure fairly well. The nonmagnetic property could therefore simply be a consequence of the uncertainty principle. Since the shell lives in a well-defined Hund's rule state for a limited time only and then makes an (apparently fast) transition into the other state with different total spin, and so on, a divergence of the susceptibility at low temperature is impossible—if this motion is sufficiently incoherent, that is, if the emitted electron is not replaced by one with exactly the same momentum and spin direction at exactly the right moment. Such a coherent mixing motion seems unlikely.

2. The Hirst model explains naturally the different effective widths of the magnetic and the nonmagnetic local state.

3. The Hirst model predicts for any ion several magnetic phases separated by nonmagnetic phases and gives a good insight for why pressure can induce magnetic-nonmagnetic transitions, and why the band structure is so much less important than traditionally believed.

It appears that this model should be explored very carefully in the future.

ACKNOWLEDGMENTS

One of us (DKW) was supported by the U.S. Air Force Office of Scientific Research under Grant No. AF-AFOSR-71-2073 during preparation of this article. He also gratefully acknowledges the hospitality of the Metal Physics group at Imperial College, London. Many of the ideas concerning nonmagnetic rare earths with intermediate valence and the effective width of the magnetic state were worked out in discussions with M. B. Maple. We have profited from discussions with many other colleagues, especially W. E. Gardner, J. W. Loram, A. Murani, and C. Rizzuto.

References

1. M. B. Maple, this Vol.
2. J. Friedel, *Nuovo Cimento Suppl.* **7**, 287 (1958).
3. P. W. Anderson, *Phys. Rev.* **124**, 41 (1961).

4. P. A. Wolff, *Phys. Rev.* **124**, 1030 (1961).
5. H. Suhl, *Phys. Rev. Lett.* **19**, 442 (1967).
6. B. Coqblin and B. Blandin, *Advan. Phys.* **17**, 281 (1968).
7. L. L. Hirst, *Phys. Kondens. Mat.* **11**, 255 (1970).
8. M. D. Daybell and W. A. Steyert, *Rev. Mod. Phys.* **40**, 380 (1968).
9. A. J. Heeger, *Solid State Phys.* **23**, 283 (1969).
10. J. E. Van Dam and G. J. Van den Berg, *Phys. Status Solidi A* **3**, 11 (1970).
11. D. Gerstenberg, *Ann. Phys. (Leipzig)* **7**, 236 (1958).
12. A. M. Clogston, B. T. Matthias, M. Peter, H. J. Williams, E. Corenzwit and R. C Sherwood, *Phys. Rev.* **125**, 541 (1962).
12a. H. Maletta and R. L. Mössbauer, *Solid State Commun.* **8**, 143 (1970).
13. N. Rivier and M. J. Zuckermann, *Phys. Rev. Lett.* **21**, 904 (1968).
14. P. Lederer and D. Mills, *Phys. Rev. Lett.* **20**, 1036 (1968).
15. A. B. Kaiser and S. Doniach, *Int. J. Magn.* **1**, 11 (1970).
16. M. B. Maple, J. G. Huber, B. R. Coles and A. C. Lawson, *J. Low Temp. Phys.* **3** 137 (1970).
17. E. C. Hirschkoff, O. G. Symko, and J. C. Wheatley, *J. Low Temp. Phys.* **5**, 155 (1971).
18. A. D. Caplin and C. Rizzuto, *Phys. Rev. Lett.* **21**, 746 (1968).
19. W. M. Star, Ph.D. Thesis, Leiden Univ., Leiden, Netherlands unpublished, 1971.
19a. W. M. Star, F. B. Basters, G. M. Nap, E. De Vroede, and C. Van Baarle, *Physica* **58**, 585 (1972).
19b. W. M. Star, *Physica* **58**, 623 (1972).
19c. W. M. Star, E. De Vroede, and C. Van Baarle, *Physica* **59**, 128 (1972).
20. L. L. Hirst, G. Williams, D. Griffiths, and B. R. Coles, *J. Appl. Phys.* **39**, 844 (1968).
21. G. Williams and L. L. Hirst, *Phys. Rev.* **185**, 407 (1969).
22. L. J. Tao, D. Davidov, R. Orbach, and E. P. Chock, *Phys. Rev. B* **4**, 5 (1971).
23. N. Kroo, *Proc. Int. Conf. Low Temp. Phys.*, *12th, Kyoto, Sept., 1970* (E. Kanda, ed.), p. 823. Academic Press of Japan, Tokyo, 1971.
23a. G. Bauer and E. Seitz, *Solid State Commun.* **11**, 179 (1972).
23b. N. Kroo and Z. Szentirmay, *Phys. Lett.* **40a**, 173 (1972).
24. G. G. Low, *Advan. Phys.* **18**, 371 (1969).
25. C. Stassis and C. G. Shull, *Phys. Rev. B* **5**, 1040 (1972).
26. R. L. Cohen, M. Eibschütz, and K. W. West, *Phys. Rev. Lett.* **24**, 393 (1970).
27. L. L. Hirst and E. R. Seidel, *J. Phys. Chem. Solids* **31**, 857 (1970).
28. I. A. Campbell, J. P. Compton, I. R. Williams, and G. V. H. Wilson, *Phys. Rev. Lett.* **19**, 1319 (1967).
29. G. N. Rao, E. Matthias, and D. A. Shirley, *Phys. Rev.* **184**, 325 (1969).
30. J. Flouquet, *Phys. Rev. Lett.* **27**, 515 (1971).
31. A. Narath, *Proc. Conf. Low Temp. Phys. 12th Kyoto, Sept., 1970* (E. Kanda, ed.), p. 675. Academic Press of Japan, Tokyo, 1971.
32. H. Alloul and H. Launois, *J. Appl. Phys.* **41**, 923 (1970).
33. J. Kondo, *Progr. Theor. Phys.* **32**, 37 (1964).
34. D. Korn, *Z. Phys.* **187**, 463 (1965).
35. D. Wohlleben, *Phys. Rev. Lett.* **21**, 1343 (1968).
36. T. A. Kitchens, W. A. Steyert, and R. D. Taylor *Phys. Rev. A* **138**, 467 (1965).
37. F. J. DuChatenier and J. DeNobel, *Physica (Utrecht)* **32**, 1097 (1966).
38. D. Wohlleben, unpublished work 1969.
39. W. B. Pearson and A. Kjekshus, *Can. J. Phys.* **40**, 98 (1961).
40. J. Souletie, *J. Low Temp. Phys.* **7**, 141 (1972).
41. W. M. Star and G. J. Nieuwenhuys, *Phys. Lett. A* **30**, 22 (1969).

42. F. T. Hedgcock and P. L. Li *Phys. Rev. B* **1**, 1342 (1970).
43. A. Narath and A. C. Gossard, *Phys. Rev.* **183**, 391 (1969).
44. B. Dreyfus, J. Souletie, J. L. Tholence, and R. Tournier *J. Appl. Phys.* **39**, 846 (1968).
45. J. Souletie and R. Tournier *J. Low Temp. Phys.* **1**, 95 (1969).
46. J. Souletie and R. Tournier, *J. Phys. Suppl.* **32**, 172 (1971).
47. J. L. Tholence and R. Tournier, *Phys. Rev. Lett.* **25**, 867 (1970).
48. B. Lecoanet and R. Tournier, *Proc. Int. Conf. Low Temp. Phys. 12th, Kyoto, Sept. 1970* (E. Kanda, ed.), p. 735, Academic Press of Japan, Tokyo, 1971.
49. L. Creveling and H. L. Luo, *Phys. Rev.* **176**, 614 (1968).
50. J. Van Dam and P. C. M. Gubbens, *Phys. Lett. A* **34**, 185 (1971).
51. V. Jaccarino and L. R. Walker, *Phys. Rev. Lett.* **15**, 258 (1965).
52. K. C. Brog and W. H. Jones, Jr., *J. Appl. Phys.* **41**, 1003 (1970).
53. K. Fischer, *Phys. Status Solidi B* **46**, 11 (1971).
54. J. H. Van Vleck, *Rev. Mod. Phys.* **25**, 220 (1953).
55. N. D. Lang and H. Ehrenreich, *Phys. Rev.* **168**, 605 (1968).
56. H. P. Myers, L. Walldén, and A. Karlsson, *Phil. Mag.* **18**, 725 (1968).
57. P. B. de Gennes, *C.R. Acad. Sci.* **247** (1958).
58. R. Brout and H. Suhl, *Phys. Rev. Lett.* **2**, 387 (1959).
59. J. R. Schrieffer and P. A. Wolff, *Phys. Rev.* **149**, 491 (1966).
60. J. R. Schrieffer, *J. Appl. Phys.* **38**, 1143 (1967).
61. G. G. Low and T. M. Holden, *Proc. Phys. Soc. London* **89**, 119 (1966).
62. B. B. Schwartz and R. B. Frankel in "Mössbauer Effect Methodology" (E. J. Gruverman, ed.), Vol VII, p. 21. Plenum, New York, 1971.
63. M. Levine and H. Suhl, *Phys. Rev.* **171**, 567 (1968).
64. W. E. Evenson, J. R. Schrieffer, and S. Q. Wang, *J. Appl. Phys.* **41**, 1199 (1970).
65. F. J. Dyson, *Phys. Rev.* **98**, 349 (1955).
66. H. Hasegawa, *Progr. Theor. Phys.* **21**, 483 (1959).
67. D. Griffiths and B. R. Coles, *Phys. Rev. Lett.* **16**, 1093 (1966).
68. R. Chui, R. Orbach, and B. L. Gehman, *J. Phys. Suppl.* **32**, C1-909 (1971).
69. A. P. Klein and A. J. Heeger, *Phys. Rev.* **144**, 458 (1966).
70. M. A. Jensen and H. Suhl, *in* "Magnetism" (G. T. Rado and H. Suhl, eds.), Vol. IIb, pp. 183–214. Academic Press, New York, 1966.
71. N. F. Berk and J. R. Schrieffer, *Phys. Rev. Lett.* **17**, 433 (1966).
72. S. Doniach and S. Engelsberg, *Phys. Rev. Lett.* **17**, 750 (1966).
73. D. L. Mills and P. Lederer, *Phys. Rev.* **160**, 590 (1967).
74. M. B. Maple, J. Wittig, and K. S. Kim, *Phys. Rev. Lett.* **23**, 1375 (1969).
75. V. Allali, P. Donzé and A. Treyvand, *Solid State Commun.* **7**, 1241 (1969).
76. B. Blandin and J. Friedel, *J. Phys. Radium* **20**, 160 (1959).
77. P. J. Ford, C. Rizuto, and E. Salamoni, to be published (1972).
78. J. W. Loram, A. D. C. Grassie, and G. A. Swallow, *Phys. Rev. B* **2**, 2761 (1970).
79. K. Kume, *J. Phys. Soc. Jap.* **23**, 1226 (1967).
80. K. Tompa, *Solid State Commun.* **10**, 1039 (1972).
81. G. Gruner and A. Janossy, to be published (1972).
82. E. Daniel, *J. Phys. Chem. Solids* **23**, 975 (1962).
83. J. W. Loram, R. J. White, and A. D. C. Grassie, *Phys. Rev.* **5B**, 3659 (1972).
84. G. J. Van den Berg, *Proc. Int. Conf. Low Temp. Phys., 12th Kyoto, Sept. 1970* (E. Kanda, ed.), p. 671. Academic Press of Japan, Tokyo, 1971.
85. B. T. Matthias, M. Peter, H. J. Williams, A. M. Clogston, E. Corenzwit, and R. C. Sherwood, *Phys. Rev. Lett.* **5**, 542 (1960).

86. A. I. Schindler and B. R. Coles, *J. Appl. Phys.* **39**, 956 (1968).
87. R. D. Blaugher, B. S. Chandrasekhar, J. K. Hulm, E. Corenzwit, and B. T. Matthias, *J. Phys. Chem. Solids* **21**, 252 (1961).
88. J. A. Cape, *Phys. Rev.* **132**, 1483 (1963).
89. B. R. Coles, *Phil. Mag.* **8**, 335 (1963).
90. T. Moriya *in* "Theory of Magnetism in Transition Metals" (W. Marshal, ed.), Enrico Fermi Course 37. Academic Press, New York, 1967.
91. G. Chouteau, R. Fourneaux, and R. Tournier, *Proc. Int. Conf. Low Temp. Phys. 12th*, *Kyoto, Sept. 1970* (E. Kanda, ed.), p. 769. Academic Press of Japan, Tokyo, 1971.
92. J. W. Loram, G. Williams, and G. A. Swallow, *Phys. Rev. B* **3**, 3060. (1971).
93. J. A. Rowlands, D. Greig, and P. Blood, *J. Phys. F* **1**, L29 (1971).
94. A. Schindler, Naval Res. Lab. Rep. No. 7057. Washington, C. C., 1970.
95. A. Tari and B. R. Coles, *J. Phys. F* **1**, L69 (1971).
96. B. R. Coles, *Phys. Lett.* **8**, 243 (1962).
97. M. P. Sarachik, *Phys. Rev.* **170**, 679 (1968).
98. G. S. Knapp *J. Appl. Phys.* **38**, 1267 (1967).
99. J. W. Loram, private communication, 1972.
100. B. R. Coles, S. Mozumder, and R. Rusby, *Proc. Int. Conf. Low Temp. Phys. 12th Kyoto, 1970* (E. Kanda, ed.), p. 737. Academic Press of Japan, Tokyo, 1971.
101. K. Fischer, *Phys. Rev.* **158**, 613 (1967).
102. J. Kondo, *Phys. Rev.* **169**, 437 (1968).
103. O. Laborde and P. Radhakrishna, *Phys. Lett. A* **37**, 209 (1971).
104. R. Rusby and B. R. Coles, *Proc. Int. Conf. Low Temp. Phys. 13th, Boulder, Colorado, Aug. 1972,* to be published.
105. A. W. Lawson and T. Y. Tang, *Phys. Rev.* **76**, 301 (1949).
106. J. M. Lock, *Proc. Phys. Soc. London Sec. B* **70**, 566 (1957).
107. B. T. Matthias, H. Suhl, and E. Corenzwit, *J. Phys. Chem. Solids* **13**, 156 (1960).
108. T. F. Smith, *Phys. Rev. Lett.* **17**, 386 (1966).
109. T. Sugawara and H. Eguchi, *J. Phys. Soc. Jap.* **21**, 725 (1966).
110. H. J. Van Daal and K. H. J. Buschow, *Phys. Status Solidi A* **3** 853 (1970).
111. K. A. Gschneidner, "Rare Earth Alloys." Van Nostrand–Reinhold, Princeton, New Jersey, 1961.
112. E. King, J. A. Lee, I. R. Harris, and T. F. Smith, *Phys. Rev. B* **1**, 1380 (1970).
113. A. Jayaraman, V. Narayanamurti, E. Bucher, and R. G. Maines, *Phys. Rev. Lett.* **25**, 368 (1970).
114. M. B. Maple and D. Wohlleben, *Phys. Rev. Lett.* **27**, 511 (1971).
115. A. Jayaraman, V. Narayanamurti, E. Bucher, and R. G. Maines, *Phys. Rev. Lett.* **25**, 1430 (1970).
116. D. Wohlleben, J. G. Huber, and M. B. Maple, *Proc. Conf. Magn. Magn. Mater, 17th, 1972* (H. C. Wolfe, ed.), p. 1478. Amer. Inst. Phys., New York 1970.
117. A. Chatterjee, A. K. Singh, and A. Jayaraman, *Phys. Rev.* **B6**, 2285 (1972).
117a. I. R. Harris, M. Norman, and W. E. Gardner, *J. Less Common Metals* **29**, 299 (1972).
118. J. C. Nickerson, R. M. White, K. N. Lee, R. Bachmann, T. H. Geballe, and G. W. Hull, *Phys. Rev. B* **3**, 2030 (1971).
119. B. Sales, D. Wohlleben, and M. B. Maple, to be published.
120. W. E. Gardner, J. Penfold, T. F. Smith, and I. R. Harris, *J. Phys. F* **2**, 133 (1972).
121. E. Bucher, A. C. Gossard, J. P. Maita, and A. S. Cooper, *Proc. Conf. Rare Eearth Res. 8th,* (T. A. Henry and R. E. Lindstrom, eds.), U. S. Government Printing Office, *Washington, D.C.*, 1, 74 (1970).

122. T. Tsuchida and W. E. Wallace, *J. Chem. Phys.* **43**, 3811 (1965).
123. M. R. MacPherson, G. E. Everett, D. Wohlleben, and M. B. Maple, *Phys. Rev. Lett.* **26**, 20 (1971).
124. A. Iandelli, *in* "Rare Earth Research" (E. V. Kleber, ed.). Macmillan, New York, 1961.
125. E. Franceschi and G. L. Olcesse, *Phys. Rev. Lett.* **22**, 1299 (1969).
126. V. U. S. Rao, R. D. Hutchens, and J. E. Greedan, *J. Phys. Chem. Solids* **32**, 2755 (1971).
127. A. Jayaraman, E. Bucher, and D. McWhan, *Proc. Conf. Rare Earth Res. 8th*, (T. A. Henry and R. E. Lindstrom, eds.), U. S. Government Printing Office, *Washington, D.C.*, **1**, 333 (1970).
128. E. E. Vainshtein, S. M. Blokhim, and Yu. B. Paderno, *Sov. Phys. Solid State* **6**, 2318 (1965).
129. J. Boes, A. J. Van Dam, and J. Bijvoet, *Phys. Lett. A* **28**, 101 (1968).
130. A. Murani, *Solid State Commun.* **12**, 295 (1973).
131. M. B. Maple and J. Wittig, *Solid State Commun.* **8**, 1611 (1971).
132. F. E. Maranzana, *Phys. Rev. Lett.* **25**, 239 (1970).
133. M. Nicolas-Francillon and D. Jerome, *Solid State Commun.* **12**, 523 (1973).
134. R. D. Hutchens, V. U. S. Rao, J. E. Greedan, W. E. Wallace, and R. S. Craig, *J. Appl. Phys.* **42**, 1293 (1971).
135. D. McWhan, private communication, 1971.
136. G. K. Shenoy, B. D. Dunlap, G. M. Kalvius, A. M. Toxen, and R. J. Gambino, *J. Appl. Phys.* **41**, 1317 (1970).
137. D. E. Cox, L. Passell, R. G. Birgeneau, and E. Bucher, private communication, 1971.
138. S. Bader, N. E. Philips, and D. Mc Whan, *Phys. Rev. B*, May 1973, in press.
139. C. Rizzuto, *Reps. Prog. Phys.*, to be published.

2. Formation of Local Magnetic Moments: Hartree–Fock Theory

A. Blandin

Université de Paris-Sud, Centre d'Orsay
Laboratoire de Physique des Solides
Orsay, France

I. Introduction	58
II. Friedel's Approach: Resonance Scattering and Virtual Bound States	59
1. Scattering Theory. Friedel's Sum Rule	59
2. Resonance Scattering and Virtual Bound States	60
3. Spin-Dependent Potential	61
4. Applications to Transitional Impurities	61
III. Anderson's Approach	64
1. The Anderson Hamiltonian	65
2. The Hartree–Fock Approximation	66
3. Comparison of the Approaches of Friedel and Anderson—Orbital Degeneracy	68
IV. Discussion of the Hartree–Fock Solutions	70
1. The Phase Diagram and the "Phase Transition"	70
2. Magnetic Moments and Susceptibilities at Finite T	71
V. The Antiferromagnetic Coupling between Localized and Conduction Electrons	73
1. The Effective Antiferromagnetic Exchange	73
2. Connections with the Kondo Problem. The Phase Diagram	75
3. The Effective Exchange for Transition and Rare Earth Impurities	76
4. Various Applications	78
VI. Transition Impurities in Transition Metals	78
1. The Scattering Problem in Narrow Bands	79
2. The Hartree–Fock Approximation in Narrow Bands (Nonmagnetic Matrices)	82
3. The Hartree–Fock Approximation for Impurities in Ferromagnetic Matrices	83
VII. Conclusions	86
References	87

I. Introduction

The problem of transitional impurities in metals has been a puzzle for a long time. Experimental results on the resistivities and on the occurrence of magnetic moments seemed to indicate very different behaviors. It took a long time to understand theoretically the key points of the problem. This paper discusses one of these points, the Hartree-Fock (HF) description of local states. In a way, it gives an historical review of this approach.

The problem of a magnetic impurity in metal (Mn in Cu, for example) was first treated with the exchange Hamiltonian [1]: the impurity is described as an ion with spin S interacting with the conduction electrons through a spin-dependent potential, the so-called s–d exchange potential:

$$V_{\text{ex}} = -J_{sd}\, \mathbf{s} \cdot \mathbf{S}. \tag{1.1}$$

This model provides a useful description of spin polarization around the impurity and indirect interactions between spins (the so-called RKKY interaction). From a theoretical point of view, it shows up some of the difficulties of the problem. Even treating J as a perturbation, V_{ex} is not a one-electron potential (the ground state of the system cannot be described by a single Slater determinant); further, even taking only the diagonal part of V_{ex}, $-J s_z S_z$, the states with different values of S_z are not orthogonal. Those difficulties had been recognized for a long time, but their importance needed a long time to emerge. As it starts from the existence of a spin (and a magnetic moment), the s–d exchange model fails evidently in explaining the occurrence of magnetic and nonmagnetic impurities.

We discuss first in section II a completely different approach which was initiated by Friedel [2], a description of the localized state (or virtual bound state) by resonance scattering theory, within an unrestricted HF theory. The application [2, 3a, b] to transitional impurities in normal metals provides an explanation to the residual resistivities and the magnetic properties of these alloys.

The third section deals with the model Hamiltonian of Anderson [4] which is the starting point for further developments. Within the HF approximation, we shall discuss the relations between both approaches.

The fourth section is devoted to a critical discussion of the HF results, particularly of the phase diagram. The inadequacy of this theory as regards the Kondo problem and the magnetic properties at finite temperature is discussed.

Section V deals with the effective antiferromagnetic coupling between localized and conduction electrons. In the strong magnetic limit, the

2. FORMATION OF LOCAL MAGNETIC MOMENTS

Schrieffer–Wolff transformation [5] shows the equivalence of the resonant scattering theory with the s–d exchange model, J being negative. This provides a useful connection with the Kondo problem [6] which is particularly interesting for antiferromagnetic coupling.

The sixth section deals with transition impurities in transitional metals. We shall discuss both cases of normal and ferromagnetic metals and apply the results to resistivities and magnetic moments of alloys.

In the last section, we show how the HF phase diagram is the starting point for further theories. We make some comments on the difficulties within the HF theory and how they can be resolved in more elaborate theories.

II. Friedel's Approach: Resonance Scattering and Virtual Bound States

This approach is based on the theory of scattering, which we shall just summarize in the case of a spin-independent potential. This situation describes the case of nonmagnetic impurities in normal metals.

1. Scattering Theory. Friedel's Sum Rule[†]

The metal being described by free electrons, an impurity in the Hartree or Hartree–Fock method, creates a self-consistent spherical potential $v(r)$. This potential gives rise to a change in the electron density $\Delta\rho(r)$. One can demonstrate that

$$\int_0^\infty \Delta\rho(r)\, d^3r = \frac{2}{\pi} \sum_l (2l+1)\, \delta_l(E_F), \qquad (2.1)$$

$\delta_l(E_F)$ being the phase shifts at the Fermi level.

If the impurity has an excess charge Z compared to the matrix, it follows the necessary condition, or Friedel's sum rule:

$$Z = \frac{2}{\pi} \sum_l (2l+1)\, \delta_l(E_F). \qquad (2.2)$$

The residual resistivity ρ_0 for a concentration x of impurities is

$$\rho_0 = \frac{4\pi x}{z k_F} \sum l \sin^2(\delta_l - \delta_{l-1}), \qquad (2.3)$$

where z is the valency of the matrix and k_F its Fermi wave vector.

[†] See [2].

The excess charge $\Delta\rho(r)$ exhibits, for large r, Friedel's oscillations:

$$\Delta\rho(r) \simeq -\frac{\alpha}{2\pi^2 r^3} \cos(2k_F r + \phi), \tag{2.4}$$

where

$$\alpha \sin \phi = \sum_l (-)^l (2l+1) \sin^2 \delta_l(E_F)$$

$$\alpha \cos \phi = \sum_l (-)^l (2l+1) \sin \delta_l(E_F) \cos \delta_l(E_F). \tag{2.5}$$

Applications of Eqs. (2.3) to (2.5) give good results for the description of normal impurities in normal metals.

2. Resonance Scattering and Virtual Bound States[†]

When the potential $v(r)$ is not strong enough, but nearly strong enough to create a bound state with quantum number l and $(2l+1)$ degeneracy, we are in a situation of resonance scattering. This situation is well known in atomic and nuclear physics. The "virtual" bound state resonates with the l spherical components of the plane waves and broadens in a region of energy Δ, the average energy being E_d. The phase shift $\delta_l(E)$ behaves as shown in Fig. 1. This corresponds to a density of localized states $\rho_l(E)$ per spin direction as given in Fig. 2:

$$\rho_l(E) = \frac{2l+1}{\pi} \frac{d\delta_l(E)}{dE}. \tag{2.6}$$

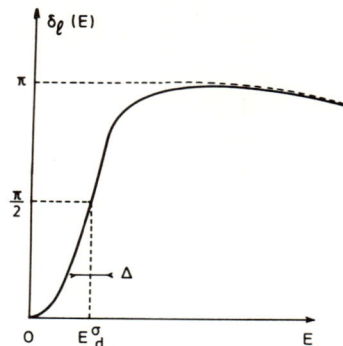

FIG. 1. Phase shift $\delta(E)$ for a real bound state (dashed lines) and for a virtual bound state (solid line).

[†] See [2].

2. FORMATION OF LOCAL MAGNETIC MOMENTS

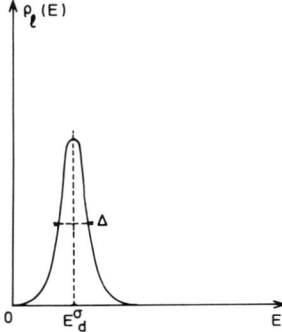

FIG. 2. Localized density of electrons for a virtual bound state.

3. Spin-Dependent Potential[†]

To describe the case of an impurity sustaining a magnetic moment one may use the unrestricted HF theory, characterized by a spin-dependent potential $v_\sigma(r)$. Focusing our attention on the l scattering, we shall neglect the other phase shifts. Writing the exchange energy $\Delta E_{ex} = -\frac{1}{2}p^2\varepsilon_{ex}$ where p is the number of unpaired d electrons and ε_{ex} is an average value of the exchange parameter, Blandin and Friedel [3a] derived a condition for the instability of the nonmagnetic state:

$$2\varepsilon_{ex}\rho_l(E_F) > 1. \tag{2.7}$$

This condition and its derivation highly resemble the Stoner condition for band ferromagnetism. A detailed discussion implies the knowledge of the shape of the density of states $\rho_l(E)$, an estimation of the width Δ and of the average position E_d^σ. The width can be estimated by a model potential calculation. It increases with increasing energy E_d^σ. Having the width and using a model density of states, one can calculate E_d^σ by application of the Friedel rule; qualitatively, the results do not depend on the exact shape of the density $\rho_l(E)$. The results can be summarized by the "phase diagram" of Fig. 3 for $l = 2$, where the curve separates two regions where the impurities are magnetic (M) or nonmagnetic (NM).

4. Applications to Transitional Impurities[‡]

The magnetic or nonmagnetic behavior of impurities is controlled by the values of the exchange parameter $\Delta\varepsilon_{ex}$ and of the width Δ.

[†] See [3a, b].
[‡] See [2, 3a, b].

The term $\Delta\varepsilon_{\mathrm{ex}}$ may be taken as constant for d electrons, whereas Δ increases with increasing Fermi energy of the matrix.

For the case of the transitional series in aluminum, experimental results suggest that $\Delta/\varepsilon_{\mathrm{ex}} > \alpha_c$, α_c being the critical value of Fig. 3.

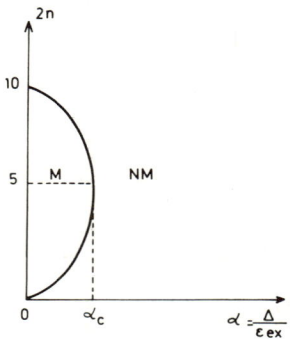

FIG. 3. Phase diagram for the HF discussion; $2n$ is the number of electrons.

Impurities are thus nonmagnetic. This results from the high value of the Fermi energy (13 eV) which gives a large width of the order of electron volts; the condition (2.7) is never fulfilled.

The contribution to the resistivity from $l = 2$ scattering is

$$\rho_0 = \frac{20\pi x}{zk_F} \sin^2 \delta_2(E_F).$$

The percentage of residual resistivities in the series should exhibit a maximum. This peak corresponds to the virtual bound state crossing down the Fermi level as the d level gets filled. The experimental results of Fig. 4 verify these conclusions.

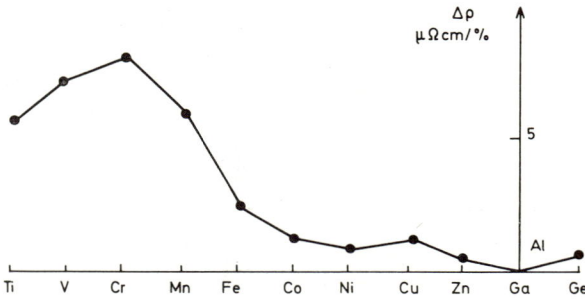

FIG. 4. Residual resistivities of transitional impurities in aluminum [2].

2. FORMATION OF LOCAL MAGNETIC MOMENTS

In noble metals, the Fermi energy is lower (5 to 7 eV), the width smaller, $\Delta/\varepsilon_{\text{ex}} < \alpha_c$, and the condition (2.7) for the occurrence of magnetism can be fulfilled in the middle of the series. Experimentally, Cu, Mn, and Fe are magnetic.[†] With magnetic decoupling, the $l = 2$ term in the residual resistivity is now

$$\rho_0 = \frac{10\pi x}{zk_F} [\sin^2 \delta_\uparrow(E_F) + \sin^2 \delta_\downarrow(E_F)].$$

Residual resistivities[†] in the series should exhibit a double peak with a minimum in the middle. The two peaks are related to the virtual states with up and down spins crossing the Fermi level as the d levels get filled. The experiments given on Fig. 5 exhibit this behavior.

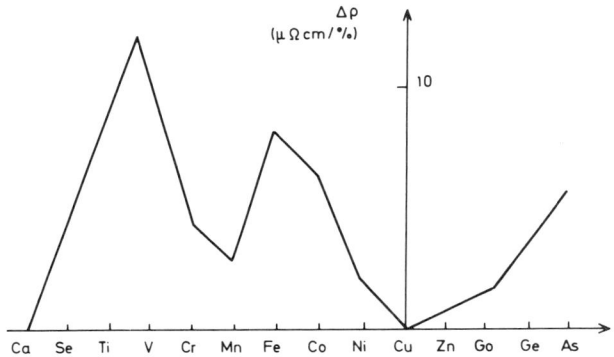

FIG. 5. Residual resistivities of transitional impurities in copper [2].

From this discussion Blandin and Friedel were able to give the data listed in Table I showing the occurrence of magnetic and nonmagnetic impurities. This table has been completed since then.

[†] Impurities are said to be magnetic if the susceptibility obeys a Curie–Weiss law in a large range of temperatures. This means that T is larger than the Kondo temperature T_K and larger than the Néel temperature T_N, below which magnetic glassy order sets in; T_N being proportional to the concentration, one needs experiments at low concentration and $T > T_K$. Along the same lines, residual resistivities are resistivities taken at low concentrations and low temperatures but above T_K and T_N.

TABLE I

Magnetic (+) and Nonmagnetic (—) Impurties in Various Matrices[a]

Matrix	Au	Cu	Mg	Zn	Al
E- in eV	5, 5	7	7, 5	9, 5	13
Cr	+	+			—
Mn	+	+	+	+	—
Fe	+	+		—	—
Co	+				—
Ni	—	—		—	—

[a] From Blandin and Friedel [3a].

III. Anderson's Approach

Anderson [4] started from the experimental results for the occurrence of localized magnetic moments (iron for example) in various transition metals and alloys [7]. As a function of the d-electron concentration, one finds sharply defined regions where localized moments are absent or present. An example is given in Fig. 6. Though introduced for the case of transitional matrices, which we shall discuss in Section V, the Anderson Hamiltonian has become a useful model Hamiltonian, widely used for theoretical investigations of the properties of local magnetic moments and applied mainly to the case of impurities in normal metals.

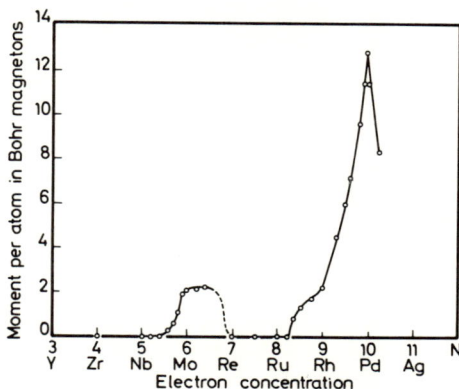

Fig. 6. Magnetic moments of Fe atoms dissolved in various second-row transition metals and alloys as a function of d-electron concentration [7].

1. The Anderson Hamiltonian

The picture suggested by Anderson is characterized by an inner local (d) state in which the Coulomb correlation integral U plays a major part. For the case of orbital nondegeneracy, U is given by

$$U = \int |\phi_{\text{loc}}(r_1)|^2 \frac{e^2}{r_{12}} |\phi_{\text{loc}}(r_2)|^2 \, d\tau_1 \, d\tau_2 . \tag{3.1}$$

Though U is a Coulomb integral, it is in a sense also an exchange integral—the exchange self-energy of the localized state. If E_0 is the energy of the "d" state for single occupancy, the Hamiltonian of the local state is

$$H_d = \sum_\sigma E_d n_{d\sigma} + U n_{d\uparrow} n_{d\downarrow} , \tag{3.2}$$

where $n_{d\sigma}$ is the occupation number operator: $n_{d\sigma} = c_{d\sigma}^+ c_{d\sigma}$. Thus the state with one electron (and spin up or down) has the energy E_d, the state with two electrons $2E_d + U$. The addition of a second electron introduces the Coulomb repulsion between the two electrons in the same orbital but with opposite spins. The interesting case will be when $E_d < E_\text{F}$ and $E_d + U > E_\text{F}$.

The conduction band of the matrix is represented in one-electron Hamiltonian in the usual band picture. Coulomb interactions in the conduction band are not taken into account. Their importance is certainly much smaller than in the d state.[†] The band Hamiltonian is

$$H_{\text{band}} = \sum_{k\sigma} \varepsilon_k n_{k\sigma} . \tag{3.3}$$

The Hamiltonians (3.2) and (3.3) represent simplified examples of the two different and opposite approaches to the study of electrons in molecules and solids. On one hand, the molecular orbital and band theory (H band) which is a one-electron theory of electrons moving in the average field of all the atoms; on the other hand, the atomic theory (H_d) where one treats only the electrons on one atom, but taking into account the Coulomb interaction (for example, with the diagonalization of the complete atomic Hamiltonian within a given shell). Magnetism on one atom needs the second approach, delocalization of electron states the first one. The interesting nature of the study of localized magnetic

---

[†] This has some importance for the screening of the impurity, which is not automatically done. One may use E_d as a parameter which is fixed by the screening charge condition, as in the approach of Friedel.

states in metals is that this problem lies in the middle: We want local magnetic properties built by nonlocalized electron states. This shows why the problem of local magnetic states in metals is difficult.

The interaction between conduction and d states, H_{sd}, is provided by a covalent admixture of s and d states:

$$H_{sd} = \sum_{k\sigma} V_{dk}(c^*_{k\sigma}c_{d\sigma} + c^*_{d\sigma}c_{k\sigma}). \tag{3.4}$$

This type of s–d interaction is a purely one-electron effect, completely different from the s–d exchange interaction.

$$V_{dk} = \langle \phi_{\text{loc}}(r) | \mathscr{H} | \phi_k(r) \rangle$$

$$= \frac{1}{\sqrt{N}} \sum_{R_n} \exp(ik \cdot R_n) \langle \phi_{\text{loc}}(r) | \mathscr{H} | W(r - R_n) \rangle$$

where $W(r - R_n)$ are the Wannier functions of the conduction band and $\mathscr{H}$ is the best one-electron Hamiltonian (usually HF) of the problem; V_{dk} is a matrix element of admixture which is very similar to the terms which are used for the admixture of atomic electronic states on different atoms in the tight binding method (LCAO).

The localized d function must usually have a different symmetry from the Wannier function. (We shall come back to this point later.) Thus the contribution to V_{dk} of the term $R_n = 0$ is zero and V_{dk} is not too large.

The total Anderson Hamiltonian is thus.

$$H = \sum_{k\sigma} \epsilon_k n_{k\sigma} + \sum_\sigma E_d n_{d\sigma} + U n_{d\uparrow} n_{d\downarrow} + \sum_{k\sigma} V_{dk}(c^*_{k\sigma}c_{d\sigma} + c^*_{d\sigma}c_{k\sigma}). \tag{3.5}$$

2. The Hartree–Fock Approximation

The Anderson Hamiltonian describes interacting particles (through the Coulomb term). It has to be treated by many-body theory. The simplest approximation is the Hartree–Fock approximation; to describe magnetic properties, one needs an unrestricted HF theory, allowing for different one-electron HF hamiltonians $\tilde{H}_\sigma$ for up and down spins. In the HF approach, one replaces the interaction term

$$U n_{d\uparrow} n_{d\downarrow} \quad \text{by} \quad U n_{d\uparrow} \langle n_{d\downarrow} \rangle + U \langle n_{d\uparrow} \rangle n_{d\downarrow} .$$

The one-electron HF Hamiltonians are

$$\tilde{H}_\sigma = H_{\text{band}} + \sum_\sigma (E_d + U \langle n_{d-\sigma} \rangle) n_{d\sigma} + H_{s-d} . \tag{3.6}$$

2. FORMATION OF LOCAL MAGNETIC MOMENTS

The problem has been treated by Anderson with the use of Green's functions. Let G be the function

$$G(E + is) = \frac{1}{E + is - \tilde{H}}.$$

The density of localized d electrons is given by

$$\rho_{d\sigma}(E) = -(1/\pi) \operatorname{Im} G_{dd}^{\sigma}(E). \tag{3.7}$$

The equation for the matrix elements of G can be easily derived. Writing $E_{d\sigma} = E_d + U\langle n_{d-\sigma}\rangle$, one has by easy calculations

$$G_{dd}^{\sigma}(E + is) = \frac{1}{E + is - E_{d\sigma} - \sum_k |V_{dk}|^2 (E + is - \varepsilon_k)^{-1}}. \tag{3.8}$$

The sum in (3.8) can be evaluated as

$$\sum_k (\cdot) = P \cdot P \cdot \sum_k \frac{|V_{dk}|^2}{E - \varepsilon_k} - i\pi \sum_k |V_{dk}|^2 \delta(E - \varepsilon_k).$$

The first term represents an energy shift which may be taken into account simply by shifting the value of E_d. The second can be written

$$-i\pi \langle |V_{dk}|^2 \rangle_{\text{av}} \rho(E)$$

where $\rho(E)$ is the density of s electrons. One usually assumes that this term is constant, introducing a "width parameter" Δ defined by:

$$\Delta = \pi \langle |V_{dk}|^2 \rangle_{\text{av}} \rho(E). \tag{3.9}$$

With this assumption, one finds a Lorentzian density

$$\rho_{d\sigma}(E) = \frac{1}{\pi} \frac{\Delta}{(E - E_{d\sigma})^2 + \Delta^2}. \tag{3.10}$$

One has then to discuss self-consistency. The average values $\langle n_{d\sigma}\rangle$ are given at $T = 0$ by

$$\langle n_{d\sigma}\rangle = \int_{-\infty}^{E_F} \rho_{d\sigma}(E) \, dE.$$

The self-consistent equations are then

$$\begin{aligned} E_d - E_F + U\langle n_{d\uparrow}\rangle &= \Delta \cot(\pi \langle n_{d\downarrow}\rangle) \\ E_d - E_F + U\langle n_{d\downarrow}\rangle &= \Delta \cot(\pi \langle n_{d\uparrow}\rangle). \end{aligned} \tag{3.11}$$

We shall not discuss the details of the solution of these self-consistent equations but only summarize the results.

There exist two regimes:

a. a nonmagnetic regime (NM) where $\langle n_{d\uparrow}\rangle = \langle n_{d\downarrow}\rangle$,
b. a magnetic regime (M) where $\langle n_{d\uparrow}\rangle \neq \langle n_{d\downarrow}\rangle$.

The occurrence of the two regimes is shown in Fig. 7, which is a "phase diagram" for the problem. The transition curve is given by

$$U\rho_d(E_F) = 1, \tag{3.12}$$

where $\rho_d(E)$ is the density of states for the nonmagnetic case. The magnetic and nonmagnetic regions are characterized by $U\rho_d(E_F) > 1$ or $U\rho_d(E_F) < 1$. In the nonmagnetic region, the susceptibility due to the d state is given by

$$\chi = \frac{g^2\mu_B^2}{2} \frac{\rho_d(E_F)}{1 - U\rho_d(E_F)}. \tag{3.13}$$

Equation (3.13) shows that as the system approaches the critical condition (3.12), the zero temperature susceptibility increases and becomes infinite.

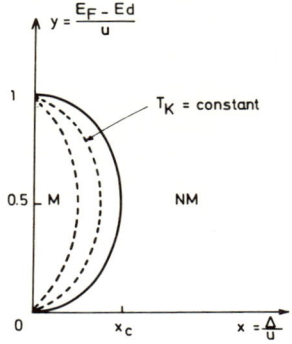

FIG. 7. Regions of magnetic (M) and nonmagnetic (NM) behavior: transition line in HF theory (solid line), and curves $T_K =$ constant (dashed lines).

3. Comparison of the Approaches of Friedel and Anderson–Orbital Degeneracy

From the HF solution of the Anderson model, one comes back to the Friedel description for orbitally nondegenerate states ($l = 1$) with the following correspondences:

$$2\varepsilon_{\text{ex}} = U, \qquad \rho_l(E) = \rho_d(E).$$

2. FORMATION OF LOCAL MAGNETIC MOMENTS

In terms of phase shifts, one can write

$$\delta_{l\sigma}(E) = \tan^{-1}\frac{\Delta}{E_{d\sigma}-E}. \tag{3.14}$$

The approximation of a constant value for Δ implies a Lorentzian shape for the density of localized states. In the scattering description, this gives the well-known Breit–Wigner formula. The phase diagrams of Figs. 3 and 7 are similar, having the same horizontal coordinates, the vertical axes being directly related, the total number $2n$ of d electrons being determined by the value of $(E_d - E_F)/\Delta$ in the magnetic and nonmagnetic regions.

Thus both approaches seem to be the same, although there are still some differences. In the Anderson model, the localized state is nondegenerate. In fact, we need orthogonality between ϕ_{ex} and the band states ϕ_k: the symmetry of ϕ_{ex} and of the Wannier states $W(r)$ of the band must be different. This ensures also a small enough width Δ in order to have magnetic solutions. There is no virtual s state within an s band, no virtual d state within a d band. If we want to describe a transitional impurity in normal metals, one should take into account the orbital degeneracy and have a virtual d state with $2l + 1 = 5$ degeneracy within a conduction band of essentially s character.

The Anderson model can be extended to orbital degenerate states. In this case, one has to consider Coulomb integrals $U_{mm'}$ and exchange integral $J_{mm'}$. In the HF treatment of the orbitally degenerate case, both integrals U and J participate in the formation of local magnetic moments. The HF solutions for the case of orbital degeneracy have been widely discussed [2, 8–10]. In this case, one may have not only a spin magnetism but also orbital magnetism. The condition (3.12) for the occurrence of magnetism is replaced by

$$(U + 2lJ)\rho_l(E_F) > 1 \quad \text{for spin magnetism}$$
$$(U - J)\rho_l(E_F) > 1 \quad \text{for orbital magnetism.} \tag{3.15}$$

For a given value of U, the existence of the intra-atomic exchange integral J helps the formation of a spin moment: It has the opposite effect for the orbital moment. Transitional impurities show spin-only magnetism, with quenched orbital moment [2, 8]. Rare earth impurities exhibit orbital magnetism. In this last case a good example of virtual bound state is cerium [9]. With orbital moment, one may have crystalline-field effects. In the approach of Friedel, only exchange was taken into account. In atoms U is much larger than J, so that one might be tempted to associate the existence of magnetic moments with the Coulomb term [11] acting as exchange self-energy. But in fact, due to correlation effects, the value of U is reduced to an effective value U_{eff}:

$$U_{\text{eff}} \simeq \frac{U}{1 + (U/\pi_\beta) \tan^{-1} \beta}, \qquad (3.16)$$

where $\beta = (E_d - E_F)/\Delta$.

Now, $U_{\text{eff}} \rho_d(E_F)$ is always smaller than 1 and the condition for a magnetic moment is no longer fulfilled in the nondegenerate case. It would thus seem necessary to have orbital degeneracy, and magnetic moment would appear owing to the existence of exchange integrals J. This conclusion is not certain. The calculations of Schrieffer and Mattis [11] at $T = 0$ may very well reflect in a way the fact that for $T < T_K$, the exact solution of the problem gives a decreasing magnetic moment with decreasing T, which disappears at $T = 0$.[†]

In conclusion, one may say that the approaches of Friedel and Anderson are very similar. For a detailed discussion, one should take care of the orbital degeneracy and use effective values of U and J. The Anderson Hamiltonian is a very good model Hamiltonian, particularly useful for a description beyond the HF approximation.

IV. Discussion of the Hartree–Fock Solutions

Let us now discuss the HF solutions somewhat critically.

1. The Phase Diagram and the "Phase Transition"

The phase diagrams of Figs. 3 and 7 show a "phase transition." For fixed values of Δ and U and varying E_d, for example, one may cross the transition line. In the magnetic region, the magnetic moment starts from zero on the critical transition line and we have a second-order transition. It can become a first-order transition with orbital degeneracy [9]. Have these transitions any real meaning?

From a theoretical point of view, the passage from a magnetic behavior to a nonmagnetic behavior cannot be a true phase transition [12] because we deal with a microscopic system which is localized in real space around the impurity; there is therefore a small number of degrees of freedom involved in the transition. On the contrary, a true phase transition needs a macroscopic system, and it is in the limit of an infinite

---

[†] The same question arises in connection with band magnetism. Is it necessary to have exchange integrals J for the appearance of ferromagnetism in a band picture?

system that the theoretical quantities of interest (energy, magnetization) do show mathematical singularities. In the case of a finite system, the transition is broadened and all physical quantities show smooth variations. Another way of saying the same thing is to say that a parameter of order can exist only for a macroscopic system and not for a microscopic one: Here the magnetic moment of the impurity cannot be a parameter of order. Thus the M–NM transition for an impurity should be a gradual transition which does not exhibit critical properties (a critical value of the parameter Δ/U for example).

From an experimental point of view, one does find usually a sharp difference between magnetic and nonmagnetic impurities. This behavior was in fact at the origin of the theoretical work on the subject. It is clear in the case of transitional impurities in transitional metals [7], and in normal metals [3a, b] though in that case some impurities appeared to be difficult to classify (Co in Cu for example). A more detailed experimental answer to the question is provided by the alloys **LaCe**. In this case, with increasing pressure (from 0 to 120 kbar) cerium impurities change from a magnetic state at low pressure to a nonmagnetic state at high pressure [13]. There is no critical value of the pressure, but rather a gradual modification of the state from magnetic to nonmagnetic behavior. This observation is not in contradiction with the previous ones: Increasing pressure by 1 kbar creates displacements of the energy levels in usual solids of the order of millielectron volts; the M–NM transition for cerium is linked with a variation of the energy of the resonant state of the order of 5×10^{-2} to 10^{-1} eV. This corresponds roughly to a variation of a few percent in the electron concentration of a matrix. Thus it appears that the transition may look sharp when one varies the electron concentration of the matrix (by alloying), but at a smaller scale of variation, with pressure experiments, the transition is smooth.

The experimental answer to the question about the nature of the transition is that there is no critical line in the phase diagram, but a gradual modification of the system appearing in a fairly narrow range of the values of the parameters (Δ/U and E_d/U). The HF theory, though introducing a fictitious transition line, gives a good qualitative description of the situation. This fact has to be explained by more elaborate theories.

2. Magnetic Moments and Susceptibilities at Finite T

All the discussion has been restricted to $T = 0$. When one tries to make a description of the system within the HF approximation at finite T, the following results come out. One finds that the magnetic moment in the magnetic HF state disappears sharply at a critical temperature T_c exactly as in a ferromagnet. One can use arguments [12] similar to that above to show that the transition should not be sharp but gradual. But there is a much bigger difficulty: The HF theory at finite temperature exhibits a decrease of the magnetic moment with increasing temperature as the observed moments decrease with decreasing temperature. This experimental behavior is linked with the Kondo problem and the fact that at low temperature there is a compensation of the local moment ("Kondo condensed state"). Thus it appears that the HF theory at finite T gives just the opposite of what is now well established from the points of view of theory and experience.

Another difficulty comes out when one tries to calculate the magnetic susceptibility in the magnetic region of the diagram. The magnetic moments should be free to rotate, giving rise to a Curie law for the susceptibility. This problem is analogous to the problem of nonspherical nuclei: First, one treats the nucleus within the HF approximation; thus the symmetry of the problem is broken; then the symmetry comes back, allowing the rotation of the HF field. For the magnetic impurity problem, it appears that there is no easy way to follow this procedure: The HF states describe the impurity and the conduction electrons of a large system, and it is not possible to rotate the HF field as in the case of the nucleus. The difficulty comes from the fact that HF states with magnetic moments pointing in different directions are eigenfunctions of different HF one-electron Hamiltonians and are not orthogonal. This means that the total HF wave function is not an eigenfunction of the total spin (S). This difficulty can be overcome for a finite system. It is not the case here. A further difficulty comes from the fact that the HF local spin is not an integer or half of an integer. This is in agreement with the experimental Curie constants which do not correspond to such spin values. But, it creates complications for a theoretical treatment. We shall come back to these points in the last section.

These various considerations could lead to the conclusion that the HF theory of impurities is completely wrong. This is not true. Certainly one has to abandon completely the description at finite temperature which leads to intolerable results. But the HF theory at $T = 0$ still retains some sense. It gives a good qualitative description if one reinterprets it as it is done in Section V. The phase diagram keeps some meaning and further theoretical ways of approaches are linked with various regions of the diagram:

1. The very magnetic region far from the transition line. This case is

discussed in the next section, and the Kondo s–d exchange Hamiltonian is valid in this region.

2. The completely nonmagnetic region far from the transition line where the ordinary theory of nonmagnetic impurities is valid.

3. The nearly magnetic region near the transition line where the theory of local spin fluctuations is applied.

In fact, it appears that the HF discussion is the basis of the different theoretical approaches which are discussed in this book.

V. The Antiferromagnetic Coupling between Localized and Conduction Electrons

In this section we concentrate essentially on the magnetic part of the phase diagrams, far from the transition line.

1. The Effective Antiferromagnetic Exchange

That admixture between localized and conduction leads to an antiferromagnetic effect was the fact underlying earlier theories from the beginning [3a, b, 4], though its importance was not fully recognized. This mechanism was pointed out by Anderson and Clogston [14], in comparison with the antiferromagnetic exchange in insulators, and appeared in the Anderson model in the so-called compensation theorem [4].

Let us discuss this point in the extreme magnetic region without orbital degeneracy where $\delta_\sigma \simeq 0$ or π [10]. Suppose that we have a magnetic moment pointing up. Then $\delta_\uparrow(E_F)$ is nearly π but slightly smaller than π, and $\delta_\downarrow(E_F)$ is nearly zero but slightly positive, as shown on Fig. 8. The "spin" of the state is

$$S_{dz} = \frac{1}{2\pi}[\delta_\uparrow(E_F) - \delta_\downarrow(E_F)].$$

It is nearly $\frac{1}{2}$ but slightly smaller, so that introducing admixture has lowered the value of the magnetic moment as if an antiferromagnetic interaction was taking place. This can be seen in more detail if one looks for the spin polarization $n(r)$ around the impurity and for the effective interaction E_{int} between impurities [10]:

$$n(r) \simeq \frac{\alpha}{4\pi^2} \frac{\cos 2k_F r}{r^3}$$

$$E_{int} \simeq \frac{2\alpha^2 E_F}{\pi} \frac{\cos 2k_F r}{(k_F r)^3} \mathbf{S}_1 \cdot \mathbf{S}_2,$$

(5.1)

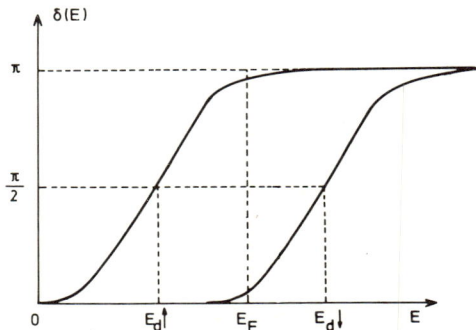

FIG. 8. Variations of the phase shifts for a magnetic solution.

where $\alpha = \delta_\downarrow(E_F) + \pi - \delta_\uparrow(E_F)$. We find results similar to those obtained with the s–d exchange interaction (1.1) if we define an effective negative exchange parameter J_{eff} :

$$J_{\text{eff}} = -\frac{2\alpha}{\pi\rho(E_F)}, \tag{5.2}$$

where $\rho(E_F)$ is the density of states of the conduction band at the Fermi level.

Here we have considered only the $S_z s_z$ term of the exchange mechanism. Schrieffer and Wolff [5] have done the correct calculation of the effective exchange Hamiltonian. They found by a perturbative canonical transformation within lowest order in the admixture term V_{kd} an effective exchange Hamiltonian $-J_{\text{eff}}\mathbf{Ss}$, the matrix elements of which are, at the Fermi level[†]:

$$J_{\text{eff}} = \frac{2\,|\,V_{kd}\,|^2\,U}{E_{dF}(E_{dF} + U)}, \tag{5.3}$$

where $E_{dF} = E_d - E_F$. This is just the value (5.2) if we use the relation (3.14) for the phase shifts. We remark than J_{eff} is negative.

In the limiting magnetic regime ($S_{dz} \simeq \tfrac{1}{2}$), we can thus describe the effective local moment by an effective exchange model Hamiltonian with $J_{\text{eff}} < 0$. One may ask whether it is possible to define a model Hamiltonian describing the various situations of the HF phase diagram [5]. The answer is no and the reason is linked with the above-mentioned

---

[†] The factor 2 in (5.3) may or may not appear, depending on the definition of the exchange Hamiltonian. With the definitions (1.1) we get this factor 2 in (5.3).

fact that the HF one-electron Hamiltonians are different, leading to wave functions which are not eigenfunctions of the total spin.

The validity of the formula (5.3) and of its variation with the position E_d of the localized level is well demonstrated by the behavior of the above-mentioned **LaCe** alloys [13]. Looking at the decrease ΔT_c of the supraconductive transition temperature of these alloys with a given amount of Ce impurities, one has the following interpretation. At low pressures E_d is far from the Fermi level ($E_F - E_d \simeq 2\Delta$ or 3Δ), J_{eff} is small, and T_c is small; when p increases, $|J_{\text{eff}}|$ increases, the total spin remaining approximately constant, and $|\Delta T_c|$ increases. This is experimentally observed. Further on when p increases, the impurity becomes nonmagnetic and $|\Delta T_c|$ decreases.

2. Connections with the Kondo Problem. The Phase Diagram†

The negative sign of J_{eff} is fundamental in connection with the Kondo problem. Starting with the exchange Hamiltonian (1.1), it is in the case of a negative J that the most interesting results occur, particularly the minimum in the variation of the resistivity with temperature. The usual direct s–d or s–f exchange integral J_{sd} or J_{sf} is positive. Thus the resonance scattering approach provides a mechanism for the sign of J.

Let us concentrate now on the Kondo temperature, which for negative J is defined as

$$T_K = D \exp[-1/\rho(E_F) |J|], \tag{5.4}$$

where D is a constant of the order of E_F. In the phase diagram of Fig. 7, the curve $T_K = $ constant is defined as $J_{\text{eff}}\rho(E_F) = $ constant. Using (5.3) one sees easily that these curves are parabolas which cut the vertical axis at $y = 0$ and $y = 1$. As T_K increases, the curves bend and come nearer to the HF transition line. There is thus a link between the HF description in the magnetic region and the Kondo description. Well inside the HF magnetic region T_K is very small. It increases rapidly when approaching the transition line.

The discussion above is valid when $\rho|J|$ is small. What does happen when $\rho|J|$ becomes large? Extrapolating (5.4) for very large values of $\rho|J|$ gives $kT_K = D$ and $(E_F - E_d) U = 0$ or 1. When $0 < (E_F - E_d) U < 1$, one would have a Kondo-like behavior: "magnetic" when $T \gg T_K$, "nonmagnetic" when $T \ll T_K$. Outside, the impurity would be nonmagnetic in the ordinary sense. In fact, again, the transition should be gradual, with no transition line from the Kondo-type behavior

† See [10, 12].

(the starting point for the discussion being an unperturbed magnetic state) to the spin fluctuation behavior [15, 16a, b] (where the starting point is a nonmagnetic state). Thus the increase of T_K (above room temperature for example) can be considered as equivalent to transition to nonmagnetism. Examples of the first type are Mn, Cr, Fe, and V impurities in Cu or Au, where T_K increases from millidegrees to hundreds of degrees. An example of the second type is Ni, Co, Fe, and Mn nonmagnetic impurities in aluminum, where the importance of spin fluctuations increases very much going from Ni to Mn.

In conclusion, it appears that the HF phase diagram with the two regions (magnetic and nonmagnetic) has something to do with the reality, if one reinterprets it. There is no transition line but a gradual transition from magnetic to nonmagnetic behavior. The transition may look sharp due to the rapid variation of T_K with the parameter J.

3. The Effective Exchange for Transition and Rare Earth Impurities

To apply the results discussed above to real impurities, one has to pay attention to the orbital degeneracy of the localized state and to the quenching (or nonquenching) of the orbital moment.

Let us look first at the case of transitional impurities where there is orbital quenching. The usual exchange Hamiltonian (1.1) describes $l = 0$ scattering. In the Kondo limit, it can describe only the effective exchange corresponding to a resonant state with $l = 0$. When $l \neq 0$, the effective exchange Hamiltonian is for S states [10, 12]:

$$-\mathscr{J}\mathbf{s}\cdot\mathbf{S},$$

the matrix elements of $\mathscr{J}$ being at the Fermi level:

$$\mathscr{J}k_F k_F' = J_{\text{eff}} P_l(\cos\theta), \tag{5.5}$$

where θ is the angle between k and k'. This scattering potential gives only phase shifts of l symmetry and J_{eff} as the same value (5.3) than for $l = 0$.

Equation (5.5) introduces differences when one calculates, for example, the resistivity within second order, following Kondo [6]:

$$\rho = ct \times J_{\text{eff}}^2[1 + J_{\text{eff}}\rho(E_F)\log(kT/D)] \quad \text{for} \quad l = 0$$

$$\rho = ct \times (2l+1)j_{\text{eff}}^2[1 + j_{\text{eff}}\rho(E_F)\log(kT/D)] \quad \text{for} \quad l \neq 0$$

where $j_{\text{eff}} = J_{\text{eff}}/(2l+1)$. The effective exchange integral which enters in the coupling constant ρJ is not J_{eff} but j_{eff}. This result is

2. FORMATION OF LOCAL MAGNETIC MOMENTS

more general and true within any order of perturbation. There are $(2l + 1)$ independent channels for the scattering.

In the case of transitional impurities, the usual s–d exchange parameter is small and the effective mechanism always prevails. In noble metals, for example [10, 17a, b, c, d], for Mn impurities, $\delta_\uparrow \simeq \pi$, $\delta_\downarrow$ is small, in order to have a virtual bound state half-full. Consequently J_{eff} is not very large, being of the order of -1.5 to -2 eV. This gives -0.3 or -0.4 eV for j_{eff}, in accordance with the value deduced by Kondo from the resistivity minimum. Thus T_K is very small. On the contrary, for Fe, Cr, or V, at least one of the phase shifts has to be fairly different from 0 or π in order to satisfy the screening and the Friedel sum rule. Here J_{eff} is fairly large and T_K may have large values, comparable to room temperature or higher.

The case of rare earth (RE) impurities is completely different for two reasons. First the width Δ is very small; secondly, there is no quenching of the orbital moment. Let us look first at the effect of the direct exchange. If **J** is the total angular momentum of the ion (**J** = **L** + **S**), and J_{sf} is the s–f atomic exchange integral, de Gennes [18] has shown that the s–f Hamiltonian to be used is

$$-2J_{sf}(g-1)\,\mathbf{s}\cdot\mathbf{J}, \tag{5.6}$$

where g is the Lande factor of the ion. The form of (5.6) leads to a puzzling result. For the first half of the rare earth series, **l** and **s** are antiparallel and $(g - 1)$ is negative; for the second half, **L** and **S** are parallel and $(g - 1)$ is positive. Thus, one should observe the Kondo effect in this first half of the series, not in the second. This is not observed. On the contrary, all RE impurities never exhibit Kondo phenomena, with the exception of Ce and Yb. Now Ce and Yb are the only cases where one expects a strong admixture of s and f electrons near the Fermi level [9] (an experimental proof is the fact that Ce and Yb impurities can be either magnetic or nonmagnetic). Thus it appears that the Kondo effect is here again directly related to the existence of a resonant state near the Fermi level. The exchange Hamiltonian (5.6) oversimplifies and distorts the situation. This is just what has been shown by Coqblin and Schrieffer [19] who studied in detail the case of cerium, within the framework of virtual bound states. They found an effective Hamiltonian which is a generalization of (1.1) and exhibits the Kondo properties [note that putting an effective negative value of J in Eq. (5.6) would forbid the Kondo effect]. Applications to various properties have widely supported this approach. This theory also explains well the properties of Yb impurities, which is a similar case (one hole in the f shell instead of one electron for the case of Ce).

4. Various Applications

We do not discuss the applications of the preceding sections here in detail. We shall simply point out some results.

For transition impurities, the HF model predicts a spin polarization $n(r)$ which is given by (5.1) for half-filled shells and more generally by using Eq. (2.4) [17a, b, c, d]:

$$n(r) = \Delta\rho^\uparrow(r) - \Delta\rho^\downarrow(r),$$

where

$$\Delta\rho^\sigma(r) \simeq -\frac{k_F^3}{4\pi^2} \sum_l (-)^l (2l+1) \sin\delta_{l\sigma} \frac{\cos(2k_F r + \delta_\rho^\sigma)}{(k_F r)^3}.$$

This result has been used with success to explain the NMR properties of dilute alloys (**Cu**Mn for example).

From the same model, one deduces the effective interaction between magnetic moments for large R, a generalization of (5.1):

$$E_{\text{int}}(R) \simeq \frac{E_F}{2\pi} \sum_{\sigma\sigma'} (2l+1) \sin\delta_{l\sigma} \sin\delta_{l\sigma'} \frac{\cos(2k_F R + \delta_{l\sigma} + \delta_{l'\sigma'})}{(k_F R)^3} \mathbf{S}_1 \cdot \mathbf{S}_2,$$

which can be written as

$$\begin{aligned} E_{\text{int}}(R) &\simeq -J(R)\, \mathbf{S}_1 \cdot \mathbf{S}_2 \\ J(R) &= \alpha\, \frac{\cos(2k_F R + \phi)}{(k_F R)^3}, \end{aligned} \quad (5.7)$$

where α and ϕ are constants depending on the phase shifts. Equation (5.7) is a generalization of the Ruderman–Kittel–Kasuya–Yosida interaction; it has been used by Caroli [17a, b, c] to explain the magnetic coupling between impurities (**Cu**Mn, for example, when $T \gg T_K$). It gives a much more effective coupling than the direct exchange. The R^{-3} dependence of the interaction is of particular interest because it gives rise to "scaling laws" for these alloys [20].

At shorter distances (nearest neighbors) the interactions between atoms have been widely discussed and applied to pure transitional metals [8, 21].

VI. Transition Impurities in Transition Metals

In this case, the matrix has uncompletely filled d bands, which are narrow bands in comparison with the conduction band. This fact has several consequences:

2. FORMATION OF LOCAL MAGNETIC MOMENTS

1. The atomic Coulomb interactions have importance in the matrix which may exhibit magnetic order at low temperatures.

2. The scattering problem has new features due to the narrow width of the band and to similar symmetries between the d states of the impurity and of the matrix.

We discuss first the scattering problem in narrow bands without Coulomb interactions. Then we shall review briefly the problem of the formation of local moments within the HF approximation in non-magnetic and in ferromagnetic matrices.

1. The Scattering Problem in Narrow Bands

The pioneer work in this field is due to Slater and Koster [22]. They consider one single band, neglecting all the other bands. The Wannier functions centered on sites R_n are $W(r - R_n)$ and the Bloch functions ψ_k with energies ε_k are

$$\psi_k = \frac{1}{\sqrt{N}} \sum_n \exp(ikR_n) W(r - R_n).$$

Thus the model is exactly similar to the band constructed in the tight binding approximation with s atomic orbitals. The perturbing potential V_p due to an impurity at the origin O is taken to be localized on the impurity site[†]:

$$\langle W(r - R_n) | V_p | W(r - R_m) \rangle = \delta_{n0} \delta_{m0} V.$$

V can be positive or negative, corresponding to repulsive or attractive potential. In the notation of second quantization, $c_{k\sigma}^*$, $c_{n\sigma}^*$ being the creation operators corresponding to the Bloch and Wannier wave functions with spin σ, the Hamiltonian is

$$H = \sum_{k\sigma} \varepsilon_k c_{k\sigma}^* c_{k\sigma} + V c_{0\sigma}^* c_{0\sigma} = H_0 + V_p. \tag{6.1}$$

The solution of (6.1) is easy and gives the following results. Let $G_0(E)$ and $G(E)$ be the unperturbed and perturbed Green's functions:

$$G_0(E) = 1/(E + is - H_0) \quad \text{and} \quad G(E) = 1/(E + is - H).$$

[†] This approximation enables an easy treatment of the scattering problem. Though it is possible in principle to treat the case of a potential extending over the neighboring atoms of the impurity site, little has been done in this direction due to technical mathematical difficulties.

One has

$$G(E) = G_0(E) + G_0(E) V_p G(E). \quad (6.2)$$

In the Wannier (atomic) representation, the matrix elements of $G_0(E)$ are

$$\langle n | G_0(E) | m \rangle = \frac{1}{N} \sum_k \frac{\exp[ik(R_m - R_n)]}{E + is - \varepsilon_k}.$$

Particularly

$$\langle 0 | G_0(E) | 0 \rangle = F(E) - i\pi n(E), \quad (6.3)$$

where $n(E)$ is the density of states of the band and $F(E)$ is its Hilbert transformation:

$$F(E) = P \cdot P \cdot \int \frac{n(\varepsilon') d\varepsilon'}{\varepsilon - \varepsilon'}. \quad (6.4)$$

Using (6.2) and (6.3) one gets

$$\langle 0 | G(E) | 0 \rangle = \frac{F(E) - i\pi n(E)}{1 - VF(E) + i\pi V n(E)}. \quad (6.5)$$

Let us discuss the consequences of Eq. (6.5). Suppose, for example, that V is repulsive ($V > 0$). Then E_M being the energy of the top of the band, $n(E)$ and $F(E)$ are schematically given in Fig. 9. One sees

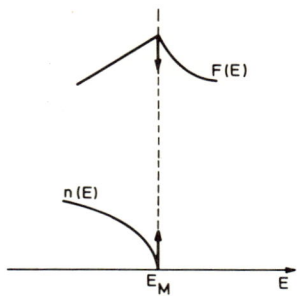

FIG. 9. Behavior of $n(E)$ and $F(E)$ near a band edge where $n(E) = (E_M - E)^{1/2}$.

immediately that if $VF(E_M) > 1$, there exists a real pole in Eq. (6.5) which gives rise to a bound state with energy given by $VF(E) = 1$. This bound state is subtracted from the band upward. If $VF(E_M) < 1$, there may be only an imaginary pole in Eq. (6.5). The discussion can be pursued within scattering theory [23]. With the simplified perturbing potential, there is only one phase shift (as with a δ-type potential for free electrons) which is given by:

$$\tan \delta(E) = -\pi V n(E)/[1 - VF(E)]. \tag{6.6}$$

This gives a local density of state per spin direction $(1/\pi)[d\delta(E)]/dE$. The variation of $\delta(E)$ near the top of the band for various cases is shown in Fig. 10. There is a strong analogy with the s phase shift for free

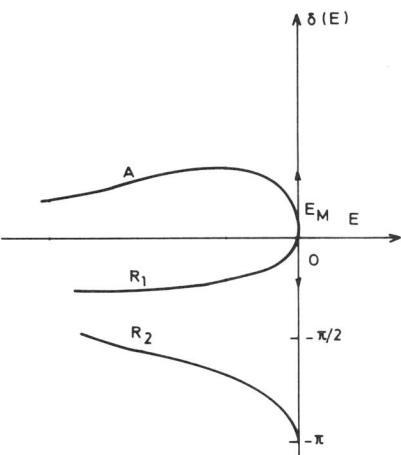

FIG. 10. Variations of the phase shift $\delta(E)$ near a band edge [25]: attractive potential (A); increasingly repulsive potentials (R_1) and (R_2).

electrons scattered by a spherical potential. But, if the potential is not sufficiently strong to push one bound state away from the band, the variation of $\delta(E)$ with energy is quite different from that of Fig. 1; it is thus difficult to speak of a virtual bound state. As it was mentioned in Section III, this is due to the similar (s) symmetry of the Wannier wave function and of the bound-state wave function, when V_p is localized on one impurity site. There is no virtual s state in an s band.

The Slater and Koster model has been generalized to d bands treated in the tight binding approximation [24, 25]. If the perturbing potential V_p is localized on the impurity site of a cubic crystal, there are two matrix elements of V_p which are different from zero:

$$\langle \Gamma_i | V_p | \Gamma_i \rangle = V(\Gamma_i).$$

Γ_i are the two classes in the irreducible representations of the point group of the lattice. The three t_{2g} states (xy, yz, zx) give the same value

$V_{t_{2g}}$ and the two e_g states ($x^2 - y^2$, $3z^2 - r^2$) give the value V_{e_g}. Then $V_{e_g} = V_{t_{2g}}$ for spherical symmetry of the potential. If one defines $n(E, \Gamma_i)$, the density of occupation of the orbitals for a given class at energy E, one can define two phase shifts which are given by a generalization of (6.6):

$$\tan \delta(E, \Gamma_i) = \frac{-\pi V(\Gamma_i) n(E, \Gamma_i)}{1 - V(\Gamma_i) F(E, \Gamma_i)}. \tag{6.7}$$

When $V(\Gamma_i) F(E, \Gamma_i)$ is large, one may have bound states of a given symmetry, and for weaker perturbations, a variation of the phase shifts which is in general analogous to that of Fig. 10. This reflects the fact that, except in very particular cases, one cannot have virtual d bound states in a d band [24, 25].

2. The Hartree–Fock Approximation in Narrow Bands (Nonmagnetic Matrices)

When one adds to the Hamiltonian (6.1), the Coulomb term $Un_{0\uparrow}n_{0\downarrow}$ as in the Anderson model, one obtains the Wolff–Clogston model [26]. Within this model, one can make the HF approximation; the local potential V is replaced by $V_\sigma = V + U\langle n_{0-\sigma}\rangle$. We shall not discuss the results in detail. They are very similar to those obtained with the Anderson model. The difference is in the shape of the localized density of states $[d\delta(E)]/dE$ which is fairly different from a Lorentzian shape.

Systematic applications to transition impurities in transition metals imply two modifications: First to extend the model to the case of five d bands; and second, to take into account the Coulomb interactions within the matrix which are comparable with the Coulomb terms of the second effect is difficult, because one deals with a perturbing potential which is no longer localized on the impurity site, and the scattering problem becomes intractable. The Coulomb terms of the matrix are particularly important to explain the formation of giant moments as shown in Fig. 6 for Fe impurities in the nearly magnetic palladium matrix. In these cases, magnetic moments can be larger than $10\mu_B$ and cannot be localized on the Fe sites; the nearest Pd atoms are also magnetized.

These difficulties explain why no detailed quantitative discussion of the overall situation has been made, particularly when giant moments occur. The Wolff–Clogston model and its extensions provide nevertheless a good qualitative basis for the discussion of the various situations which for nonmagnetic matrices are comparable to the case of normal matrices. One finds magnetic systems of the Kondo type with various Kondo temperatures [27], normal nonmagnetic impurities, and nearly magnetic cases. It is in the situation of the last type that Lederer and Mills [15, 28] first introduced the concept of localized spin fluctuations in order to explain the magnetic behavior of **Pd**Ni dilute alloys.

3. The Hartree–Fock Approximation for Impurities in Ferromagnetic Matrices

Though this problem is slightly outside the main lines of interest of this paper, it is interesting to discuss it because it confirms the validity of the model in various situations.

The average moments $\bar{\mu}$ of transitional ferromagnetic alloys are well known and are given in Fig. 11 (Slater–Pauling curve). If one can explain the behavior of some alloys (**Ni**Co, for example) within the rigid band picture (weak perturbations which can be treated within the Born approximation), the cases of **Ni**Cr or **Ni**V alloys fall evidently outside this approach. Friedel [2, 25] has explained this behavior. He supposed that the repulsive potential can be strong enough to push bound states through the Fermi level as shown in Fig. 12. The bound d levels with spin up empty themselves into the Fermi sea of d electrons with spin down. This effect produces a lowering of the magnetic moment, which is a characteristic feature of the Slater–Pauling curve. Due to the presence of the conduction band, there is admixture between the bound states and the conduction electrons. The d states are in fact virtual d states in the s conduction band, as shown in Fig. 12.

These qualitatively simple ideas have been worked out quantitatively [29, 30] in order to explain the Slater–Pauling curve and the magnetic

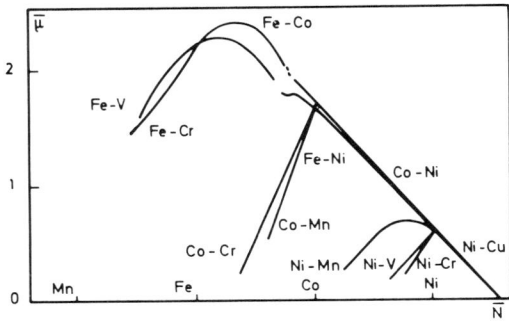

FIG. 11. The Slater–Pauling curve giving the average magnetic moment $\bar{\mu}$ per atom in Bohr magnetons versus the average number $\bar{N}$ of d electrons for binary alloys of the first transitional series [25].

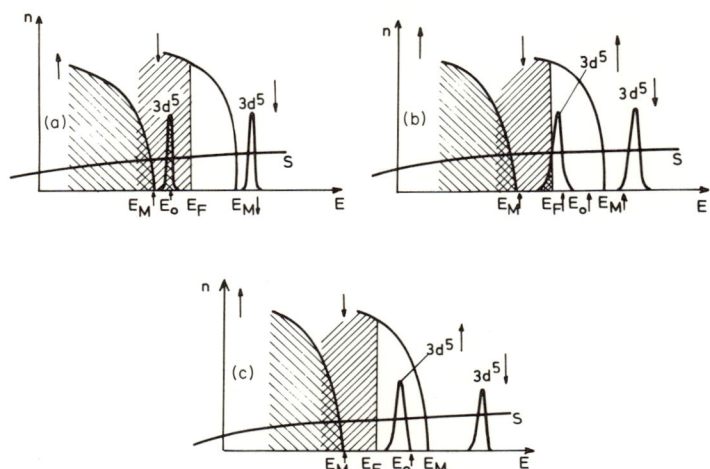

FIG. 12. Virtual bound states $3d^5$ with spin up being pushed through the Fermi level in Nickel-based alloys [25]: (a) NiMn; (b) NiCr; (c) NiV.

form factors as deduced from neutron scattering experiments [31]. In order to simplify the question, the perturbing potential V_p is taken as localized spherical ($V_{e_g} = V_{t_{2g}}$) and $n(E, \Gamma_i)$ as independent of Γ_i.

The same model may be applied to explain the observed residual resistivities of impurities in ferromagnetic alloys [32] (Fig. 13). Mott proposed a model where the carriers are the electrons of the conduction band, the resistivity being controlled by s–d transitions [33]. In a ferromagnetic metal, these transitions are different for spin up and spin down, and this fact leads to a model with two currents in parallel. If $\rho_\uparrow$ and $\rho_\downarrow$ are the resistivities for spin up and spin down, the total resistivity is $\rho = \rho_\downarrow \rho_\uparrow / (\rho_\downarrow + \rho_\downarrow)$. The observed resistivities (Fig. 13) clearly exhibit the crossing of the Fermi level by the virtual d state. Detailed experiments with one or two kinds of impurities which show deviations from Matthiessen's rule are well explained within the model of two currents [33], taking for ρ_σ the values calculated with the model Hamiltonian described above. Examples for ρ_σ are given in Fig. 14.

In conclusion, one can say that the formation of local magnetic moments in ferromagnetic metals is well explained by the HF theory. If we compare this case to the case of normal metals, there is an important reason for this result: in ferromagnetic metals, elastic spin-flip scattering is forbidden; thus there does not exist difficult (and interesting) problems of the Kondo type. Impurities in ferromagnetic metals are good examples where, at least at low temperature, the unrestricted HF theory has full

2. FORMATION OF LOCAL MAGNETIC MOMENTS

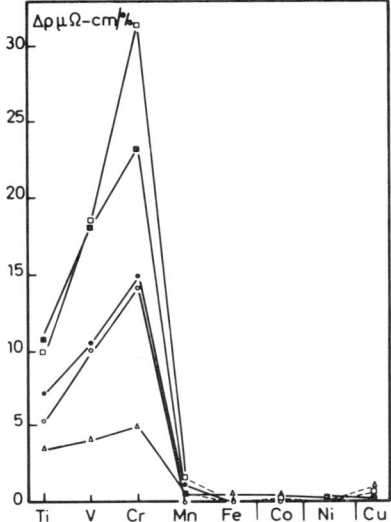

FIG. 13. Residual resistivities of impurities of the first transitional series dissolved in nickel and nickel-based alloys [25, 34]: △ Ni; □ FeNi$_3$; ■ FeCo; ● CoNi; ○ Ni$_{0.6}$, Cu$_{0.4}$.

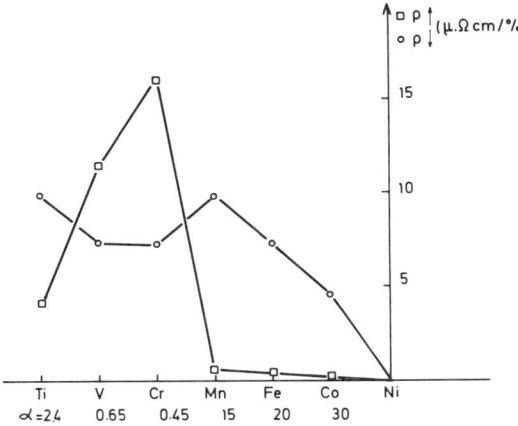

FIG. 14. Partial resistivities for s carriers of spin up and spin down in nickel-based alloys. The values of $\alpha = \rho_\downarrow/\rho_\uparrow$ are deduced from experiment [34].

validity. Extension to the case of an antiferromagnetic matrix (chromium for example) could be done in principle, though the existence of new Brillouin zones would certainly give complicated computations.

VII. Conclusions

The HF description of magnetic moments at $T = 0$ gives a good qualitative description for the occurrence of magnetic moments in normal and magnetic metals. For impurities in normal metals, the HF phase diagram makes sense if one understands it as it was done in Section V. The HF description fails to explain the Kondo effect and the gradual appearance of magnetic moments with temperature. But, the phase diagram provides a basis for further developments.

In the *magnetic region*, the above-mentioned difficulty of having different one-electron potentials for magnetic moments pointing up and down has been a major point for the understanding of the Kondo problem. Anderson [34] demonstrated that the matrix element between two determinantal wave functions constructed from different scattering states (different one-electron perturbing potentials) is vanishingly small when the phase shifts at the Fermi surface of the two scattering states are greatly different (infrared catastrophe). This fact has applications in order to explain x-ray absorption and emission edges in metals [35]. This was also a key point in solving the Kondo problem with *s*–*d* exchange: Suppose that we treat first the diagonal part of the *s*–*d* exchange potential, $-Js_zS_z$. The two states with magnetic moments pointing up and down correspond to different one-electron potentials and different phase shifts. Treating at any order in perturbation theory the nondiagonal part $-\frac{1}{2}J(s^-S^+ + s^+S^-)$ describes the flipping up and down of the magnetic moment. The problem is directly related to infrared catastrophe. This provides a new method of handling the Kondo problem successfully, as it has been done by Anderson and Yuval [36]. Outside the extreme magnetic region where there is equivalence between the *s*–*d* model and the virtual bound-state model, less has been done. Anderson tried a $T = 0$ solution of the problem [37]. In order to introduce "communications" between states with magnetic moments up and down, he choose a trial wave function where the two phase shifts δ_σ change quickly near the Fermi level and are equal to $\pi/2$ at $E = E_F$. The scattering cross section of the impurity attains the "unitarity limit" which was first discussed by Suhl [38]. No theoretical extension of this work has been made at $T \neq 0$. But it is interesting to mention the phenomenological theories [39] where one introduces temperature-dependent phase shifts $\delta_\sigma(E_F, T)$. At high temperature, $\delta_\sigma(E_F, T) = \delta_\sigma(E_F)$, the phase shift of the magnetic HF solution; at low temperatures, $\delta_\sigma(E_F, T)$ is equal to the phase shift of the nonmagnetic HF solution. We have a picture which provides a link between magnetic and nonmagnetic impurities as it was first introduced by Schrieffer [12] and which gives a fairly good fit with experiments, at least for problems where spin-flip scattering is not dominant.

In the *nonmagnetic region of the HF phase diagram*, hope had been raised by the theory of local spin fluctuations [15, 16] starting from the nonmagnetic HF solution. In fact, for impurities which are not very magnetic (Mn in Al for instance), such a theory is fairly successful in explaining, for example, the variations of the magnetic susceptibility and of the resistivity which vary as T^2 at low temperature [40]. But, approaching the transition line of the HF phase diagram, we find the situation less desirable: the transition line is washed out by the fluctuations, but the theory has not been able to give back the Kondo anomaly and temperature [41].

It remained to find a theory which could describe the complete situation from nonmagnetic cases to magnetic cases. A good approach in this direction is given by the functional integral method of Schrieffer and Hamann [42]. With this method, the first approximation (where one takes into account only static effects) gives back the two regions of the HF method. Further, taking into account dynamical effects, one finds the spin fluctuation regime and the spin-flip regime of the Kondo type. Thus a step has been made in the direction of a theory which would cover continuously the whole phase diagram of the HF theory.

References

1. C. Zener, *Phys. Rev.* **81**, 440 (1951); M. A. Ruderman and C. Kittel, *Phys. Rev.* **96**, 99 (1954); T. Kasuya, *Progr. Theor. Phys.* **16**, 45 (1956); K. Yosida, *Phys. Rev.* **106**, 893 (1957).
2. J. Friedel, *Can. J. Phys.* **34**, 1190 (1956); *Nuovo Cimento Suppl.* **7**, 287 (1958); P. de Fajet de Calteljau and J. Friedel, *J. Phys. Rad.* **17**, 27 (1956).
3a. A. Blandin and J. Friedel, *J. Phys. Radium* **20**, 160 (1959);
3b. A. Blandin, Thesis, Univ. of Paris, Paris, France, unpublished, 1961.
4. P. W. Anderson, *Phys. Rev.* **124**, 41 (1961).
5. J. R. Schrieffer and P. A. Wolff, *Phys. Rev.* **149**, 491 (1966).
6. J. Kondo *Progr. Theor. Phys.* **32**, 37 (1964).
7. B. T. Matthias, M. Peter, H. J. Williams, A. M. Clogston, E. Corenzwit, and R. C. Sherwood, *Phys. Rev.* **125**, 541 (1962); *Phys. Rev. Lett.* **5**, 542 (1960).
8. T. Moriya, *Proc. Varenna School*, 1966 (W. Marshall, ed.), p. 206. Academic Press, New York, 1967.
9. B. Coqblin and A. Blandin, *Advan. Phys.* **17**, 281 (1968).
10. A. Blandin, *J. Appl. Phys.* **38**, 1143 (1967).
11. J. R. Schrieffer and D. C. Mattis, *Phys. Rev. A* **140**, 1412 (1965).
12. J. R. Schrieffer, *J. Appl. Phys.* **38**, 1143 (1967).
13. T. F. Smith, *Phys. Rev. Lett.* **17**, 386 (1966); B. Coqblin and C. F. Ratto, *Phys. Rev. Lett.* **21**, 1065 (1968); M. B. Maple, J. Wittig, and K. S. Kim, *Phys. Rev. Lett.* **23**, 118 (1969); B. Coqblin, M. B. Maple, and G. Toulouse, *Int. J. Magn.* **1**, 333 (1971).

14. P. W. Anderson and A. M. Clogston, *Bull. Am. Phys. Soc.* **6**, 124 (1961).
15. P. W. Anderson, *Comments Solid State Phys.* **1**, 190 (1969).
16a. P. Lederer and D. L. Mills, *Solid State Commun* **5**, 131 (1967).
16b. N. Rivier and M. Zuckermann, *Phys. Rev. Lett.* **21**, 904 (1968); H. Suhl, *Phys. Rev. Lett.* **19**, 442 (1967); *Phys. Rev.* **171**, 567 (1968).
17a. B. Caroli, Thesis, Univ. of Paris, Orsay, 1967.
17b. B. Caroli and A. Blandin, *J. Phys. Chem. Solids* **27**, 503 (1966).
17c. B. Caroli, *J. Phys. Chem. Solids* **28**, 1427.
17d. A. Blandin *Proc. Varenna School*, 1966, (W. Marshall, ed.), p. 393. Academic Press, New York, 1967.
18. P. G. de Gennes, *J. Phys. Radium* **23**, 510 (1962).
19. B. Coqblin and J. R. Schrieffer, *Phys. Rev.* **185**, 847 (1969).
20. J. Souletie and R. Tournier, *J. Phys. C* **1**, 172 (1971).
21. S. Alexander and P. W. Anderson, *Phys. Rev. A* **133**, 1594 (1964).
22. G. F. Koster and J. C. Slater, *Phys. Rev.* **95**, 1167, **96**, 1208 (1954); G. F. Koster, *Phys. Rev.* **95**, 1436 (1954).
23. A. M. Clogston, *Phys. Rev.* **125**, 439 (1961).
24. F. Gauthier, *Ann. Phys. (Paris)* **10**, 275 (1965).
25. J. Friedel, *Proc. Varenna School*, 1966 (W. Marshall, ed.), p. 293. Academic Press, New York, 1967.
26. P. A. Wolff, *Phys. Rev.* **124**, 1030 (1961).
27. B. R. Coles, *Phys. Lett.* **8**, 243 (1964).
28. P. Lederer and D. L. Mills, *Phys. Rev.* **165**, 837 (1967).
29. J. Kanamori, *J. Appl. Phys.* **36**, 929 (1965).
30. A. A. Gomès, *J. Phys. Chem. Solids* **27**, 451 (1966); A. A. Gomès and I. A. Campbell *Proc. Phys. Soc. London* **89**, 319 (1967).
31. C. G. E. Low and M. F. Collins, *J. Appl. Phys.* **34**, 1195 (1963); *Proc. Phys. Soc.* **86**, 535 (1965).
32. C. W. Chen, *Phil. Mag.* **7**, 1753 (1962); *Phys. Rev.* **129**, 121 (1962).
33. N. F. Mott, *Advan. Phys.* **13**, 325, (1964); I. A. Campbell, A. Fert, and A. R. Pomeroy, *Phil. Mag.* **15**, 977 (1967); A. Fert and I. A. Campbell, *Phys. Rev. Lett.* **21**, 1190 (1968); T. Farrell and D. Greig, *J. Phys. C* **2**, 1359 (1968); P. Leonard, M. C. Cadeville, J. Durand, and F. Gautier, *J. Phys. Chem. Solids* **30**, 2169 (1969); J. Durand and F. Gautier, *J. Phys. Chem. Solids* **31**, 2773 (1970); A. Fert and I. A. Campbell, *J. Phys. Paris* **32**, 1, 46, (1971).
34. P. W. Anderson, *Phys. Rev. Lett.* **18**, 1049 (1967).
35. P. Nozières and C. T. de Dominicis, *Phys. Rev.* **178**, 1097 (1969).
36. G. Yuval and P. W. Anderson, *Phys. Rev. B* **1**, 1552 (1970); P. W. Anderson, G. Yuval, and D. R. Hamann, *Ibid.* **1**, 4464 (1970).
37. P. W. Anderson, *Phys. Rev.* **164**, 352 (1968).
38. H. Suhl, *Phys. Rev. A* **138**, 1112 (1965); *Proc. Varenna School 1966* (W. Marshall, ed.), p. 116. Academic Press, New York, 1967.
39. J. Loram, *Commun. European Meeting, Florence*, 1970; J. Souletie, *J. Low Temp. Phys.* **7**, 141 (1972).
40. A. D. Caplin and C. Rizzutto, *Phys. Rev. Lett.* **21**, 746 (1968).
41. H. Suhl, *Phys. Rev. Lett.* **19**, 442 (1967); M. Levine and H. Suhl, *Phys. Rev.* **171**, 567 (1968); D. R. Hamann, *Ibid.* **186**, 549 (1969).
42. S. Q. Wang, W. E. Evenson, and J. R. Schrieffer, *Phys. Rev. Lett.* **23**, 9 (1969); W. E. Evenson, J. R. Schrieffer, and S. Q. Wang, *J. Appl. Phys.* **41**, 1199 (1970); D. R. Hamann, *Phys. Rev. B* **2**, 1373 (1970).

3. Spin Fluctuations around Impurities: Magnetic and Nonmagnetic Cases

D. L. Mills[†]

Department of Physics
University of California
Irvine, California

M. T. Béal-Monod and P. Lederer

Laboratoire de Physique des Solides[‡]
Faculte des Sciences
Orsay, France

I. Introduction	89
II. Theory of Local Spin Fluctuations Associated with Paramagnetic Impurities in Metals; Mean Field Description	92
III. Renormalized Theories of Local Spin Fluctuations: Nonmagnetic and Magnetic Cases	100
1. Introduction	100
2. Breakdown of Mean Field Theory in the Transition Region	101
3. The "Renormalized Random Phase Approximation" (RRPA) of Suhl and Co-Workers	102
4. The Functional Integral Method of Wang et al.	103
5. Importance of Vertex Corrections	104
6. Failure of the RPA' in the Functional Integral Approach except in the Weak Coupling Regime	107
7. Phenomenological Renormalization	108
8. Conclusion	108
IV. The Effect of Local Spin Fluctuations on the Properties of Dilute Alloys	109
1. The Static Susceptibility of the Alloy	109
2. The Change in Specific Heat upon Alloying	110
3. Nuclear Magnetic Resonance at the Impurity Site	111
4. Transport Properties	113
References	116

I. Introduction

Many of the physical properties of dilute alloys may be successfully explained by a theory based on the one-electron picture of metals. The

[†] Supported in part by the Air Force Office of Scientific Research, Office of Aerospace Research, USAF, under Grant No. AFOSR 70-1936.
[‡] Laboratoire associé au C.N.R.S.

fundamental concepts that form the basis of this theory have been developed and successfully applied to a wide variety of phenomena by Friedel and his collaborators. For a detailed discussion of this approach, we refer the reader to Chapter 2 by Blandin. For our purposes, it will be useful to recall the underlying assumptions of the theory.

Suppose one places a transition metal impurity in a noble metal or transition metal host. Then the electrons in the metal are perturbed by the change in crystal potential produced by the presence of the impurity. The change in crystal potential is confined principally to the unit cell which contains the impurity in these systems, since the electrons of the host screen the perturbation efficiently. In that limit one may obtain analytic expressions for the wave functions and energy levels of the electrons in the alloy. The crucial feature of the theory is a condition of self-consistency that is imposed by the Friedel sum rule: The change ΔN in the number of electrons produced by altering the crystal potential must precisely equal the difference Z in the effective nuclear charge between the impurity and the host ion. The Friedel sum rule thus allows one to determine the strength of the change in crystal potential once the relative position of the host atom and impurity atom in the periodic table is noted. The ability of this theory to explain the systematic variation of the properties of dilute alloys constructed by doping a given noble metal or transition metal host matrix with a sequence of impurities is most impressive.

In this article, we discuss the theory of certain many-body effects that have proved to produce striking deviations in the properties of a number of alloys from the behavior expected on the basis of one-electron theory. We also review a number of experimental studies which provide illustrations of the sort of behavior produced by these many-body effects.

Suppose we apply a spatially uniform, static magnetic field to a pure paramagnetic metal. Then the magnetic field induces a spin polarization in the gas of conduction electrons. Considerations of translational invariance demand that the spin moment assumes the same value in each unit cell of the crystal. If we ignore the effect of interactions between the electrons, then the contribution χ_s of the spin paramagnetism to the total susceptibility is given by an expression due to Pauli [1]; χ_s is proportional to the density of states at the Fermi level. Now consider the effect on the spin susceptibility of interactions between the electrons. The Coulomb interaction between electrons in consort with the exclusion principle leads to an enhancement of χ_s to a value greater than that given by the one-electron theory. The physical origin of this enhancement effect may be appreciated by considering a single, nondegenerate band of electrons. Suppose an external field creates a net spin polarization in the conduction band in the $+z$ direction. An electron with spin down interacts principally with spin-up electrons. This is so because the exclusion principle does not allow electrons of parallel spin to be in the same unit cell, while no such restriction obtains for electrons of opposite spin. Since there are more electrons with spin up than spin down in the presence of an external magnetic field, the electrostatic contribution to the energy of a spin-down electron is greater than that for an electron with spin up. Thus, in the presence of Coulomb interactions between electrons, the splitting in energy between up- and down-spin bands is greater than the splitting expected from the free electron model; the magnetic moment induced by the field and also the Pauli spin susceptibility are therefore enhanced by the interaction. This exchange enhancement of the spin susceptibility was first discussed by Stoner [2]. The theory of paramagnon effects in metals with strongly enhanced susceptibilities [3a,b] follows by generalizing Stoner's approach to obtain the response of the pure paramagnetic matrix to an applied magnetic field that varies in space and time.

Now suppose that a single impurity is added to the host matrix described above. It will be convenient to presume that the electrons move in a band of the tight binding form; the d bands of transition metals have the character of tight binding bands, so this approximation is a reasonable one. In this case, the intra-atomic Coulomb interaction U_I between two electrons in the impurity cell will differ from the value U_0 appropriate to the host. Suppose that $\Delta U = U_I - U_0 > 0$. Then the reasoning in the previous paragraphs leads one to expect that a uniform magnetic field will induce a spin polarization in the impurity cell that is larger than in the host matrix, because the enhancement effect is larger when the electrons are in the impurity cell. This is indeed the case, as we shall see. In fact, if the host is paramagnetic, and ΔU exceeds a certain critical value ΔU_c, then Hartree–Fock (HF) theory predicts that the paramagnetic state becomes locally unstable [4a,b,c], and a local moment forms. When $\Delta U < \Delta U_c$, the ground state is paramagnetic, but the spin density induced in the impurity cell can be very much larger than that in the host. (Actually, the enhancement effect is not confined only to the impurity cell, but extends into the host matrix with a certain range. The spatial extent of the enhanced polarization can become considerable if the exchange enhancement of the host susceptibility is large.) When the local enhancement is large, then the susceptibility of the alloy exhibits a concentration dependence much stronger than expected on the basis of one-electron theory. A direct measure of the size of the local enhancement effect may be obtained if the core polarization contribution to the Knight shift of the impurity nuclear magnetic resonance (NMR) is studied.

So far, we have confined our attention to the enhancement of the spin moment near the impurity cell induced by a static external field. This enhanced response implies that the fluctuations in spin density in the conduction-electron gas have an amplitude near the impurity cell that is larger than in the host matrix. The enhanced spin density fluctuations lead to a striking concentration dependence of the contribution from electron–electron scattering to the transport coefficients of the alloy. In addition, the enhanced spin fluctuations lead to very rapid longitudinal relaxation of the impurity nucleus, to a large contribution to the specific heat, and to other striking effects.

If the exchange enhancement in the host matrix is appreciable, and $\Delta U < 0$ (i.e., $U_I < U_0$), then the Coulomb interactions lead to a suppression of the response of the host matrix to an applied field, and the enhanced spin fluctuations ("paramagnons") characteristic of the host will be suppressed near the impurity cell.

The purpose of this article is to present a discussion of the theory of local spin fluctuations, and to review experimental data on certain alloy systems in which locally enhanced spin fluctuations appear to play an important role. In Section II we discuss a theory of the inhomogeneous response to an external field of a metal with a single impurity present using a simple model. This discussion is based on a molecular field theory, and is equivalent to the random phase approximation (RPA) of many-body theory. In Section III we describe a number of attempts to improve the RPA description of the response, with the goal of obtaining from first principles a theory accurate in the limit of large local enhancement. In the final section, we discuss some experimental data which provide illustrations of the effect of enhanced local response, and locally enhanced spin fluctuations on a number of properties of alloys. Section IV is not a complete review of all the existing data. We prefer to confine our attention to a few specific examples which provide illustrations of these effects.

II. Theory of Local Spin Fluctuations Associated with Paramagnetic Impurities in Metals: Mean Field Description

In this section, we present a brief derivation of the response to an external magnetic field of a paramagnetic metal which contains a single paramagnetic impurity. We consider a highly simplified model of the metal similar to that used in the early papers [5, 6] on the subject.

After a discussion of the properties of this model, we conclude with some brief remarks on some extensions of the theory to other situations that have proved useful.

In a metal of the transition metal series, in which the Fermi level lies within the energy bands formed from the d levels of the free atoms, in many cases the dominant contribution to the density of states at the Fermi level comes from regions of the Fermi surface in which the wave functions have strong d character. This is particularly true near the right end of the transition metal series, where pockets of d-like holes near the zone boundary give the dominant contribution to the density of states. A qualitatively correct description of the wave functions and energy bands in this region of the zone results from the tight binding approach to the band structure problem. For the moment, we confine our attention to host materials for which this description is a reasonable one. Furthermore, effects associated with the presence of orbital degeneracy will also be neglected. Later in this section, we comment on the extension of the theory to the case of noble metal host matrices, where the conduction-electron wave functions and conduction bands have nearly free electron character, and on the effect of including orbital degeneracy in the theory.

In the one-electron theory of metals, the Hamiltonian H of the pure metal is equal to the kinetic energy T of the electrons. Let $C_{\mathbf{k}\sigma}$, $C_{\mathbf{k}\sigma}^+$ be the annihilation and creation operator associated with the Bloch state of wave vector $\mathbf{k}$ and spin σ. The operators $C_{i\sigma}$, $C_{i\sigma}^+$ destroy and create electrons in the Wannier state of spin σ, in the ith unit cell. Then

$$T = \sum_{\mathbf{k}} E(\mathbf{k}) C_{\mathbf{k}\sigma}^+ C_{\mathbf{k}\sigma} = \sum_{ij} T_{ij} C^+(i\sigma) C(j\sigma), \qquad (2.1)$$

where $E(\mathbf{k})$ describes the shape of the energy band, and T_{ij} is the matrix element for hopping of an electron from site i to site j of the lattice.

To describe the interacting band of electrons we add to Eq. (2.1) the Coulomb interactions between the electrons. Following Wolff [4c] and Hubbard [7], we presume that in the tight binding limit, two electrons interact only when they are in the same unit cell. Thus we retain in the interaction term only the intra-atomic Coulomb interaction U_0. In the one-band model, the Hamiltonian then assumes the form

$$H_0 = T + U_0 \sum_i n_{i\uparrow} n_{i\downarrow}, \qquad (2.2)$$

where $n_{i\sigma} = C_{i\sigma}^+ C_{i\sigma}$ is the number operator for electrons in the Wannier function centered at the ith site.

Now add a single impurity at site I of the matrix. The impurity has two effects. First of all, the impurity produces a change in the crystal potential. Since this potential is efficiently screened, the disturbances in the periodic potential will be confined principally to the impurity cell. We describe this effect by adding the term $V \sum_\sigma n_{I\sigma}$ to H_0. Secondly, the strength of the intra-atomic Coulomb potential U will assume a value U_I in the impurity cell that differs from the value U_0 appropriate to the host. Thus the Hamiltonian for the host matrix with one impurity added becomes

$$H = H_0 + V \sum_\sigma n_{I\sigma} + \Delta U n_{I\uparrow} n_{I\downarrow}, \qquad (2.3)$$

where $\Delta U = U_I - U_0$.

Now suppose an external magnetic field is applied to the system. The magnetic field is parallel to the z axis, varies in space, and varies in time like $e^{i\Omega t}$. If h_i is the amplitude of the magnetic field in cell i, then if h_i is measured in appropriate units, the effect of **h** is to add to Eq. (2.3) the Zeeman energy

$$H_Z = -\tfrac{1}{2} \sum_i (n_{i\uparrow} - n_{i\downarrow}) h_i e^{i\Omega t}, \qquad (2.4)$$

where the z component of spin $s^z(i)$ in cell i is described by the operator $\tfrac{1}{2}(n_{i\uparrow} - n_{i\downarrow})$.

In the paramagnetic state, the expectation value $\langle s^z(i) \rangle$ of the operator $s^z(i)$ is zero. In the presence of the external field, we introduce the dynamical susceptibility $\chi(i, j; \Omega)$ of the alloy as follows:

$$\langle s^z(i) \rangle = \tfrac{1}{2} \sum_j \chi(i, j; \Omega) h_j. \qquad (2.5)$$

In the pure metal, $\chi(i, j; \Omega)$ is a function only of $i - j$. This is not so in the alloy, since the presence of the impurity breaks down translational invariance.

To find the response of a gas of electrons described by the Hamiltonian of Eq. (2.3) leads us into a very complex many-body problem. We shall employ a very simple dynamic molecular field approximation to obtain the form of this response. The present approach, while it involves an approximation that is hard to justify from first principles, leads to a workable expression for the response, and gives a physical picture of the origin of the contribution to the inhomogeneity in the response of the alloy to the perturbation. Attempts to improve the result obtained from this procedure are discussed in Section III. We begin by noting that in the presence of the external field,

$$\langle n_{i\uparrow} \rangle = \bar{n}_i + \langle s_i^z \rangle e^{i\Omega t}, \qquad \langle n_{i\downarrow} \rangle = \bar{n}_i - \langle s_i^z \rangle e^{i\Omega t},$$

3. SPIN FLUCTUATIONS AROUND IMPURITIES

where $\bar{n}_i$ is time independent. We now replace the electron–electron interaction term in the Hamiltonian by a (time-dependent) single-particle Coulomb potential

$$\sum_i U_i \langle n_{i\uparrow} \rangle n_{i\downarrow} + \sum_i U_i n_{i\uparrow} \langle n_{i\downarrow} \rangle. \tag{2.6}$$

When this approximate form for the interaction term is inserted into Eq. (2.3), the Hamiltonian (including the Zeeman term) becomes

$$H_{\text{eff}} = T + V_{\text{HF}} \sum_\sigma n_{I\sigma} + \tfrac{1}{2} e^{i\Omega t} \sum_i (n_{i\uparrow} - n_{i\downarrow}) h_i^{(\text{eff})},$$

where $V_{\text{HF}} = V + \Delta U \bar{n}_I$ is the Hartree–Fock approximation to the change in crystal potential that the impurity produces, and $h_i^{(\text{eff})}$ is an effective magnetic field seen by an electron in cell i:

$$h_i^{(\text{eff})} = h_i + 2 U_0 \langle s_i^z \rangle + 2 \Delta U \langle s_I^z \rangle \delta_{i,I}. \tag{2.7}$$

The subsequent discussion shall be simplified by ignoring the effect of potential scattering; that is, we set $V_{\text{HF}} = 0$. The theory that results strictly speaking then applies only to isoelectronic alloys where the effect of potential scattering may be small in many cases. For the case where $U_0 = 0$, the form of the response when $V_{\text{HF}} \neq 0$ has been discussed in detail elsewhere [5]. Even if $V_{\text{HF}} = 0$, the response of the electrons to a uniform applied field is inhomogeneous because of the term in Eq. (2.7) proportional to ΔU. In addition to this effect, the electrons "see" an additional spatially inhomogeneous field if $\langle s_i^z \rangle$ is different from the value appropriate to the host in the vicinity of the impurity, because of the term in Eq. (2.7) proportional to U_0.

We must now solve for $\langle s_i^z \rangle$ in a self-consistent manner. After the neglect of V_{HF}, examination of the effective Hamiltonian shows that one must consider the response of the noninteracting gas of electrons in the pure matrix to the spatially inhomogeneous magnetic field exhibited in Eq. (2.7). Let $\chi_0(i - j; \Omega)$ denote the response of the noninteracting conduction electrons of the pure metal in cell i to a time-dependent field of frequency Ω applied to cell j. Then when $U_0, \Delta U \neq 0$ one obtains the following equation from which $\langle s_i^z \rangle$ may be determined:

$$\langle s_i^z \rangle = \tfrac{1}{2} \sum_j \chi_0(i - j; \Omega) h_j + U_0 \sum_j \chi_0(i - j; \Omega) \langle s_j^z \rangle + \Delta U \chi_0(i - I) \langle s_I^z \rangle. \tag{2.8}$$

The solution to Eq. (2.8) may be obtained by straightforward means.

It will be useful to introduce some new definitions before the form of the solution is exhibited. Let

$$\chi_0(\mathbf{q}, \Omega) = N^{-1} \sum_i \exp(i\mathbf{q} \cdot \mathbf{x}_i) \chi_0(i, \Omega), \quad (2.9a)$$

$$\tilde{\chi}(\mathbf{q}, \Omega) = \chi_0(\mathbf{q}, \Omega)/[1 - U_0 \chi_0(\mathbf{q}, \Omega)] \quad (2.9b)$$

$$\tilde{\chi}(i - j; \Omega) = N^{-1} \sum_\mathbf{q} \exp[i\mathbf{q} \cdot (\mathbf{x}_i - \mathbf{x}_j)] \tilde{\chi}(\mathbf{q}, \Omega) \quad (2.9c)$$

and

$$\bar{\chi}(\Omega) = N^{-1} \sum_\mathbf{q} \tilde{\chi}(\mathbf{q}, \Omega). \quad (2.9d)$$

In Eqs. (2.9), $\chi_0(\mathbf{q}, \Omega)$ is the dynamic susceptibility of the d electrons of the host in the absence of interaction, and $\tilde{\chi}(\mathbf{q}, \Omega)$ is the response of the host d electrons in the presence of U_0. It is the function $\tilde{\chi}(\mathbf{q}, \Omega)$ that plays the central role in the theory of enhanced spin fluctuations (paramagnons) in nearly ferromagnetic pure metals [3a,b]. The quantity $\tilde{\chi}(i - j, \Omega)$ describes the response of cell i of the pure metal to a time-dependent field (of frequency Ω) applied to the jth cell. Finally, $\bar{\chi}(\Omega) = \tilde{\chi}(i - j, \Omega)|_{i=j}$. With these definitions, the response function $\chi(i, j; \Omega)$ of the alloy is

$$\chi(i, j; \Omega) = \tilde{\chi}(i - j; \Omega) + \Delta U \frac{\tilde{\chi}(i - I; \Omega) \tilde{\chi}(I - j; \Omega)}{1 - \Delta U \bar{\chi}(\Omega)}. \quad (2.10)$$

Equation (2.10) is a central result. We now examine its properties. We denote the static limit ($\Omega = 0$) of the response functions in Eq. (2.10) by appending a subscript s to the response function, that is, we denote $\bar{\chi}(0)$ by $\bar{\chi}_s$.

Consider the spin density induced by a spatially uniform, static magnetic field h_0. Let $\alpha = [1 - \Delta U \bar{\chi}_s]^{-1}$. Then from Eq. (2.10),

$$\langle s_i^z \rangle = \tfrac{1}{2} h_0 \tilde{\chi}_s (1 + \alpha \Delta U \tilde{\chi}_s(i - I)). \quad (2.11)$$

Note that $\tilde{\chi}_s$ is the static (exchange-enhanced) susceptibility of the host.

Far from the impurity, $\tilde{\chi}_s(i - I)$ is small, and $\langle s_i^z \rangle$ assumes the value appropriate to the host. If $\Delta U > 0$, then the static response near the impurity is enhanced to a value greater than that appropriate to the host. In particular, note that

$$\langle s_I^z \rangle = \tfrac{1}{2} h_0 \tilde{\chi}_s \alpha. \quad (2.12)$$

3. SPIN FLUCTUATIONS AROUND IMPURITIES

The quantity α is called the local exchange enhancement factor [5, 6]; α provides a measure of how much greater the response is in the impurity cell, when compared to the host response.

The range of the inhomogeneity in the spin response is determined by the range of $\tilde{\chi}_s(i - I)$. If the magnetic response of the matrix is well described by the free electron model (no exchange enhancement; $U_0 = 0$), then the range of $\tilde{\chi}_s(i - I)$ is of the order of a lattice constant. If U_0 is so large that the metal is nearly ferromagnetic, then the range of $\tilde{\chi}_s(i - I)$ can be quite large [8], and the disturbance in the spin response produced by the impurity will have a considerable spatial extent.

It will be useful to exhibit an expression for the static susceptibility $\chi_s^{(A)}$ of the alloy:

$$\langle s_{\text{tot}}^z \rangle = \sum_i \langle s_i^z \rangle = \frac{N}{2} h_0 \chi_s^{(A)} = \frac{Nh_0}{2} \left[\tilde{\chi}_s + \alpha \frac{\Delta U}{N} \tilde{\chi}_s^2 \right]. \quad (2.13)$$

The term proportional to ΔU is the contribution from the impurity.

If we imagine increasing ΔU by increasing U_I, then α increases and eventually becomes infinite. The point where α becomes infinite corresponds to the value of $U_I = U_I^{(c)}$ for which HF theory predicts that a local moment will form.[†] Thus, for $U_I > U_I^{(c)}$, our assumption that the ground state is paramagnetic breaks down. In Section IV, we shall see that in a number of nonmagnetic alloy systems, the data indicate that α is large compared to unity, that is, the impurity is "nearly magnetic," in much the same sense as a pure metal with a strongly exchange-enhanced susceptibility is "nearly ferromagnetic."

If $U_I < U_0$, then $\Delta U < 0$ and $\alpha < 1$. In this case, the impurity acts to *suppress* the magnetic response of the matrix locally.

So far we have confined our attention to the response of the matrix to an applied field. When $\alpha \gg 1$, the amplitude of the (intrinsic) spin density fluctuations near the impurity is strongly enhanced over the value of the fluctuations far from the impurity. The response function $\chi(i, j; \Omega)$ may be used to obtain the amplitude of the spin density fluctuations in the alloy. This follows from the fluctuation–dissipation theorem, which in this particular case takes the following form. Let

$$S_{ij}(\Omega) = \int_{-\infty}^{+\infty} \frac{dt}{2\pi} e^{i\Omega t} \langle s_i^z(0) s_j^z(t) \rangle.$$

[†] When $U_0 = 0$, this point has been discussed explicitly by Mills and Lederer [9].

The quantity $S_{ii}(\Omega)$ measures the amplitude of the spin density fluctuation in cell i with frequency Ω. Then

$$S_{ij}(\Omega) = \frac{1}{2\pi i} n(\Omega) \mathscr{I}m[\chi(i,j;\Omega)], \qquad (2.14)$$

where $n(\Omega) = [\exp(\hbar\Omega/k_B T) - 1]^{-1}$ is the Bose–Einstein factor.

From Eq. (2.14), one finds that the low-frequency spin density fluctuations in the alloy can become very large near the impurity. To see this, note that for small Ω, $\mathscr{I}m[\tilde{\chi}(\Omega)] = i\tilde{\chi}'\Omega$, where $\tilde{\chi}'$ is a parameter characteristic of the host. Then far from the impurity,

$$S_{ii}(\Omega) = \frac{\tilde{\chi}'}{2\pi} n(\Omega)\Omega \qquad (2.15)$$

for small frequencies. At the impurity site

$$S_{II}(\Omega) = \frac{n(\Omega)}{2\pi} \frac{\chi'\Omega}{\alpha^{-2} + (\Delta U \tilde{\chi}')^2 \Omega^2} \qquad (2.16a)$$

$$= \frac{n(\Omega)}{2\pi} \alpha^2 \chi' \Omega \qquad \text{for small } \Omega. \qquad (2.16b)$$

Equation (2.16b) shows that for small Ω, the spin density fluctuations in the impurity cell can be very strongly enhanced indeed. The low-frequency spin fluctuations at the impurity site are larger than those in the bulk by the factor α^2. In Section IV, we shall see that these large-amplitude, low-frequency fluctuations lead to a very large enhancement of the core polarization contribution to the longitudinal nuclear relaxation rate T_1^{-1}, strong concentration-dependent contributions to the portion of the electrical and thermal resistivity that arises from electron–electron scattering, and large enhancements of the specific heat.

Equation (2.15) gives only the low-frequency limit of $S_{ii}(\Omega)$. In the host, $S_{ii}(\Omega)$ will exhibit a broad maximum at a frequency of the order of $E_F/\hbar$. [This is true even in strongly exchange-enhanced materials since $S_{ii}(\Omega)$ obtains its largest contribution from fluctuations of the large wave vector, which are only weakly enhanced.] On the other hand, Eq. (2.16a) shows that when $\alpha \gg 1$, $S_{II}(\Omega)$ contains a strong peak at the local spin fluctuation frequency $\Omega_{\rm sf} = (\Delta U \tilde{\chi}' \alpha)^{-1}$. One expects $(\Delta \tilde{\chi}' \Delta U)^{-1}$ to be typically of the order of $E_F/\hbar$ in magnitude. Thus when $\alpha \gg 1$, $\Omega_{\rm sf} \ll E_F/\hbar$, and $S_{II}(\Omega)$ contains a strong low-frequency peak.

We conclude this section with brief remarks about certain features of the above-mentioned derivation, as well as extensions of the theory that have appeared.

First, it should be noted that the introduction of the dynamic effective field to compute the response function is equivalent to the random phase approximation (RPA) of many-body theory. A derivation of the response function by a Green's function decoupling scheme equivalent to the RPA has been carried out for a closely related problem [9]. As mentioned above, the effect of potential scattering ($V_{HF} \neq 0$) on the properties of a locally enhanced paramagnetic metal have been discussed elsewhere for the case where $U_0 = 0$ [5].

If a transition metal impurity is placed in a noble metal host, the Friedel–Anderson Hamiltonian [4a,b,c] provides an appropriate description of the system. Subsequent to the work of Lederer and Mills [5, 6, 10], Rivier and co-workers [11] have presented a discussion of local spin fluctuations in the Friedel–Anderson model. The physical picture obtained by these authors is identical to the one described above, and differs only in detail. The expression for Ω_{sf} and the susceptibility of the impurity cell at low temperatures are identical for the two models, and the frequency dependencies of the self-energy of the electron in the impurity cell are very similar. However, different mechanisms have been proposed to explain the temperature dependence of the electrical resistivity in these two cases: Rivier and co-workers have also applied the theory to the case where $kT \gg \hbar\Omega_{sf}$, while the earlier work confined its attention to the low-temperature region $kT \ll \hbar\Omega_{sf}$. When $kT \gtrsim \hbar\Omega_{sf}$, it is found [11] that some predictions of the local spin fluctuation theory assume a form qualitatively similar to the behavior expected from the Kondo effect. It has been argued [11] that the theory of the Kondo effect and of local spin fluctuation effects are equivalent. While it may be true that certain properties of the alloy are sensitive only to the existence of a long correlation time for spin fluctuations in the impurity cell, and are insensitive to the precise mechanism responsible for the long correlation time, the physical processes currently believed responsible for the Kondo anomaly are not included in the simple forms of the local spin fluctuation theory discussed in this section.

Our discussion has neglected all effects associated with orbital degeneracy. The effect of orbital degeneracy in the impurity cell has been discussed within the framework of the Friedel–Anderson model by Caroli *et al.* [12], and also by Dworin and Narath [13]. We discuss the principal features of this work in Section IV.

A more careful examination of the theory of local spin fluctuations shows that when $\alpha \gg 1$, the validity of the RPA description breaks down. A number of attempts to provide a complete description of the phenomena have been carried out. These theories are reviewed in Section III. While the investigations employ a variety of powerful and sophisticated approaches, at the time of this writing, no complete and satisfactory description of local spin fluctuations has appeared valid when $\alpha \gg 1$. It is therefore useful to note that the structure obtained for $\chi(i, j; \Omega)$ from the RPA [Eq. (2.10)] is quite general. One may show [14] that the dynamic susceptibility has a form identical to Eq. (2.10) even if the RPA is not valid, provided the response functions $\tilde{\chi}(i - j, \Omega)$, $\tilde{\chi}(\Omega)$ are replaced by suitably defined irreducible particle–hole propagators. Thus many of the results derived from the RPA form of the theory may have more general validity, provided the local spin fluctuation frequency is regarded as a phenomenological parameter that may differ significantly from the value provided by the RPA.

III. Renormalized Theories of Local Spin Fluctuations: Nonmagnetic and Magnetic Cases

1. Introduction

In this section we outline the major difficulties encountered once one realizes that actually the RPA description of local spin fluctuations breaks down in the moment formation region, so that it appears necessary to build up a renormalized theory including higher order corrections in the perturbation expansion, beyond RPA. Indeed, Hartree–Fock (HF) [4a,b,c] and improved HF theories [15] of local moment formation in metals lead to an unphysically sharp transition from the nonmagnetic to the magnetic state. At the same time, rotational invariance of the theory is not maintained, so that it is not possible to discuss a number of fundamental properties of the alloy, such as the magnetic susceptibility, contribution of the impurity to the specific heat, and so on. A renormalized theory should therefore be constructed in a rotationally invariant manner, and should in principle describe a more realistic smooth transition from the magnetic to the nonmagnetic regime. Unfortunately the same difficulties as those encountered in real phase transition problems arise here. Starting from the HF criterion for formation of a local moment ($\alpha \to \infty$), all the corrections to the RPA happen to be equally important and to diverge to infinite order. The theory of the transition appears to be unrenormalizable, despite considerable efforts displayed by several authors using sophisticated methods. Outside the transition region (or intermediate coupling regime), when the impurity is not very close to becoming magnetic (weak coupling regime), or, on the other side, when a local moment is well formed (strong coupling regime), then perturbation theory holds again and it is possible to select the leading corrections to a simple zero order theory. But the description of the transition region remains so far an open matter. A possible means of constructing a semiphenomenological theory in the transition region is to take for granted the existence of a renormalized local enhancement α_{ren} (different from the HF one and thus not infinite), to extract a measured value of α_{ren} from experiments, and to calculate the quantities of interest within RPA but using α_{ren} instead of α. As we shall see in the next section, this procedure has been successfully employed in the theory of transport phenomena in systems with strong local enhancement.

2. Breakdown of Mean Field Theory in the Transition Region

Let us first briefly show that the validity of the RPA and, more generally, that perturbation theory breaks down when $\alpha \to \infty$. For simplicity, we use the Wolff Hamiltonian (2.3) with $V = 0$. Let us consider the actual (renormalized) local susceptibility $\bar{\chi}_{\text{ren}}(\Omega)$ which may be written as follows [14]:

$$\bar{\chi}_{\text{ren}}^{\text{RPA}}(\Omega) = \frac{\bar{\chi}^i(\Omega)}{1 - \Delta U \bar{\chi}^i(\Omega)}. \qquad (3.1)$$

$\bar{\chi}^i(\Omega)$ is the renormalized irreducible particle–hole bubble (in the sense defined in ref. 4 of Béal-Monod and Mills [14]); $\{1 - \Delta U \bar{\chi}^i(0)\}^{-1} = \alpha_{\text{ren}}$ will be the renormalized local enhancement factor. Note that when $\bar{\chi}^i(\Omega)$ is reduced to the bare HF bubble $\bar{\chi}(\Omega)$, α_{ren} is reduced to $\alpha = \{1 - \Delta U \bar{\chi}(0)\}^{-1} = \{1 - \Delta U \bar{\chi}_s\}^{-1}$, and (3.1) leads back to the bare RPA local susceptibility $\bar{\chi}(\Omega)/\{1 - \Delta U \bar{\chi}(\Omega)\} = \bar{\chi}^{\text{RPA}}(\Omega)$, as used in the previous section.

In order to prove that perturbation theory holds and the RPA is valid to describe the local susceptibility, we must show that the first few corrections to $\bar{\chi}(\Omega)$ in $\bar{\chi}^i(\Omega)$ are small:

$$\bar{\chi}^i(\Omega) = \bar{\chi}(\Omega) + \bar{\chi}^{(1)}(\Omega) + \bar{\chi}^{(2)}(\Omega) + \cdots. \qquad (3.2)$$

One way of calculating the corrections $\bar{\chi}^{(n)}(\Omega)$ to $\bar{\chi}(\Omega)$ is to dress the bare particle–hole bubble representing $\bar{\chi}(\Omega)$ with all possible insertions of local spin fluctuation propagators (i.e., local paramagnon lines), so that $\bar{\chi}^{(1)}(\Omega)$ involves one insertion (in the form of either a self-energy or a vertex correction; see, for instance, Béal-Monod and Mills [14, Fig. 1]), $\bar{\chi}^{(2)}(\Omega)$ involves two insertions, and so on. For the details, we refer the reader to the paper by Béal-Monod and Mills [14]. It turns out that the corrections $\bar{\chi}^{(n)}(\Omega)$ are not small and diverge like powers of $\ln(1/\alpha)$ and powers of α. This proves that the RPA description of local spin fluctuations is actually not valid. Furthermore the coefficients in front of the different divergent "corrections" are all of order 1 $\{(1 - 1/\alpha)^n \sim 1$ when $\alpha \to \infty)$, so that the perturbation series as a whole breaks down when the impurity is very close to becoming magnetic in the HF sense.

3. The "Renormalized Random Phase Approximation" (RRPA) of Suhl and Co-Workers

The first who tried to solve the difficulties raised by the improbable sharp HF transition were Suhl [16a], Levine and Suhl [16b], and Levine et al. [16c]. They considered a band of interacting electrons and a local impurity potential, which, in the single-band version of the Wolff [4a,b,c] model, leads to the sharp instability in HF theory. Suhl et al. reasoned that the large local spin fluctuations actually existing near this threshold, if taken into account, would modify the effective impurity potential so that the instability threshold would never quite be reached. The analog of the Wolff instability in many-body theory is that the Bethe–Salpeter equation for a certain reducible vertex Γ should have a singular solution; but Suhl and co-workers have argued that such an instability will occur only if uncorrected single-particle propagators are used. When Γ becomes large, so does the self-energy, so that the propagators diminish, inhibiting the instability. Therefore, they wrote down a set of two coupled nonlinear integral equations for the local susceptibility and the electron propagator including self-energy (and only self-energy) corrections to the electron lines. The integral equations are described graphically in Fig. 1. These equations were solved numerically. Hamann [17], in order to explore this work in more detail, succeeded in giving an approximate analytic solution of the integral equations of Suhl et al. in the Anderson model. An appealing result of all these works was that, as the strength of the local interaction was

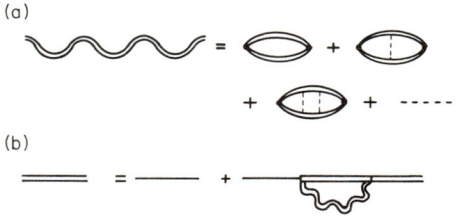

FIG. 1. A graphical depiction of the integral equations studied by Suhl, Levine, Ramakrishnan and Weiner, where (a) describes the local paramagnon propagator (double wiggly line), which is approximated by a series of ladder graphs with renormalized single-particle propagators inserted (double curved lines), and (b) is the Dyson equation for the single-particle Green's function.

increased, the transition was found to occur gradually, as it is expected to be physically, with a susceptibility going smoothly from a Pauli type to a Curie one, flattening out to a constant at low temperatures [16a,b,c]. Furthermore, in the case of strong coupling, Suhl *et al.* found a Kondo-type resistivity in ln T over a wide range of temperature. Unfortunately their treatment is not valid in the strong coupling limit although they extrapolated it into this region. In particular in this region, where a local moment is formed, Hamann [17] showed that the characteristic temperature T_{sf} involved in the work of Suhl *et al.*, which is expected to be identified with the Kondo temperature T_K, is actually exponentially smaller than the value one expects for T_K from elementary considerations. Moreover the Curie constant obtained at high temperatures [16a,b,c] is much smaller than the value appropriate to spin $\frac{1}{2}$. Finally Suhl *et al.* could verify the consistency of their approximations with the requirement of spin conservation only in the weak coupling limit. One then can reasonably suppose that the diagrams considered by Suhl and his co-workers are not sufficient and that some important class of diagrams is missing in their work. In fact, as is explained in Subsection 5, vertex corrections, which were omitted by Suhl and co-workers, turn out to be at least as important as self-energy corrections only retained in the theory discussed above.

4. The Functional Integral Method of Wang et al.

The functional integral approach of Wang *et al.* [18] is mentioned just briefly here, since Chapter 8 is devoted to the details of this method. This approach replaces the system of interacting electrons by a system of noninteracting electrons but moving in a local but time-dependent external magnetic field, the field being averaged over with a Gaussian weight function. Wang *et al.* [18], in the Anderson model, then found the proper values for the limiting Pauli and Curie susceptibilities for weak and strong coupling, respectively, and, in this last case they found a Curie law without starting from a zero-order approximation which already contains a moment. For the intermediate coupling, in the transition range from magnetism to nonmagnetism, they obtained, numerically, a smooth interpolation between the two regimes. However, some aspects of this work have been criticized [19], as is shown in Subsection 6. In particular the validity of the so-called RPA' that results from this approach is open to question together with the possibility of obtaining an actual meaningful description of the transition region with the functional integral method applied to the intermediate coupling regime. Although the functional integral method seems *a priori* much more powerful than the diagrammatic method, Béal-Monod et al. [19] as well as Keiter [20] argue that this method does not turn out to be better than other manybody approaches.

5. Importance of Vertex Corrections

a. *Nearly magnetic case.* We noted in Subsection 3 that the theory of Suhl et al. most likely failed because it dropped an important class of diagrams. Béal-Monod and Mills [14] showed that vertex corrections in the local susceptibility irreducible diagrams are at least as important as self-energy corrections, so that in the diagrams of Suhl et al., the first diagram on the right-hand side of Fig. 1a should actually be replaced by

instead of

For instance, in Eq. (3.2), the first correction $\bar{\chi}^{(1)}(\Omega)$ to $\bar{\chi}(\Omega)$ contains vertex corrections (b) as well as self-energy corrections (a):

(a)   (b)

Both are equally divergent, for intermediate coupling ($\alpha \to \infty$), so that

$$\bar{\chi}^{(1)}(0) \sim \ln \alpha. \tag{3.3}$$

In the same manner, self-energy and vertex corrections contribute to give

$$\bar{\chi}^{(2)}(0) \sim \ln \alpha + (\ln \alpha)^2. \tag{3.4}$$

Moreover with three local paramagnon insertions, a new type of divergence arises from vertex corrections and one gets

$$\bar{\chi}^{(3)}(0) \sim \ln \alpha + (\ln \alpha)^2 + (\ln \alpha)^3 + \alpha \tag{3.5}$$

with all coefficients of order 1. At this point, it is clear that the class of diagrams involving vertex corrections must be considered as well as self-energy corrections. Another important remark results from the analysis above: For a given order, say (n), the next to leading divergent diagrams in the next higher order, say (n + 1), give contributions comparable to the leading divergence in the order (n) [cf. (ln α)² in $\bar{\chi}^{(3)}(0)$ and in $\bar{\chi}^{(2)}(0)$]. So, at that point, it already looks impossible to obtain accurate results by performing a partial summation of the most divergent diagram in each order. Béal-Monod *et al.* [19] actually reformulated the problem in a rotationally invariant way. Then they could calculate, with the help of a form of dimensional analysis well known in quantum field theory [21], the leading divergence in any diagram of "order" n (i.e., with n local paramagnon insertions):

$$\bar{\chi}^{(n)}(0) \sim (\ln \alpha)^{n_0} (\alpha)^{n_1-1}, \qquad (3.6)$$

where n_0 is the number of completely independent paramagnon frequencies (those paramagnons with both ends attached to the same electron line), and n_1 is the number of closed electron loops [so that there exists $(n_1 - 1)$ energy conservation relations among the $(n - n_0)$ remaining paramagnons, both of whose ends are attached to different electron lines]. Considering (3.6) leads to a convenient redefinition of the "order" of a diagram which can be more useful and avoids the difficulties arising when two (or more) paramagnon frequencies are identical. Instead of referring to the number n of inserted paramagnons, the "order" of a given diagram would then be equal to the number m of independent paramagnon frequencies in that diagram [22a].[†] In the following, we adopt this new definition. The quantity $\bar{\chi}^{(m)}(0)$ is defined as the correction to $\chi(0)$ which contains m "independent" paramagnons. With this new definition, as $\alpha \to \infty$, one obtains

$$\bar{\chi}^{(m)}(0) \sim (\ln \alpha)^m + \sum_{p=0}^{m-1-q} \sum_{q=0}^{m-1} \{(\ln \alpha)^p \alpha^q\}, \qquad (3.6a)$$

with again, coefficients all of order 1, in the intermediate coupling regime. Then in a classification of the diagrams into skeleton [23] [corresponding to $n_0 = 0$ in (3.6)] and nonskeleton $(n_0 \neq 0)$ types, Béal-Monod *et al.* [19] noted that this series of nonskeleton-type diagrams can be summed up leading to a partially renormalized fluctuation lifetime analogous to the RRPA of Suhl *et al.* and Hamman, and consistent with it. But the series of skeleton diagrams remains impossible to sum up although it contains, at least as far as static properties are concerned, crucial divergences involved in the vertices of interacting spin fluctuations as indicated below:

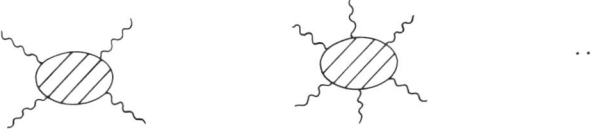

[†] See also a similar classification of diagrams in the theory of superconductivity [22b].

They conclude that, in the intermediate coupling regime one should consider altogether the renormalized quantities represented by the infinite series of fluctuation–fluctuation vertices set above, and the renormalized electron and fluctuation self-energies and the electron–fluctuation vertex (see diagrams a, b, and c below).

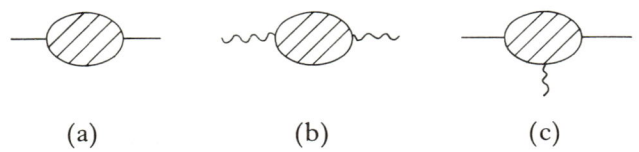

(a)   (b)   (c)

In Subsection 6 we show that an answer to that problem can be formulated with the help of the functional integral method but only for a much weaker coupling regime. Therefore in the intermediate coupling regime, the problem is essentially unrenormalizable and no simple consistent renormalization procedure appears to exist.

b. *Magnetic case.* In the case of strong coupling, Weiner [24] tried to take care of some important vertex corrections in Suhl's equations by using parquet diagrams. He found a value for the Curie constant of the susceptibility which is better than in the nonparquet results of Suhl *et al.* [16a,b,c] but still too small. The interesting point of that work is the appearance of a two-peaked frequency-dependent single-particle density of states reminiscent of those found in HF by Anderson [4b], on each side of the resonance state close to the Fermi level, although the background between the two peaks is too strong to obey normalization conditions. This may be due to some of the approximations, in particular, to the fact that Weiner [24], as in the first papers of Suhl *et al.* [16a,b,c], used for the bare paramagnon mode a δ function peaked at zero frequency while the actual mode has a finite frequency width.

More recently, Weiner [25], still using a δ-function mode for the bare fluctuation, tried to improve his numerical calculations by including some vertex corrections from the self-consistent parquets which were left out of his previous work [24]. The result was that he could not find any Curie law behavior over the whole range of temperature: At low T, the susceptibility turns out to behave like $T^{-2/3}$ and the power of T is even weaker at higher temperature. He also mentioned [25] some speculations which could be thought of in the case where one could keep real frequency dependence of the fluctuation into the parquets. The parquet renormalizations of the vertex are expected to be weaker, because one could not get perfect overlap of self-consistent fluctuation propagators; however, and this is an important point to be noted, this renormalization should still be strong enough to increase the self-consistent spin fluctuation temperature value, maybe up to the correct range as predicted in the Anderson model for the Kondo temperature; it might also be that the conserving susceptibility would not diverge at $T > 0$ (as it did in all the preceding works using Suhl's theory [16a,b,c, 24, 25]) although the effective Curie-type constant may be increased drastically. Unfortunately it seems that any attempt to get the real frequency dependence of the paramagnon propagator into the parquet calculation would turn out to be an incredibly difficult computing task. At any rate, Weiner [25] does not expect the real frequency dependence of the vertex to give anything like a Curie law behavior, so that if this is correct the parquet method does not look like a powerful and adequate treatment of the local moment formation problem.

6. Failure of the RPA' in the Functional Integral Approach except in the Weak Coupling Regime

In the weak coupling regime, that is, when the impurity is not very close to becoming magnetic, Béal-Monod et al. [19] showed that the renormalization described in Subsection 5a can be solved through the functional integral method. More specifically they showed that, for the static properties, the set of renormalized quantities to be considered may be restricted, so that one only needed to study the system of interacting fluctuations. To do so they used a functional Lagrangian which, in the weak coupling limit, reduces to two terms: one describing a free fluctuation, and the other, two fluctuations interacting through a bare electron loop (see a and b below, respectively).

(a)   (b)

They could then write the renormalized local spin fluctuation propagator in a closed form. However, the practical evaluation of the functional integral, even in this simplified weak coupling limit, is very difficult. In the "static" approximation (see Schrieffer and Hamann [25a]) Béal-Monod et al. [19] could obtain an explicit result only at very low temperature with the saddle-point method. The next approximation, the RPA' (see Schrieffer–Hamann), either does not give a sensible improvement over the "static" approximation or, if it does, the higher order corrections are still very important. That is why we mentioned in Subsection 4 that there is some question about the results obtained by Schrieffer *et al.* [18]. Actually Béal-Monod *et al.* [19] have found that when the "static" approximation is valid (i.e., RPA' and higher order approximations may be neglected) the functional integral approach is equivalent to just retaining the first few terms in the diagrammatic series expansion, so that this can be obtained as well diagrammatically. They pointed out that the same conclusion must be drawn in the Anderson model studied by Schrieffer *et al.* [18] so that in their case the "static" approximation, evaluated with the saddle-point method, leads back to the early Anderson result [4a,b,c]. An important remark was noted [[19]: The interactions between fluctuations actually prevent the formation of a local moment, so that the description of the transition to magnetism, although it should take into account interactions between fluctuations, is made at the same time more difficult. In a more recent paper, Schrieffer *et al.* [26] took into account the system of interacting fluctuations but arrived also at the same conclusion that the intermediate coupling region is still poorly described either by existing diagrammatic or by the functional integral methods.

7. Phenomenological Renormalization

Since it is clear that serious difficulties arise in the calculation of the actual renormalized local susceptibility in the intermediate coupling regime, what can be done as far as the interpretation of experiments is concerned? For physical reasons we have shown that $\alpha_{\text{ren}} \neq \alpha$ so that $\alpha_{\text{ren}} \nrightarrow \infty$. One can then take that condition for granted together with the observation that the renormalized local susceptibility has the form (3.1). Then one may use, for practical purposes, this formula with a phenomenological α_{ren} which one does not know how to calculate, but which may be extracted from experiments. Then, knowing this phenomenologically renormalized local susceptibility, one could, in principle, proceed to calculate the susceptibility of the alloy. (See, for instance, Béal-Monod and Mills [14, formula (2)].)

8. Conclusion

As a conclusion of this section one should say that the intermediate coupling regime and the transition from nonmagnetic to magnetic cases remain an open question. But any theory dealing with that case should definitely take into account interactions between local fluctuations which have been proved [14, 19] to play a crucial role. In the strong coupling limit recent developments by Hamann [27] look promising. In the weak coupling regime the functional integral method turns out to be not more powerful than the perturbational diagrammatic method. Both give as well the first few corrections to the bare RPA result. Note that, in the weak coupling regime, the recent functional integral general formulation of Hamann [27] reduces to the result obtained previously by Béal-Monod et al. [19] as described in Subsection 6.

IV. The Effect of Local Spin Fluctuations on the Properties of Dilute Alloys

In this section, we present a brief, qualitative discussion of the effect of strong local enhancement, and the associated large-amplitude local spin fluctuations on some properties of the dilute alloy. We also mention a number of experimental observations which appear consistent with the local enhancement model. As we mentioned at the end of Section I, we shall cite only a limited number of experimental results. Our aim is to provide a sketch of the kind of effects one may expect to observe when the local enhancement is strong rather than to present a complete review of existing data. We note that in a recent article [28], one of us has presented a more extensive review of a number of the points discussed below. We consider various properties of the alloy in turn. The discussion is based primarily on the results of the mean field theory of the dynamic susceptibility presented in Section II.

1. The Static Susceptibility of the Alloy

In the low-concentration limit, where the change in susceptibility varies linearly with impurity concentration c, from Eq. (2.13) one has for the isoelectronic alloy[†]

$$\frac{1}{\chi}\frac{d\chi}{dc} = \Delta U \tilde{\chi}_s \alpha, \tag{4.1}$$

where $\tilde{\chi}_s$ is the static susceptibility of the host. Thus when the local enhancement is strong ($\alpha \gg 1$), $\chi^{-1}(d\chi/dc)$ becomes very large. The effect is particularly pronounced in hosts with large exchange enhancement, such as Pd, since in such a material $\tilde{\chi}_s$ is anomalously large.

A striking example of an alloy for which $\chi^{-1}(d\chi/dc)$ is large is provided

---

[†] When $U_0 = 0$ (in Lederer and Mills [5]), it is demonstrated that the effect of potential scattering is to add a term $\Delta n/n_0$ to Eq. (4.1) where Δn and n_0 are the change in local density of states at E_F, and the density of states of the host at E_F.

by dilute alloys of Ni in Pd. From the data of Shaltiel et al. [29], one estimates $\chi^{-1}(d\chi/dc) \approx 100$. From this value, and data on the resistivity of these alloys (discussed below), an analysis indicates that $\alpha \approx 20$ for this impurity. We shall see that a number of properties of dilute **PdNi** alloys appear strongly influenced by local enhancement effects. For dilute **PtNi** alloys [30], $\chi^{-1}(d\chi/dc) \approx 10$ or 12. A large portion of the difference between these two systems has its origin in the large value of χ_s produced by the strong exchange enhancement of the Pd matrix; the exchange enhancement is nearly absent from Pt. From the rigid band model, one expects $\chi^{-1}(d\chi/dc)$ to be close to unity for an unenhanced transition metal host such as Pt. Thus the large value of $\chi^{-1}(d\chi/dc)$ suggests that $\alpha \gg 1$ in **PtNi** also.

As we saw in Section III, one can say little from the theory about the temperature dependence of the impurity contribution χ_{imp} to χ, when $\alpha \gg 1$. If one can make an analogy between locally enhanced alloys, and nearly ferromagnetic pure materials,[†] then one would expect $(d\chi/dc)$ to decrease in magnitude when $T \gtrsim T_{\text{sf}} = \hbar\Omega_{\text{sf}}/k_B$. Rivier and Zukermann [11] have suggested that $\chi_{\text{imp}} \propto (T + T_{\text{sf}})^{-1}$.

As remarked in Section II, Rivier and Zukermann have pointed out that an impurity in a noble metal host may also exhibit local enhancement effects [11]. They suggest that **AlMn** is such a system, with $T_{\text{sf}} \sim 500°\text{K}$.

2. The Change in Specific Heat upon Alloying

The large-amplitude spin fluctuations present in systems with strong local enhancement lead to a strong concentration dependence of the coefficient γ of the term linear in temperature in the electronic specific heat. For the case where exchange enhancement is unimportant in the host matrix, a theory of this phenomena for the one-band model has been presented by Lederer and Mills [10]. An extension of the theory of the one-band model to host metals with strong exchange enhancement has been developed by Brinkman et al. [31]. For this more general case,

$$\frac{1}{\gamma}\frac{d\gamma}{dc} = \frac{3}{2}\frac{m}{m^*}\frac{1}{\chi}\frac{d\chi}{dc}\left[(1-\bar{I}) + \frac{\bar{I}}{3}\sigma\right]^{-1}.$$

In this expression, m and m^* are the band structure and specific heat effective mass of the host, respectively, $(1 - \bar{I})^{-1}$ is the exchange enhancement parameter of the host, and σ measures the range in the wave vector **q** of static susceptibility $\chi_0(\mathbf{q}, 0)$ of the host.

[†] See the work of Béal-Monod et al. [3b].

For **Pd**Ni alloys, experimental data show that $\gamma^{-1}(d\gamma/dc) \approx 15$ [32, 33]. With the value for $\chi^{-1}(d\chi/dc)$ quoted earlier, $m^*/m = 1.8$ [34], $\bar{I} = 0.9$, and $\sigma = 6$, the theoretical result gives $\gamma^{-1}(d\gamma/dc)$ too large by roughly a factor of 3. Thus, while it is true that the data show a very large concentration dependence for γ, the agreement between the data and theory of the specific heat enhancement for a one-band model is only qualitative. The theory predicts that $\gamma^{-1}(d\gamma/dc)$ for dilute **Pt**Ni alloys should be very much smaller than in **Pd**Ni, in accord with the data [37], but the theory of the one-band model still overestimates the value of $\gamma^{-1}(d\gamma/dc)$, unless a rather large value of σ is presumed [35]. A theoretical study of $\gamma^{-1}(d\gamma/dc)$ for the Anderson model in the presence of orbital degeneracy in the impurity cell has been carried out by Caroli et al. [12]. These authors find that in the limit of strong local enhancement, the theoretical value of $\gamma^{-1}(d\gamma/dc)$ can be reduced from the value predicted by the one-band model by a factor as large as $1/p$, where p is the degeneracy of the impurity orbital. Thus, while no theoretical study of $\gamma^{-1}(d\gamma/dc)$ has been completed in a host with both orbital degeneracy and strong exchange enhancement, it is clear from the work of Caroli et al. that the one-band model overestimates the contribution of local spin fluctuations to γ because the effect of orbital degeneracy plays a particularly important role when $\alpha \gg 1$.

In dilute **Pd**Ni alloys, Chouteau et al. [36] have observed a large concentration dependence of the T^3 term in the specific heat (from the data, it is difficult to differentiate between a T^3 or $T^3 \ln T$ temperature variation for this contribution). A large T^3 term is expected from the impurity when $\alpha \gg 1$, and a $T^3 \ln T$ term arises when host enhancement is present. The size of the observed effect in **Pd**Ni is in rough accord with theory [38].

We conclude by pointing out that if an isolated impurity is characterized by a large local enhancement factor, then a pair of such impurities may have an effective α large compared with that of the isolated impurity. In this instance, one may observe contributions to γ and χ from pairs, even at low concentrations. This has been pointed out by Tournier and Blandin [37]. These authors have analyzed specific heat and susceptibility data on **Cu**Co, and demonstrate that nearest neighbor pairs of Co impurities are characterized by a local spin fluctuation temperature small compared to that associated with an isolated Co.

3. Nuclear Magnetic Resonance at the Impurity Site

The technique of nuclear magnetic resonance provides a powerful means of probing the magnetic properties of the impurity cell and its vicinity. The contribution K_c to the impurity Knight shift from the core polarization part of the hyperfine field is proportional to the magnetic moment induced in the impurity cell by the external field. From Eq. (2.12), we see that the magnetic moment in the impurity cell is proportional to $\alpha\tilde{\chi}_s$. Thus, when K_c can be extracted from the total shift, one obtains a direct measure of α. (Notice that the total moment induced by the field is $\Delta U \tilde{\chi}_s^2 \alpha$, while the moment induced in the impurity cell is $\tilde{\chi}_s \alpha$. These two quantities differ because of the presence of spin polarization in the matrix outside of the impurity cell. In exchange-enhanced hosts, $\Delta U \tilde{\chi}_s \gg 1$, and this difference can be large.)

The core polarization part of the hyperfine interaction allows the impurity nucleus to relax by coupling to the low-frequency spin fluctuations in the d band. In the limit $\alpha \gg 1$, we have seen in Section II that the amplitude of the low-frequency part of the spin fluctuation spectrum is proportional to α^2 [Eq. (2.16b)]. Thus, when α is large, one observes a large contribution T_{1c}^{-1} to the longitudinal relaxation rate T_1^{-1} from this source. A theory of the core polarization part of the Knight shift and T_{1c} in the presence of local exchange enhancement effects was first presented by Lederer and Mills [5]. For the case where exchange enhancement in the host may be ignored, these authors point out that even though K_c and T_{1c} may be greatly enhanced when $\alpha \gg 1$, the Korringa product $K_c^2 T_{1c} T$ remains unaffected, and assumes a value expected from the one electron picture. The value of $K_c^2 T_{1c} T$ remains equal to the one electron value when $\alpha \gg 1$ even when strong potential scattering is present, provided exchange enhancement in the host may be neglected. This relation appears well satisfied in a number of systems in which local enhancement effects are observed. We refer the reader to a recent paper by Narath [38], which reviews the experimental data. The one electron value for the Korringa product obtains only by virtue of the intra-atomic character of the electron–electron interaction. An interaction with an extended spatial range, such as the form suggested by Clogston [8], would lead to a deviation from the free electron value for the Korringa product.

When exchange enhancement is present in the host, Moriya [39] has shown that for the host nucleus in the pure metal the Korringa product is reduced from the free electron value by a factor R that is related to an average of the exchange-enhanced, wave-vector-dependent susceptibility of the host. If an impurity is added to the exchange-enhanced host, then Lederer [40] has pointed out that in the absence of potential scattering, the Korringa product $K_c^2 T_{1c} T$ is reduced from the free electron value by the same factor R, independent of the local enhancement factor.

We conclude by noting that Dworin and Narath [13] find that when

3. SPIN FLUCTUATIONS AROUND IMPURITIES

$\alpha \gg 1$, the orbital contribution to T_1^{-1} and the Knight shift also experience a considerable enhancement.

4. Transport Properties

Striking evidence for the existence of large-amplitude local spin fluctuations in systems with strong local enhancement comes from data on the temperature dependence of the electrical resistivity ρ and thermal resistivity w of transition metal alloys.

The scattering of electrons by spin fluctuations leads to a term in ρ proportional to T^2 [41]. In materials with strong exchange enhancement, the contribution to ρ from electron–spin fluctuation scattering may be expected to dominate the T^2 contribution from Coulomb (Baber) scattering [42] of the electrons. Indeed, in pure Pd, one observes a very large contribution to ρ of the form AT^2 [43]. Schindler and Rice [43] have studied the dependence on concentration c of A in a series of dilute **Pd**Ni alloys. For $c < 1\%$, A varies linearly with c, while $A^{-1}(dA/dc)$ assumes the extraordinarily large value of 750. Lederer and Mills [6] have suggested that this strong concentration dependence of A is associated with the scattering of electrons from the very large amplitude spin fluctuations which occur in the vicinity of the impurity cell. As we have seen, the susceptibility data on this system indicate that $\alpha \gg 1$. A calculation of the concentration dependence of A that results from this mechanism produces a value of $A^{-1}(dA/dc)$ in good accord with the data; the strong concentration dependence of both χ and A can be interrelated in a quantitative manner. Recently, a study of $A^{-1}(dA/dc)$ in **Pt**Ni alloys has been reported [35]. It is found that $A^{-1}(dA/dc) \approx 28$. This value is also consistent with the result obtained upon assuming that the concentration dependence of A results from the scattering of the electrons from locally enhanced spin fluctuations.

The study of the thermal resistivity w provides an important test of the model. The scattering of electrons from spin fluctuations produces a term linear in T in w. One expects a strong concentration dependence of this term in systems with $\alpha \gg 1$. Of particular importance is the concentration dependence of the Lorenz number L formed from the AT^2 term in ρ, and the term linear in T in w. If the enhancement of these terms in ρ and w is associated with scattering from local spin fluctuations, one expects L to approach a constant value as c increases [44]. If the enhancement comes from scattering of the electrons from spatially extended spin fluctuations (paramagnons) with a character very similar to paramagnons in a pure material, as Schindler and Rice have suggested [43], then L should decrease rapidly as c increases [44, 45], as

Rice first pointed out. Experimental data [44] on the thermal resistivity of **Pd**Ni alloys indicate that L remains constant as c increases, with a value of roughly one-half of the classical Sommerfeld value L_S. This gives further support to the suggestion that local spin fluctuations are responsible for the strong concentration dependence of A, and the term linear in T in the thermal resistivity.

The scattering of electrons from local spin fluctuations produces terms in ρ and w with the temperature dependence cited above only in the limit that the temperature $T \ll T_{sf}$, the local spin fluctuation temperature. Kaiser and Doniach [46] have recently extended the theory of the contribution to ρ from this source to higher temperatures. They employ a phenomenological method mentioned briefly earlier: The expression for dynamic susceptibility obtained from the mean field theory (Section II) is utilized in the calculation. Then the local spin fluctuation temperature T_{sf} is treated as a temperature-dependent parameter that may be obtained from the data in an empirical fashion. Kaiser and Doniach obtain $T_{sf}(T)$ from resistivity data on dilute **Ir**Fe and **Rh**Fe alloys, and compare this quantity with the observed temperature dependence of the impurity contribution to χ. The argument is quite good in both cases.

A partial justification of the procedure employed by Kaiser and Doniach follows from the observation [14] that generally the dynamic susceptibility of a host with one impurity present can be written in a form quite identical to the mean field result. Thus, computation of the parameters that appear in the general form (i.e., the local spin fluctuation temperature) presents difficulties. However, we may use the mean field form for the dynamic susceptibility, and treat T_{sf} as a phenomenological parameter. One must also presume that for small Ω, the exact form for $\mathscr{I}m(\bar{\chi}(\Omega))$ is proportional to Ω. This assumption is difficult to justify in theory. However, if $\mathscr{I}m(\bar{\chi}(\Omega))$ is nonanalytic for small Ω, the system would not behave like a "normal Fermi gas," that is, the impurity contribution to the specific heat would not be proportional to T, ρ and w would not vary like T^2 and T, respectively, when $T \ll T_{sf}$, and the Korringa relation would be altered. Thus the data provide support for this assumption. In many instances, one may thus use mean field theory when $\alpha \gg 1$ where the RPA breaks down, provided one regards the parameters as phenomenological quantities poorly approximated by the mean field theory.

The procedure employed by Kaiser and Doniach has been extended to the thermal resistivity by Kaiser [47]. Kaiser predicts that as T increases through T_{sf}, the Lorenz number L found from the local spin fluctuation contributions to ρ and w should increase from the value of

3. SPIN FLUCTUATIONS AROUND IMPURITIES

approximately $\approx \frac{1}{2} L_S$ characteristic of the low-temperature region [44] $T < T_{sf}$ to L_S when $T > T_{sf}$. Thus a measurement of L can provide a direct measure of T_{sf}. At this writing, no data on the thermal resistivity are available for an alloy in the temperature region where $T \sim T_{sf}$.

The discussion of transport properties has been confined to systems with transition metal impurities in transition metal hosts. Large T^2 terms in ρ have also been observed in simple and noble metals doped with transition metal impurities. Rivier and Zukermann [11] have argued that in **AlMn**, such a term comes from interaction of the conduction electrons with local spin fluctuations associated with the Mn impurity. If the energy level of the Mn impurity lies at the Fermi level, the theory produces a T^2 term in ρ with a negative coefficient. The mechanism discussed by Rivier and Zukermann is quite different than the scattering process considered dominant in **PdNi**. In a system such as **AlMn**, the conduction electron interacts directly with the local spin fluctuations by means of the s–d mixing term (the "V_{dk}" term) in the Anderson Hamiltonian. In the limit of small T_{sf}, the local spin fluctuations produce a resonance at the Fermi energy in the conduction-electron t matrix. The system behaves as if a virtual level of width of about T_{sf} is centered at the Fermi energy; a negative T^2 term in ρ obtains as a consequence, when the temperature dependence of ρ is examined. For a host material like Pd, the s–d model of transition metals is appropriate. The primary contribution to the electrical current comes from the s electrons (more precisely, from the electron surface centered at Γ; their wave functions are actually a strong admixture of s and d character), while the magnetic properties and the specific heat are dominated by the pockets of heavy d holes at the zone faces. Large-amplitude local spin fluctuations produce structure at the Fermi energy in the t matrix of the d electrons that is very similar to that in the conduction-electron t matrix for the Anderson model. The large values of $\gamma^{-1}(d\gamma/dc)$ in **PdNi** are a reflection of the presence of this structure [10]. Since the d-electron contribution to the electric (and thermal) currents is small in a host such as Pd, one does not observe the analog of the negative T^2 contribution to ρ that appears in **AlMn**. However, the s electrons can sample the spin fluctuations in the d band directly via the s–d exchange interaction $J\mathbf{S} \cdot \mathbf{s}$ [41]. This produces a T^2 term in ρ; the coefficient of the T^2 term associated with this mechanism is necessarily positive. In the presence of an impurity with $\alpha \gg 1$, the T^2 term may exhibit a very strong concentration dependence. Also, if $\Delta U < 0$, the effect of the impurity is to decrease the magnitude of the host T^2 term ($A^{-1} dA/dc < 0$); however, the coefficient of the T^2 term associated with this scattering process must be positive. A negative T^2 term in a transition metal alloy could result if the *d*-electron contribution to the electrical current is appreciable, or if the *s–d* mixing parameter is altered locally by the impurity.

ACKNOWLEDGMENT

One of us (M. T. Béal-Monod) is grateful to Dr. K. Maki and Dr. R. Weiner for many useful comments on the material in Section III.

References

1. C. Kittel, "Introduction to Solid State Physics," 4th ed., p. 518. Wiley, New York, 1971.
2. E. C. Stoner, *Phil. Mag.* **19**, 565 (1935).
3a. S. Doniach and S. Engelsberg, *Phys. Rev. Lett.* **17**, 750 (1966); N. Berk and J. R. Schrieffer, *Ibid.* **17**, 433 (1966); N. Berk, Thesis, Univ. of Pennsylvania (1966) unpublished; W. Brinkman and S. Engelsberg, *Phys. Rev.* **169**, 417 (1968).
3b. M. T. Béal-Monod, S. K. Ma, and D. Fredkin, *Phys. Rev. Lett.* **20**, 929 (1968).
4a. J. Friedel, *Nuovo Cimento Suppl.* **7**, 287 (1958).
4b. P. W. Anderson, *Phys. Rev.* **124**, 41 (1961).
4c. P. A. Wolff, *Phys. Rev.* **124**, 1030 (1961).
5. P. Lederer and D. L. Mills, *Solid State Commun.* **5**, 131 (1967).
6. P. Lederer and D. L. Mills, *Phys. Rev.* **165**, 837 (1968).
7. J. Hubbard, *Proc. Roy. Soc. Ser. A* **276**, 238 (1963); **281**, 401 (1964).
8. A. M. Clogston, *Phys. Rev. Lett.* **19**, 583 (1967).
9. D. L. Mills and P. Lederer, *Phys. Rev.* **160**, 590 (1967).
10. P. Lederer and D. L. Mills, *Phys. Rev. Lett.* **20**, 1036 (1968).
11. N. Rivier and M. Zukermann, *Phys. Rev. Lett.* **21**, 904 (1968); N. Rivier, M. Sunjic, and M. Zukermann, *Phys. Lett. A* **28**, 492 (1968).
12. B. Caroli, P. Lederer, and D. St. James, *Phys. Rev. Lett.* **23**, 700 (1969).
13. L. Dworin and A. Narath, *Phys. Rev. Lett.* **25**, 1287 (1970).
14. M. T. Béal-Monod and D. L. Mills, *Phys. Rev. Lett.* **24**, 225 (1970).
15. J. R. Schrieffer and D. C. Mattis, *Phys. Rev. A* **140**, 1412 (1965).
16a. H. Suhl, *Phys. Rev. Lett.* **19**, 442 (1967).
16b. M. Levine and H. Suhl, *Phys. Rev.* **171**, 567 (1968).
16c. M. Levine, T. V. Ramakrishnan, and R. A. Weiner, *Phys. Rev. Lett.* **20**, 1370 (1968).
17. D. R. Hamann, *Phys. Rev.* **186**, 549 (1969).
18. S. Q. Wang, W. E. Evenson, and J. R. Schrieffer, *Phys. Rev. Lett.* **23**, 92 (1969); J. R. Schrieffer, *C.A.P. Summer School, Baniff, Canada,* 1969, to be published; W. E. Evenson, J. R. Schrieffer, and S. Q. Wang, *J. Appl. Phys.* **41**, 1199 (1970).
19. M. T. Béal-Monod, J. P. Hurault, and K. Maki, *Proc. Int. Conf. Low Temp. Phys. 12th, Kyoto, 1970,* (E. Kanda, ed.). Academic Press of Japan, Tokyo, 1971; and to be published.
20. H. Keiter, *Phys. Rev. B* **2**, 3777 (1970).
21. N. N. Bogoliubov and D. V. Shirkov, "Introduction to the Theory of Quantized Fields," p. 357. Wiley (Interscience), New York, 1959.

22a. K. Maki, private communication.
22b. H. Tagayama and K. Maki, *Progr. Theor. Phys. Jap.* **46**, 42 (1971).
23. F. J. Dyson, *Phys. Rev.* **75**, 1736 (1949).
24. R. A. Weiner, *Phys. Rev. Lett.* **24**, 1071 (1970).
25. R. A. Weiner, *Phys. Rev. B* **4**, 3165 (1971).
25a. J. R. Schrieffer and D. R. Hamann, this Vol.
26. J. R. Schrieffer, W. E. Evenson, and S. Q. Wang, Int. Conf. of Magnetism, 7th, 1970, *J. Phys. Suppl.* n° 2.3, **32** C1-19 (1971).
27. D. R. Hamann, *Phys. Rev. B* **2**, 1373 (1970); D. R. Hamann and J. R. Schrieffer, this Vol.
28. P. Lederer, *Proc. Summer School, La Colle sur Loup*, 1970, to be published.
29. D. Shaltiel, J. H. Wernick, and H. J. Williams, *Phys. Rev.* **135**, 1346 (1964).
30. H. Launois, Thesis, Univ. of Paris, Paris, France, 1969, unpublished; A. Thorpe and S. Sullivan, unpublished.
31. W. Brinkman, S. Englesberg, and S. Doniach, *Phys. Rev. Lett.* **20**, 1040 (1967)
32. A. I. Schindler and C. A. Mackliet, *Phys. Rev. Lett.* **20**, 15 (1968).
33. G. Chouteau, R. Fourneaux, K. Gobrecht, and R. Tournier, *Phys. Rev. Lett.* **20**, 5 (1968).
34. F. M. Mueller, A. J. Freeman, J. O. Dimnock, and A. M. Furdyna, *Phys. Rev. B* **1**, 4617 (1970).
35. C. A. Mackliet, A. I. Schindler, and D. J. Gillespie, *Phys. Rev. B* **1**, 3283 (1970).
36. G. Chouteau, R. Fourneaux, R. Tournier, and P. Lederer, *Phys. Rev. Lett.* **21**, 1082 (1968).
37. R. Tournier and A. Blandin, *Phys. Rev. Lett.* **24**, 397 (1970).
38. A. Narath, *J. Appl. Phys.* **41**, 1122 (1970).
39. T. Moriya, *J. Phys. Soc. Jap.* **18**, 516 (1963).
40. P. Lederer, *Solid State Commun.* **7**, 209 (1969).
41. D. L. Mills and P. Lederer, *J. Phys. Chem. Solids* **27**, 1805 (1966).
42. W. G. Baber, *Proc. Roy. Soc. Ser. A* **153**, 699 (1936).
43. A. I. Schindler and M. J. Rice, *Phys. Rev.* **164**, 759 (1967).
44. A. I. Schindler, T. J. Schriempf, and D. L. Mills, *Phys. Rev.* **187**, 959 (1969).
45. M. J. Rice, *Phys. Lett. A* **26**, 86 (1967).
46. A. B. Kaiser and S. Doniach, *Int. J. Magn.* **1**, 11 (1970).
47. A. B. Kaiser, *Phys. Rev. B* **3**, 3040 (1971).

Part II
THE s–d MODEL

4. The s–d Model and the Kondo Effect: Thermal and Transport Properties

Melvin D. Daybell

Department of Physics
University of Southern California
Los Angeles, California

I. Introduction	121
II. Thermal and Transport Properties	125
1. Kondo Temperature	126
2. Susceptibility	128
3. Resistivity	131
4. Specific Heat	137
5. Thermoelectric Power	138
6. Other Measurements	140
III. Interactions	140
1. Strong Interactions	141
2. Weak Interactions	143
IV. Review Articles	144
References	144

I. Introduction

The properties of a dilute alloy of a transition metal like vanadium or iron (or a few of the rare earths) dissolved in a simple metal solvent like copper or gold (or certain transition metals like molybdenum) are dominated by the interaction of the localized magnetic moment of the impurity with the conduction electrons of the host metal. At high temperatures, the impurity behaves much like a free magnetic moment, and contributes a Curie- (or more properly Curie–Weiss-) like term to the susceptibility of the alloy, which is, therefore, paramagnetic. Below a temperature T_K specific to each alloy, the Kondo temperature, the impurity becomes almost completely nonmagnetic in the sense that although the susceptibility may greatly exceed the normal Pauli paramagnetism of the host, little or no temperature-dependent magnetism is present. As this gradual transition from the paramagnetic to the nonmagnetic regime takes place, the contribution of the impurity to most of the bulk properties of the alloy undergoes a dramatic change, resulting in characteristic Kondo anomalies in the susceptibility, the resistivity, the specific heat, the thermopower, and so on. These changes can be used to estimate the Kondo temperature of a given alloy, and when this is done [1] for the 3d-shell transition series as impurities in gold or copper, for example, a systematic variation of T_K with the position of the impurity in the periodic table is revealed (Fig. 1). Above the curve, which represents a prediction based on a simple $s-d$ model [1], the impurities are magnetic, while below it they are nonmagnetic. It is seen that ln T_K varies inversely with the number of unpaired 3d electrons on

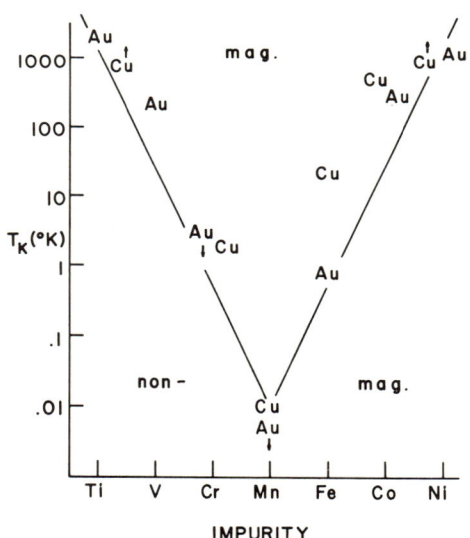

FIG. 1. The Kondo temperature varies in a systematic way for the 3d transition metals as local moments in gold and copper hosts. Log T_K/T_F varies roughly as $-n_{ud}$.

the impurity, n_{ud}. Even in systems where "T_K" is too high to be observable, and the local moment does not exist long enough for the $s-d$ model to apply, Kondo anomalies similar to those found in the nonmagnetic regime of alloys like the gold- and copper-based systems of Fig. 1 may be observed. For example, when the same 3d impurities are dissolved in aluminum, Mn is more nearly "magnetic" than Fe or Cr, which are, in turn, more nearly magnetic than their other neighbors in the periodic table [2].

To establish the existence of one of these Kondo anomalies, it is necessary to work with alloys which are sufficiently dilute, which means in practice that the impurity concentration c should certainly be less than 50 ppm times T_K, and in many cases impurity–impurity interactions are apparent at concentrations an order of magnitude lower.

Much of the theoretical work on these alloys is based on the s–d exchange model of Kondo [3], which is, in turn, based on the hope that the interaction of the conduction electrons with the paramagnetic d electrons localized on the transition metal impurity in the Friedel–Anderson [4, 5] picture would be weak enough that the problem of the formation of a local moment could be treated separately from the problem of its interaction with the conduction electrons. In this model, the conduction-electron spin $\mathbf{s}$ is coupled to the local moment $\mathbf{S}$ with an effective antiferromagnetic exchange coupling J, leading to a Hamiltonian $\mathscr{H}_{sd} = -J\mathbf{S}\cdot\mathbf{s} + V$, where $J < 0$ and V is included to reflect any ordinary potential scattering from the ion carrying the local moment. Any orbital effects are ignored. The result is that below the Kondo temperature [given by $T_K = D\exp\{-[1/|\tilde{J}|N(0)]\}$ where D is of the order of E_{Fermi}, $N(0)$ is the conduction-electron density of states per spin, per host atom at the Fermi surface, and $\tilde{J} \simeq J$, but depends on V and S] a many-body "quasibound state" is formed by the local moment and the conduction electrons, severely inhibiting the ability of the local moment to respond to external fields and pushing the conduction-electron scattering cross section up to the limit imposed by unitarity for the scattering partial wave considered, usually the d-wave portion of the conduction-electron wave function [6]. This state, which represents the ground state of the s–d interaction model, is not a bound state in the usual sense but is a convenient label to use in classifying the properties of Kondo alloys. The broad transition to the quasibound state is accompanied by a decrease in the entropy and magnetic moment of the impurity system as well as by an increase in the resistivity and a peak in the thermopower of the alloy (Fig. 2). Except for this peak [7], whose sign and magnitude depend on those of V, the influence of J and V appears only indirectly, through the expression for T_K, while T_K functions as a scaling factor for the temperature. The only other important parameter in the theory is the impurity spin, S, which determines the magnitude of many of the quantities of experimental interest, as well as, in principle, the rate at which the quasibound state forms [8] with decreasing temperature and magnetic field.

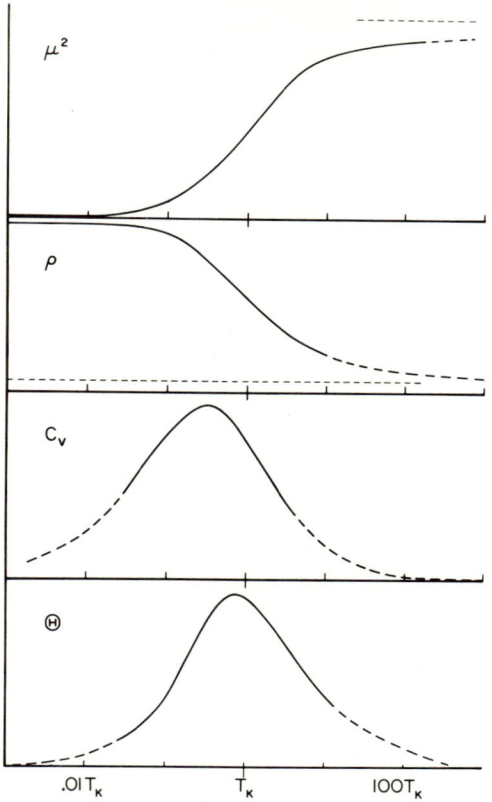

FIG. 2. Typical Kondo anomalies in the magnetic moment $\mu^2 = (3k_B T)\chi$ (dashed line is the free ion moment); the resistivity ρ, including a temperature-independent contribution from the potential scattering V (dashed line); the specific heat C_v; and the thermopower Θ, with V assumed negative and the potential scattering phase shift $\delta_v < \pi/4$. Solid curves are based on experimental behavior, the dashed portions of the curves have not been examined experimentally, and represent predictions based largely on the Suhl–Nagaoka theory. (The properties of the pure host material have been subtracted out.)

Many of the predictions of this model have been borne out experimentally, but it suffers from the fact that quite large values of the "small" coupling constant $\tilde{\gamma} \equiv -|\tilde{J} \cdot N(0)|$ are needed to yield the values of T_K seen in the experiments, and it is then no longer obvious that the s–d interaction problem can, in fact, be separated from that of the formation of the local moment.

If the lifetime of the local moment itself is severely limited by its interaction with the conduction electrons, the Kondo model does not apply, and the local spin must be considered to fluctuate [9] at a rate τ_{sf}^{-1}, where τ_{sf} is the lifetime of the spin fluctuation.† When this rate is much greater than the rate that would be produced by thermal fluctuations, the impurity appears nonmagnetic, but at higher temperatures where many thermal fluctuations occur in the time occupied by one spin fluctuation, the impurity behaves like a well-defined local moment. The boundary $T_{sf} = \hbar/k_B\tau_{sf}$ between the two regimes is the analog of T_K in the Kondo model, and is, in this picture, represented by a curve similar to that in Fig. 1. The predictions of the two models are very similar [9], and it would be a difficult task to choose between them on experimental grounds, so that it is fortunate that recent theoretical work [10, 11] has indicated that they are two aspects of a single model.

In what follows, "Kondo temperature" is used to mean the temperature near which a Kondo anomaly appears, even though the Kondo model is not appropriate if T_K is large.

When the host itself is a transition metal, the simple s–d interaction concept is modified by the exchange enhancement of the conduction-electron susceptibility in the neighborhood of the local moment [12]. While it is possible to consider Kondo anomalies in alloys [13] like MoCo, where the exchange enhancement is weak, the properties of local moments in such materials as palladium are dominated by exchange enhancement, and lie outside the scope of the present chapter. Kondo anomalies have also been seen [14] in nonmagnetic intermetallic compounds such as CoAl when one of the constituents is present in a concentration slightly in excess of the stoichiometric value. In this case, an isolated local moment appears to lie on the *excess* ions, vanishing at the exact stoichiometric composition.

Attempts to test the expression $T_K = D \exp\{-[1/|\tilde{J}|N(0)]\}$ by varying the average density of states at the Fermi surface [15, 16] have been inconclusive, possibly because $\tilde{J}$ and $N(0)$ vary together, and also because the local value of $N(0)$ near the impurity can differ considerably from its average value. There is some evidence that increasing the lattice constant of the host by low-temperature deposition of the alloy [17] lowers T_K.

II. Thermal and Transport Properties

Comparisons between the measured properties of systems exhibiting the Kondo anomalies and the various theories fall naturally into two

---

† Mention should be made of a third possibility, wherein $\hbar\tau_{sf}^{-1} \ll k_B T_K$, and a distinction has to be made between these two energies. Then $T_{sf} \neq T_K$.

segments. First, in the region where the temperature or $\mu_B H_K/k_B$, its external magnetic field equivalent, or both are far above T_K, the impurity is manifestly magnetic, with a well-defined magnetic moment μ more or less equal to its free ion value, μ_e. Here, spin–spin correlation effects are small, and the fluctuation effects responsible for the theoretical difficulty of the quasibound state at low temperatures and magnetic fields are relatively unimportant. In this "perturbation" region, which has, in effect, been extended to about a decade below T_K by the success of the Suhl–Nagaoka approach [7, 8, 18, 19], it is possible to account for nearly all of the results obtained experimentally (the gross features of Fig. 2) by calculations equivalent to the application of a rather sophisticated form of perturbation theory to the original $\mathcal{H}_{sd}$ used by Kondo. [Because $\tilde{\gamma}$ is large in actual dilute alloys, truncating the perturbation series after the first divergent term, as Kondo did, is not expected to be valid except at temperatures $T/T_K (\equiv \tau) \geqslant 10^4$, which is above the experimental range.] The major exception is that it is necessary to choose the impurity spin $S \simeq \frac{1}{2}$ in the expressions that predict the temperature dependence of the Kondo anomalies, rather than using the value determined by the high-temperature susceptibility. The ordinary potential V is often unknown, but its major effect can be absorbed into the definition of $\tilde{J}$ [20].

These calculations appear to break down as T approaches zero (for $H = 0$), although the increasing importance of impurity–impurity interactions at very low temperatures makes this difficult to verify experimentally. In this region, where the quasibound state dominates the behavior of the properties of the alloy, the predominant theoretical approach has been to apply variational methods to an assumed class of ground-state wave functions (usually a singlet) and attempt to calculate thermal and transport properties of the ground state near $T = 0$ [21]. In this case, the interest is focused on the functional dependence on temperature of the experimental parameter in question, and its value at $T = 0$.

1. Kondo Temperature

In most systems where the Kondo temperature is considered to be known, it has been estimated from the temperatures characteristic of one or more of the Kondo anomalies associated with either the susceptibility, χ, the resistivity, ρ, the specific heat, C_v, or the thermoelectric power, Y, of the added impurity (Fig. 2). Only in a few systems, such as **Cu**Fe, **Cu**Cr, **Au**V, **Au**Fe, **Y**Ce, and **Mo**Co, has it been possible to obtain a clear picture of a majority of these Kondo anomalies in a single alloy. More often it is necessary to extract an estimate of T_K from a single experimental curve, or even from the $T = 0$ limit of χ or C_v/T, if T_K is quite high. If T_K is low, impurity–impurity interactions limit the accuracy with which it can be measured. When T_K is high, on the other hand, though interaction effects are less important, the total effect of the local moment is small, at least in the low-temperature range where it is often necessary to work to avoid phonons. These problems, together with the typical logarithmic temperature dependence of Kondo

TABLE I

CHARACTERISTIC TEMPERATURES, T_K, FOR "KONDO" ALLOYS

Host	Impurity	$T_K{}^a$ (°K)	Ref.
Mg	Mn	<1	[1]
Al	Cr	1200	[22]
	Mn	500	[22]
Cu	Ti	>500	[1]
	Cr	2	[text]
	Mn	0.01	[23]
	Fe	30	[text]
	Co	·500	[24]
	Ni	>1000	[1]
Zn	Fe	>500	[1]
	Mn	1	[25]
	Cr	·4?	[1]
Y	Ce	30	[26]
Mo	Fe	1	[27]
	Co	25	[13]
Pd	Cr	40–100	[28]
Ag	Mn	<0.4	[1]
	Fe	10?	[1]
Cd	Mn	0.1	[29]
La	Ce	0.5	[30]
W	Co	10	[31]
Pt	Cr	200	[28]
Au	Ti	>2000	[1]
	V	300	[32]
	Cr	<4	[33, 34]
	Mn	~0	[35]
	Fe	0.8	[text]
	Co	300–700	[1, 36]
	Ni	1200	[1]
Th	U	100	[37]

^{a} Most of these numbers are known to no better than a factor of 2.

anomalies, usually lead to uncertainties of T_K of factors of 2 or more. However, since it is often only $\log(T_K)$ that is of interest, available estimates of T_K are frequently adequate. Estimates of T_K for some systems known to exhibit at least one Kondo anomaly are tabulated in Table I. Many of these numbers are taken directly from the original papers, others have been obtained using prescriptions described in the following sections.

2. Susceptibility

In perhaps no other way is the effect of the conduction electrons on the impurity spin more strikingly demonstrated than in the modification of the magnetic field dependence of its magnetic moment at low temperatures. The major effect of the interaction of the local moment with the conduction electrons is to prevent it from lining up with the applied field at low temperatures as a free spin would [38]. The approximate Curie–Weiss form of the initial ($H \to 0$) susceptibility $\chi \propto (T + \theta)^{-1}$, $\theta \sim \tau_{sf}^{-1}$, is a result of this interaction, which leads to a finite lifetime τ_{sf} for the local moment, as mentioned earlier. From the observed behavior of the low-field susceptibility as a function of temperature, it is reasonable to expect that the magnetization in the presence of a Kondo anomaly will depend on $\mu_e H/kT_K$ rather than on $\mu_e H/kT$ as for a free spin and this is, in fact, the case. As an example, at 1°K the spin of a free iron atom would be almost completely aligned in a field of a few kilooersteds. The moment of an iron impurity in copper ($T_K \simeq 30°K$), on the other hand, is still increasing toward the free ion value at fields as high as 200 kOe [39]. In systems with lower Kondo temperatures, attempts to saturate the moment at high fields have only been partly successful [40, 41], and even at the Kondo field, $H_K = kT_K/\mu_B$, the development of the full free ion moment on the impurity is not complete.[†]

Measurements of the hyperfine field at the nuclei of both the local moment [42, 43] and the host [43, 44] have indicated that the bulk of the susceptibility associated with the local moment resides on the d orbitals of the impurity, with little polarization of the conduction electrons detected beyond that associated with the RKKY result (which is adequately described by perturbation theory to first order in J).

In Fig. 3 is shown the typical behavior of the susceptibility of a localized moment in a metal for $H \ll H_K$. Far above T_K, the tempera-

---

[†] The saturation value of the moment is not expected to reach exactly the free ion value because of the small negative RKKY polarization on the surrounding conduction electrons.

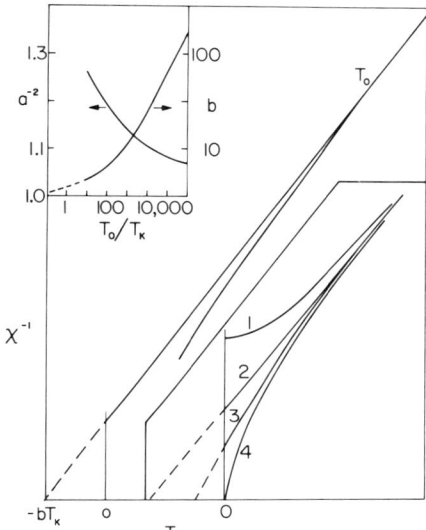

FIG. 3. Inverse susceptibility $1/\chi$ versus temperature. The different low-temperature behaviors are discussed in the text. Inset shows values of the normalized moment $a = \mu_{\text{CW}}/\mu_e$ and normalized Curie–Weiss temperature $b = \theta_{\text{CW}}/T_K$ required to generate a "Curie–Weiss law" tangent to the very slowly changing theoretical curve [Eq. (2.1)] at the point $T = T_0$. Lower inset shows various low temperature limits.

ture-dependent part of the susceptibility contributed by each impurity slowly approaches the limit $\mu_e^2/3k_B T$. Here μ_e is the effective number of Bohr magnetons, $\mu_e^2 = \mu_B^2 g^2 S(S+1)$ for an impurity of spin S, g factor g (neglecting a factor $1 - |\gamma|$ arising from the first-order perturbative conduction-electron polarization [45]). S is usually close to the spin of the free impurity ion. Letting $T/T_K \equiv \tau$, perturbation theory gives [46]

$$\frac{1}{\chi} = \left[\frac{\mu_e^2}{3kT_K}\right]^{-1} \tau \left[\frac{\ln \tau}{\ln \tau - 1 + \kappa/\ln \tau}\right], \tag{2.1}$$

for $\tau \gtrsim 3$, with $\kappa = 0$, while a higher order expansion in $\ln \tau$ gives [47] the same expression with $\kappa \simeq 0.38$. For the range of temperatures covered in a typical experiment carried out above about $2T_K$, this expression is approximated quite well by a Curie–Weiss law of the form [1] (straight line, Fig. 3) $\chi = (a\mu_e)^2/(T + bT_K)$, with $a \simeq 0.9$, $b \simeq 2.5$ to 10. The values of a and b that yield a "Curie–Weiss law" curve that is tangent to the curve of Eq. (2.1) at the point of tangency T_0 are shown in an insert of Fig. 3. As T drops below T_K, theoretical predictions differ. Most of the variational calculations would have $1/\chi$ drop to $\mu_e^2/3kT_K$ at $T = 0$ (curve 3, Fig. 3) while Suhl–Nagaoka theory [8] predicts a susceptibility diverging (approximately as $T^{-1/2}$) as $T \to 0$ (curve 4, Fig. 3). A calculation based on scaling theory [48] behaves as a Curie–Weiss law at high temperatures (with $b \simeq \frac{1}{3}$), but predicts that $1/\chi = \alpha + \beta T^2$ below T_K (cf. Rivier and Zuckermann [9]) rising about 50% above the Curie–Weiss value as $T \to 0$ (curve 1, Fig. 3). A fourth possibility is that the Curie–Weiss law continues to hold down to $T = 0$ (curve 2).

Experimentally, all four behaviors have been observed [41, 49–51]. All four can also result from impurity–impurity interactions. A few parts per million of free spins of any type will contribute to curve 4, so that host purity is an important consideration. For local moments on rare earth impurities, crystal-field effects [52] can also simulate curve 4.

The validity of the unmodified Curie–Weiss law down to $T_K/20$ has been established in copper–iron [51], with $b \simeq 1$ or 2, at least in a 60-kOe field. This is well below H_K in this alloy, and can be expected to represent the low-field susceptibility. In **Au**V, where T_K is high, and interactions are relatively unimportant, a Curie–Weiss law with $b \simeq 1$ is observed at high temperatures [32] but in the region below 5°K, a T^2 law similar to curve 3 is seen [50]. In copper–manganese, a Curie–Weiss law is also obeyed [23], with $bT_K \simeq 10\text{m}°\text{K}$. (If $b \simeq 1$, these data extend from near T_K to about $35T_K$. In **Au**Fe, $b \simeq 1$ to 2 at temperatures [41] up to $\simeq 10T_K$, but is much larger [53] above $100T_K$. A similar result is found [54] in **Cu**Mn. Some of this difference may be due to impurity interactions, and the validity of Eq. (2.1) is not well established experimentally.) In the tabulation of Table I, b was taken as 1.

The susceptibility of a rare earth local moment like cerium in a nonmagnetic host can be more complicated than the simple picture just presented, since crystal-field effects can be important [55], as can the orbital moment on the impurity [56], although the results are not affected very much if the crystal-field splitting is large. A well-defined $T^{-1/2}$ term in the susceptibility occurs in some of these materials [57] which may be a single-impurity effect although the concentrations are rather high, as are the values of τ where it is seen.

In a system of alloys like the $3d$ transition metals as impurities in aluminum, the impurity susceptibility is nearly temperature independent, but it increases systematically with n_{ud} with a maximum value at manganese [2]. This behavior would be expected if Kondo anomalies were present in each alloy, all with very high T_K values. In this case, T_K would be lowest at manganese, and increase as we moved out in either direction along the 3d series (compare with Fig. 1). Here, T_K is far too high to apply the simple Kondo picture, of course, but resistivity [22], thermopower [58], and other data [2] permit one to view this system as a limiting case of the s–d model of the local moment problem.

3. Resistivity

It was the striking logarithmic rise in resistivity as $T \rightarrow 0$ which led Kondo to a divergence in the third-order perturbation term in the simple s–d exchange model [3] and generated the subsequent interest in the local moment problem. It is this $-\log(T/T_K)$ behavior, coupled with the minimum produced in the total resistivity of the alloy by the increasing scattering of conduction electrons by phonons at higher temperatures, that has become known as the Kondo effect, although the term is often used in a more general sense. The logarithmic increase of the resistivity as T is lowered reflects the presence of a large peak at or near E_F in the conduction-electron scattering rate as a function of energy. This peak appears because the scattering center has an internal degree of freedom, its spin direction, unlike an ordinary nonmagnetic scatterer. As $T \rightarrow 0$, only electrons with energy near E_F carry current, and these electrons are strongly scattered by the local moment. Because the s–d exchange Hamiltonian has its origin in a resonant interaction between the conduction electrons and d-valence electron orbitals on the local moment [5], only the $l = 2$ (or d) partial wave of the incoming conduction electron is scattered strongly [6]. (For rare earth impurities, the f partial wave is scattered, since the local moment is on an $l = 3$ level in the rare earths.)

As T decreases below T_K, the divergence must saturate when all of the d wave is being scattered, and the resistivity cannot increase above the value $\rho_{du} = 5\rho_{su}$, where ρ_{du} represents the limit imposed on the total amount of d-wave scattering by the unitarity of the S matrix.† This d-wave unitarity limit is $2l + 1$ times the unitarity limit for s-wave ($l = 0$) scattering, ρ_{su}; $\rho_{su} = 2m^*c/\pi z e^2 N(0)$ (m^* is the conduction-electron effective mass, c is the impurity concentration, z is the number of conduction electrons per host atom, and e is the electronic charge). For copper, ρ_{su} is about 0.38 nΩ-cm/ppm of impurity. This saturation behavior was first seen in **Cu**Fe [59], although $\rho(T = 0)$ was less than ρ_{du}. Saturation at the d-wave unitarity limit ρ_{du} as $T \rightarrow 0$ was subsequently observed [40] in **Cu**Cr. In addition, a temperature-independent

---

† If there is appreciable scattering in more than one partial wave, the d-wave resistivity cannot reach even the unitarity limit because of restrictions imposed by the Friedel sum rule [4], which ensures charge neutrality for the impurity site.

plateau in the resistivity at temperatures below the $-\log(T/T_K)$ region has been seen in **AuV** [32], **AuFe** [60], **AuCo** [61], **CuAuFe** [62], **MoCo** [13], **YCe** [55], in **AlCr**, and **AlMn** where a well-defined $-\log(T/T_K)$ term has yet to be observed [22], and also in the exchange-enhanced alloys **PtCr**, **PdCr**, and **RhCr** [28]. The gap between the logarithmic region and the low-temperature plateau is bridged by a region in which ρ varies as $\rho(T=0)[1 - (T/T^*)^2]$ where T^* is a constant equal to $\sim 0.1\, T_K$ or larger. If T_K is large, this gap may constitute the entire range of the experimental data, and the $-\log(T/_K T)$ term is then not observed.

Assuming for a minute that the phonon contribution to the resistivity above about $10°K$ could be separated from the contribution of the impurity,[†] the logarithmic part of the resistivity of the local moment should gradually drop to a limiting high-temperature value determined by the scattering by any ordinary potential V present on the impurity (Fig. 2). Largely because of a breakdown of Matthiesen's rule induced by the strong **k**-vector dependence of the scattering both by phonons and by the local moment [62], this high-temperature limit has never been observed. However, the expected curvature away from a straight $-\log(T/T_K)$ dependence is seen in **AuFe** and in a few systems where a binary host material such as copper–gold has been used expressly to make the dominant scattering mechanism less dependent on **k** and thus minimize deviations from Matthiesen's rule [62].

Ignoring the temperature-independent part of the resistivity, the features of the Kondo anomaly are represented by three parameters: the step height B between the low- and high-temperature plateaus; the slope $d\rho/d\lambda\,[\lambda = \log(T/T_K)]$ of the logarithmic region, determined by the spin S of the local moment and modified by the potential scattering V; and the temperature scale, determined by T_K. The temperature-independent part of the resistivity, A, is a fourth parameter needed to complete the characterization of the experimental data. These features have appeared in several theoretical calculations [7, 8, 18, 47]. If only d-wave scattering is considered, the Suhl–Nagaoka theory gives [8, 19], for the total resistance of the alloy in the absence of phonon effects,

$$\rho = A_d + \frac{1}{2} B \left\{ 1 - \frac{\ln(T/T_K)}{[\ln^2(T/T_K) + \pi^2 S(S+1)]^{1/2}} \right\}, \qquad (2.2)$$

where $A_d = \rho_{du} \sin^2 \delta_v$, $B = \rho_{du} \cos 2\delta_v$, $\tan \delta_v \simeq \pi N(0) V$, and $J \simeq$

[†] In a temperature range where the phonon scattering dominates the resistivity of the pure host metal, corrections must be made for significant effects due to deviations from Matthiesen's rule [62].

$J\cos^2\delta_v$ in the expression for T_K. A more exact calculation requiring a computer is also available [7]. Rewriting Eq. (2.2) in the form

$$\rho = \frac{\rho_{du}}{2}\left\{1 - \frac{\cos 2\delta_v \cdot \lambda}{[\lambda^2 + \pi^2 S(S+1)]^{1/2}}\right\}, \qquad (2.2\text{a})$$

we see that the height of the step in the resistivity is $\rho_{du}\cos 2\delta_v$, its slope $d\rho/d\lambda$ near T_K is $-\rho_{du}\cos 2\delta_v/2\pi[S(S+1)]^{1/2}$, its width is approximately proportional to S, and its center is at $T = T_K$. At $T = 0$ ($\lambda = -\infty$), $\rho(0) = \rho_{du}\cos 2\delta_v$, while $\rho(T \to \infty) = \rho_{du}\sin^2\delta_v$.

A constant A_0 must be added to Eqs. (2.2) and (2.2a) to allow for (nonresonant) scattering in other partial waves.

Because of the logarithmic nature of the Kondo effect, an experimental check of an expression such as Eq. (2.2) must extend over many decades in temperature. This is difficult to do on a single alloy, as there is a tendency to be limited at very low temperatures by interaction effects, and at high temperatures by phonons. It is possible, however, to vary T_K by changing the composition of the host, and to reduce the influence of electron–phonon scattering by using a binary host alloy. When the resistivity of the local moment of iron in gold, copper–5% gold, and copper matrices is plotted as a function of $\ln T/T_K$ choosing $T_K = 0.88$, 23, and 34°K, respectively, and using a different A ($\equiv A_d + A_0$) for each alloy [59, 62], the curve for the spin-dependent part of the total resistivity can be mapped out as a function of λ (Fig. 4, Table II).

TABLE II

PARAMETERS FOR FIGURE 4

Host	Impurity	Concentration (ppm)	A	B	A_0	δ_v	T_K (°K)	S	Ref.
				(nΩ-cm/ppm)					
Au	Fe	20	0.896	0.41	0.024	39.5°	0.88	0.45	[62]
Cu–5% Au	Fe	300	0.870	0.41	0.125	38.8°	23	0.45	[62]
Cu	Fe	60 and 400	0.765	0.41	0.020	38.8°	34	0.45	[59, 62]
Cu	Cr	15 and 30	0.680	1.19	0.32	25.7°	2.2	0.45	[40, 63]

The temperature dependence of the resistivity of Cr in copper ($T_K \simeq 2.2$°K) is almost identical [40] to that of these iron alloys (cf. Fig. 4). Above T_K, all the experimental data can be fit quite well with Eq. (2.2) if S is taken as approximately $\frac{1}{2}$ (solid curve in Fig. 4). If the free ion spin is used, the fit is unsatisfactory.

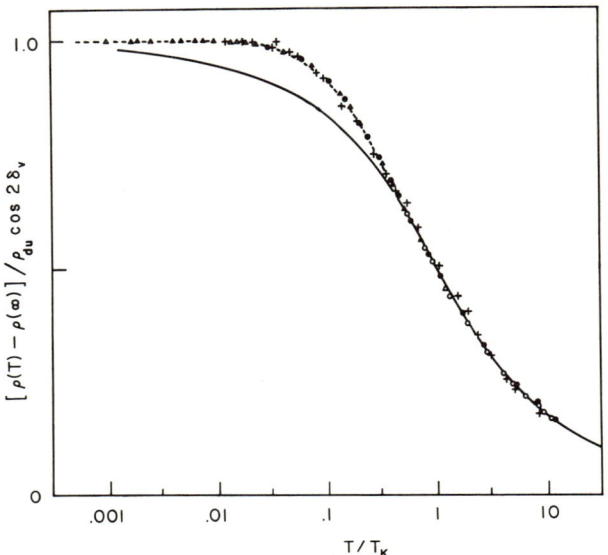

FIG. 4. Temperature-dependent part of the resistivity per impurity versus log T compared to prediction of the Suhl–Nagaoka theory (solid curve) and an empirical expression for $T \leqslant T_K$ (dashed curve). △, **Cu Fe**; ○, **Au Fe**; ●, **Cu-5% Au Fe**; and +, **CuCr**. Parameters used are given in Table II. The spin S was obtained from the fit to the data for Fe.

Not surprisingly, the fit is poor below T_K where the transition to the quasibound state is too rapid to be reproduced by the theory. The resistivity approaches saturation much faster than predicted but it is well represented from $T = 0$ to above T_K by

$$\rho'(T) \doteq A' - B' \ln[(\theta^2 + T^2)/T_K^2]^{1/2}$$
$$\{A' = (A + B/2),\ B' = B/2\pi[S(S+1)^{1/2}],\ \ln(\theta/T_K) = -\pi[S(S+1)]^{1/2}\},$$

(2.3)

where the constants have been chosen to match the value and slope of this empirical formula with those of Eq. (2.2) at $T = 0$ and $T = T_K$ (Fig. 4, dashed curve). For $S = \frac{1}{2}$, $\theta = 0.066 T_K$. If $T \lesssim \theta$,

$$\frac{\rho'(T)}{\rho'(0)} \simeq 1 - \alpha \left[\frac{T}{T_K}\right]^2, \qquad (2.3a)$$

with $\alpha \simeq 42B/(B+A)$ if $S = \frac{1}{2}$. Equation (2.3a) is a form often used successfully for empirical fits for T much less than T_K. Its curvature is opposite in sign from that of Eq. (2.2) on a linear T plot.

Expressions like Eq. (2.3a) come out of a number of calculations [64, 65] that take the finite lifetime of the impurity spin into account.[†] The essential point is that any mechanism which limits the lifetime of the local moment increases the width in energy of the peak in the electron scattering amplitude responsible for the Kondo effect so that the divergence in the resistivity as $T \to 0$ is arrested at the temperature ($\sim \theta$) where the thermal width of the conduction-electron Fermi function is exceeded by the width arising from the finite lifetime of the spin. Such effects are seen experimentally when the spin lifetime is shortened by interactions with other magnetic [63] or nonmagnetic [67] impurities, and the fact that the resistivity of isolated local moments is well described by Eq. (2.3a) may reflect an intrinsic lifetime of order $\hbar/k\theta$ for the local moment. If this is so, the step height would be less than ρ_{du}, but this decrease could not be distinguished from the decrease in B produced by a nonzero δ_v.

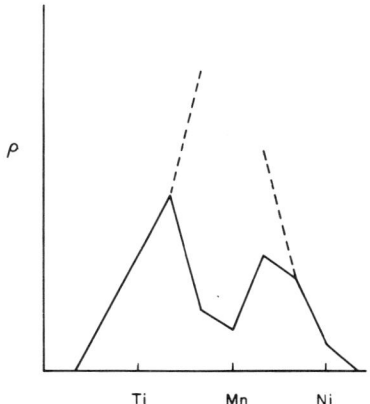

FIG. 5. Impurity resistivity at $T = 0$ versus n_d, the number of 3d electrons on the free impurity ion (dashed curve). Same at $T = 4.2°K$ (solid curve). Host is copper.

As T approaches zero, the resistivity of the copper and gold alloys containing 3d-transition local moments approaches a limiting value roughly proportional to the number of unpaired 3d electrons one would expect to find on the free transition metal ions n_{ud} (see Fig. 5). At higher temperatures the resistivity begins to decrease rapidly for ions like manganese, chromium, and iron, near the middle of the third row, and

[†] An extension of the Suhl–Nagaoka model to include the effect of spin correlations on the scattering also leads to this equation, with $\alpha \simeq 5 \ln T_F/T_K$, $\simeq 40$ in **Cu**Fe, and $A = 0$, since V was taken as 0 [66].

at 4°K, the characteristic notched "residual resistance" curve of the textbooks results (Fig. 5, solid line). This peaking of $\rho(T=0)$ near manganese also occurs for first-row transition impurities in aluminum, normally considered to be a good example of a system that is essentially nonmagnetic because the high spin-fluctuation rates imply unobservably high values of T_K. By assigning the orbital magnetic quantum numbers, m_l, of the 3d electrons in a particular way, Schrieffer predicted [6] $\rho(T=0) = n_{ud}\rho_{su}$ (if $\delta_v = 0$), in good qualitative agreement with experiment, although a similar result is obtained if the $T=0$ limit of Eq. (2.2a) is used with $\delta_v = (\pi/2)[n_d/5] - 1$ [68].

In an external magnetic field H, the anomalous resistivity of a dilute magnetic alloy decreases, giving rise to a negative magnetoresistance. (The magnetoresistance is positive for nonmagnetic scattering.) For $T \gg T_K$, the magnetoresistance is primarily the result of the partial alignment of the impurity spin, which inhibits the spin-flip scattering of the conduction electrons, and $\rho(H, T) - \rho(0, 0) \propto -M^2(H, T)$ as predicted on the basis of first-order perturbation theory [69] using $\mathscr{H}_{sd}$ (M is the bulk magnetization of the impurity). In copper–manganese ($T_K \simeq 10 \text{m}°\text{K}$) the relation above holds reasonably well [70] down to 1°K, if the *experimental* value is taken for M. Since for fixed H, M increases as $T \to 0$, $\rho(H, T)$ exhibits a peak as a function of T at fixed H if T_K is low. As the temperature is lowered below T_K, where the alignment of the local moment is no longer a function of H/T alone, the magnetoresistance, though still negative, varies more slowly with H, approximately as $\log H$. In fact, in copper–chromium [40] for example, no peak in $\rho(H, T)$ occurs, and $\rho(H/H_K, 0) = \rho(0, T/T_K)$ for $H < H_K$, so that the field removes the divergent electron scattering in the ground state of the local moment in much the same way as the increased temperature does in zero field. Perturbation theory [71, 72] reproduces this $\log H$ behavior, but predicts additional effects due to alignment of the local moment by the field which are not observed if $T < T_K$. This is not surprising since the same theory cannot predict the susceptibility resulting from this same alignment of spins, once $T \lesssim T_K$. Existing calculations with the Suhl–Nagaoka model overemphasize the effect of H [38].

Correction for the background of positive magnetoresistance (arising from the Lorentz force term in the Boltzmann equation, which is ignored here) in a real alloy is complicated by the fact that this correction depends on the scattering rate from the local moments, which itself depends on H and T. The correction can be large and uncertain, and is particularly troublesome if H_K and T_K are large, especially if the concentration is small [73].

4. Specific Heat

The Kondo anomaly in the specific heat per impurity C_v (Fig. 2) appears as a broad Schottky-like peak of a more or less symmetric "Gaussian" shape on a log T plot, centered at about $3.15 T_K$, and about two decades wide. In **Cu**Cr this peak (Fig. 6) occurs at 0.67°K, and in

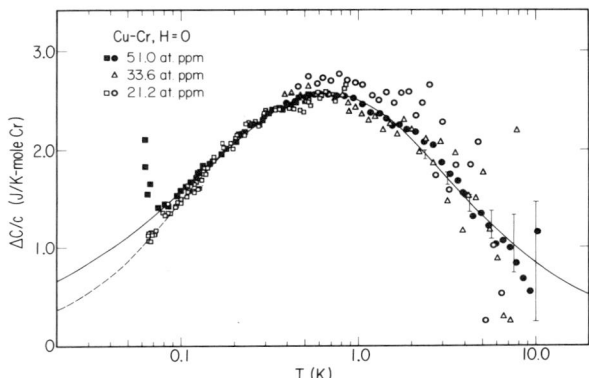

FIG. 6. Impurity specific heat in **Cu**Cr. The solid curve is from Bloomfield and Hamann, the dashed curve assumes $C_v/T =$ constant as $T \to 0$ [74].

CuFe it is near 9°K [74]. The entropy under the peak $\Delta \mathscr{S} = \int_0^\infty C_v \, d(\ln T)$ can be expressed as $k_B \ln(2S + 1)$, where the experimental value of S is close to that determined from the effective moment measured by the high-temperature susceptibility. Calculations based on the Suhl–Nagaoka approach [8, 75] reproduce this behavior well, except that the theory yields

$$\Delta \mathscr{S} = k_B[(2S + 2)\ln(2S + 2) + 2S \ln(2S) - 2(2S + 1)\ln(2S + 1)]$$

and also as $T \to 0$, C_v/T actually approaches D/T_K ($D \simeq 3 k_B$) rather than diverging slowly as in the theory. [$C_v/T \to \ln(\tau)$ as $T \to 0$ is another possibility that is not ruled out by the experiments.] This limiting behavior has also been observed [76] in dilute **Au**V alloys ($T_K \simeq 300°$K), where the peak would be at too high a temperature to be observed, and in the aluminum-based $3d$ transition impurity alloy series mentioned earlier [2]. Any effect that increases the density of states at the Fermi surface on alloying will also increase C_v/T, of course, so that a T_K determined from the value of D/T_K is only significant if T_K is already known. The agreement is good in **Cu**Fe, **Cu**Cr, and **Au**V, however.

In a magnetic field of the order of H_K, the specific heat approaches a Schottky anomaly appropriate on the free ion spin S, but with an appreciable amount of excess entropy from the Kondo anomaly remaining on the low-temperature side of the peak [74].

5. Thermoelectric Power

Several Kondo anomalies in the thermopower Θ of dilute magnetic alloys are familiar to the experimentalist, who has used them for years for low-temperature thermometry in the form of **Au**Fe ($T_K \simeq 0.8°K$) and **Au**Co ($T_K \simeq 300°K$) thermocouples. A concentration-insensitive peak occurs in Θ near T_K, with a sign opposite to that of V, and a height that depends on the magnitude of V. The thermopower of **Cu**Fe is typical [77] with a negative peak at $T \simeq 22°K$ (Fig. 7). The kink in the high-temperature side of the 300-ppm data can be removed by correcting for phonon scattering, and then the curves for 300, 750, and

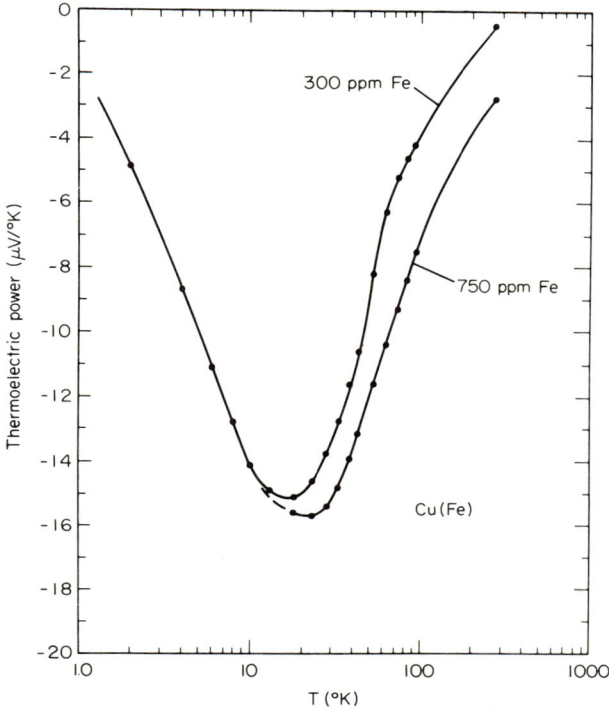

FIG. 7. Thermoelectric power Θ of Fe in copper, showing the Kondo anomaly at $T = 22°K$.

1000 ppm are quite similar, although phonon effects are small below 50°K and do not account for the slight shift in the peak temperature of the low-concentration sample. In the gold- and copper-based 3d transition series, the peak is small and positive for impurities left of manganese in the periodic table, and negative and somewhat larger for iron, cobalt, and nickel. A peak has been seen [26] in the rare earth Kondo system YCe. The thermopower of several of the dilute magnetic aluminum alloys is also anomalously large, although no peak has been found [58]. Some peak heights and temperatures are given in Table III.

TABLE III

Kondo Anomalies in the Thermopower[a]

Host	Impurity	Tempurature (°K)	Peak sign	Magnitude ($\mu V/°K$)	Ref.
Mg	Mn	<30	−	3.6	[58]
Al	Mn	>80	−	<1	[58]
Cu	Ti		+?		[78]
	V		+?		[78]
	Cr	5?	−?	0.2?	[79] but cf. [80]
	Fe	22	−	16	[77]
	Co	240	−	~15	[80]
	Ni	>1000	−	>15	[80]
Zn	Mn	<20	−	5	[58]
Y	Ce	25	+	3	[26]
Pd	Cr	35	+	7.5	[81]
Ag	Mn	<4	−?		[82]
	Fe	<10	−	7	[83]
Au	V	240	+	5	[84]
	Cr	~4	+	>2	[33]
	Mn	<9	+?		[34]
	Fe	<4	−	14	[34]
	Co	270	−	40	[85]
	Ni	550	−	16	[86]

[a] Values are quite sensitive to the quality of the alloy, and are merely indicative of the observed behaviors.

Except that the magnitude of the peak is a factor of 2 or 3 too small, the thermopower results are in good agreement with the results of the Suhl–Nagaoka theory [7, 47] if S is taken to be about $\frac{1}{2}$. The peak is expected to occur somewhat below T_K [as measured by $\rho(T)$] for phase shifts as large as those in CuFe, and this is in fact the case in this alloy.

In a magnetic field, the thermopower is expected [86a] to vary as $1/H$, and although a field dependence of the correct sign is seen in **Au**Fe, the actual behavior is considerably more complicated [87].

6. Other Measurements

The change dT_c/dc of the superconducting transition temperature per magnetic impurity added to a superconducting host like Zn or Al is greatly reduced if T_K of the impurity is greater than T_c, so that the pair breaking effect of the impurity is destroyed by the quenching of its moment. If T_K is lowered from, for example, above 300°K in **Al**Mn to about 0.25°K in **Zn**Mn (which is below T_c for either Al or Zn) by alloying Zn into **Al**Mn, dT_c/dc increases rapidly with the Zn content of the host [ref. 271 of 88].

Other impurity-related anomalies have, so far, prevented the identification of a Kondo anomaly in the Hall coefficient of the local moment system [89]. Little is known about the thermal conductivity of Kondo alloys. Microscopic properties of the local moment and the host metal nearby have been investigated using Mossbauer, NMR, EPR, nuclear orientation, and de Haas–Van Alphen techniques. Discussions are available in several review articles [1, 88, 90].

III. Interactions

A simple first-order perturbation calculation on the s–d exchange Hamiltonian $\mathcal{H}_{s-d} = -J\mathbf{s} \cdot \mathbf{S}$ leads to an oscillatory conduction-electron spin density proportional to $(\cos 2k_F r)/r^3$ near a local moment, as mentioned earlier. A spin density of this form leads to a long-range interaction between two local moments which is of the form

$$\mathcal{H}_{\text{RKKY}} \sim \frac{J^2 \mathbf{S}_1 \cdot \mathbf{S}_2 \cos(2k_F r)}{r^3}.$$

This interaction is expected to persist [91] in the presence of the Kondo effect and modifies the experimental behavior of the various Kondo anomalies down to very dilute concentrations.

The effects of interactions on the bulk properties of simple Kondo systems are usually quite apparent above a concentration of about 50 ppm times T_K but are present, occasionally only in a subtle form, in nearly all existing results on the properties of local moments in dilute alloys with noble metals, down to concentrations below 10 ppm. Many of these effects are of interest in their own right, but they have contributed a certain amount of confusion to the proper interpretation of a number of measurements at first believed to represent the properties of isolated impurities.

For descriptive purposes, interaction effects will be separated into those seen easily when the concentration c is greater than about 50 ppm times T_K: the *strong* interaction effects studied thoroughly before the Kondo effect was understood; and the *weak* interaction effects observed during the course of more recent experiments on the very dilute alloys commonly used to study Kondo anomalies. Both effects can arise from a conduction-electron spin density oscillation of the RKKY type, but the weak interaction behavior appears to reflect the presence of the ground state of the local moment.

The strong interactions affect the shape of the experimental curves as a function of temperature and concentration, whereas the weak interactions may have only a small effect on the shape of the curves, and occasionally scale so nearly with concentration as to appear to be a property of a single impurity if the concentration range investigated is not sufficiently large. While the observation of strong interactions is associated with systems in which $\mathcal{H}_{RKKY}$ is larger than T_K, interaction effects are not as well understood for alloys where T_K is greater than $\mathcal{H}_{RKKY}$, although $\mathcal{H}_{RKKY}$ is reduced to some extent as the local moment disappears below T_K. The presence of these two energies in the problem (one varying as J^2c, and the other exponentially in J) means, of course, that the criterion that an alloy be sufficiently dilute to see single-impurity effects is actually more complicated than the simple "50 ppm $\times$ T_K" rule above.

1. Strong Interactions

If T_K is much less than $\mathcal{H}_{RKKY}$, it is possible to account for most of the properties of dilute magnetic alloys above T_K by assuming that each local moment experiences a random internal effective magnetic field H_e distributed with probability distribution $P(H_e)$, produced by the presence of all the other impurities. $P(H_e)$ is symmetric in H_e with a width of the order of $J^2(S(S+1))c$, since the average of r^{-3} is c, if the impurity distribution is random. In addition, $P(H_e = 0)$ is proportional to c^{-1}, so that a finite fraction of the impurities are free of interactions. Certain simple scaling laws on H, T, and c follow directly from these properties of $P(H)$ [92]. In addition to the effects of H_e, two impurities with no other moments nearby can couple at low enough temperatures to form a "correlated" pair. This pair may be coupled ferromagnetically or antiferromagnetically, depending on the phase of $\cos(2k_F r)$ for the pair separation, and breaks up as T exceeds the coupling energy [93, 94]. Such pairs represent the simplest form of the short-range magnetic order found in more complicated forms in more concentrated "dilute" alloys of concentrations of 0.5% and higher.

In this Klein–Brout–Marshall (KBM) model [95], the local moment is assumed to obey a Curie law, the magnetization $M(H_e, T)$ is proportional to H/T, the single-impurity resistivity is taken from the third-order perturbation result of Kondo, and the specific heat arises from the response of the bare local moments to H_e.

a. *Susceptibility.* As the temperature is lowered, a Curie law appropriate to c isolated impurities of moment μ_e gives way to a maximum in χ at $T_m \equiv \Gamma_1 c$ as χ starts to drop to its $T = 0$ value, which is approximately *independent* of c. The peak represents the onset of short-range antiferromagnetic ordering, which reduces χ (T_m is typically a few tens of Kelvins per atomic percent).

b. *Resistivity.* The slope of $\rho(T)/c$ vs. log T, which would be independent of c for isolated impurities, decreases as c increases, going as c^{-1} at large c. A maximum appears in $\rho(T)$ at $T_m' \equiv \Gamma_2 c$ (with $\Gamma_1 \sim 2.3$ times Γ_2), and below T_m', ρ decreases linearly in T to $\rho(T = 0)$, which is approximately independent of c. As c is lowered, the peaks in χ and in ρ vanish, and both quantities tend to scale more nearly with concentration, first only at high temperatures, and finally, as c goes to zero, at all temperatures.

The peak is the result of competition between the Kondo divergence on the single impurity and the "freezing out" of the spin-flip scattering part of this resistivity as H_e/T increases when T is lowered toward zero. As long as T_K is vanishingly small, the fact that ρ does not drop exponentially to zero below the peak implies that $P(H_e = 0)$ is not zero; that is, there is no long-range order.

c. *Specific heat.* A peak (at T_m'') appears in $C_v(T)$ in the KBM model just as in the Kondo effect, but now T_m'' depends more or less linearly on c. In addition, C_v/T of the whole alloy increases as T goes to zero, approaching a value independent of c, while C_v/T for the Kondo anomaly is flat near $T = 0$ and proportional to concentration. At temperatures above T_m'' where interactions are dominated by thermal effects, the specific heat approaches that of the isolated impurities, although it is usually masked by the specific heat of the lattice.

d. *Thermopower.* The Kondo anomaly here is independent of concentration, as long as enough local moments are present to dominate the total electron scattering. In concentrated alloys, the peak height is decreased, while the peak is displaced to higher T, and the high-temperature side of the peak becomes much less steep as the magnitude of Θ at high temperatures more nearly equals the now lower peak value. The presence of ordering produces its own thermopower peak at a concentration-dependent temperature [96].

2. Weak Interactions

No systematic experimental or theoretical studies have been made in the regime where $\mathscr{H}_{RKKY}$ is less than T_K. At high temperatures, interactions have not been a problem in alloys of this type. The data that are available have come from attempts to study the ground state of the isolated local moment in a metal, which is inherently the most interesting part of the entire local moment problem. It is in attempting to study the $T = 0$, $c = 0$ limit, that the presence of the long-range "weak" interactions has limited the experimental understanding of the local moment. The weak interactions appear in the susceptibility [23, 97] as an extra term proportional to T^{-a}, $a = 0.5$ to 0.6, that vanishes in magnetic fields too small to have any reason to affect the Kondo susceptibility [98]. At the same time, C_v/T shows a rise [99] as $T \to 0$. Both effects vary as c^n, where n is between 1 and 2, so that in some cases (when $n \simeq 1$), the interaction results in what appears to be a single-impurity effect.

In a weak interaction alloy, there is no resistivity maximum, but the Kondo $\log(T)$ divergence terminates at a temperature θ_c that depends approximately linearly [63] on c. This temperature is of the order of the average magnitude of H_e in the KBM model. Below θ_c the resistivity is temperature independent. Thus ρ is well represented by $-\log(T^2 + \theta_c^2)^{1/2}$ plus a constant. These properties of the resistivity have a simple theoretical interpretation [100] based on the statistical effect of the random interaction of all the spins in the sample, although better agreement with experiment is obtained if a finite fraction of the moments are allowed to form pairs and cease scattering [93]. Because the shape of the $\rho(T)$ curve in this case is nearly identical to that of an isolated impurity near the temperature region where the Kondo divergence saturates at the unitarity limit, it is important to vary c to determine whether the Kondo divergence is saturating at unitarity, or merely at θ_c. In the language of Eq. (2.3) it is necessary to demonstrate that θ is independent of c.

Only interaction effects for alloys in which the spatial distribution of impurities is random have been considered. Spatial agglomeration and precipitation of impurities is known to occur in most alloys under some

144   MELVIN D. DAYBELL

conditions, and this agglomeration shows up strongly in the interaction-related properties of dilute alloys.

Table IV lists the lowest concentrations at which the $T = 0$ limit of various properties of several Kondo alloys have been investigated, along with an indication of whether or not the single-impurity limit appears to have been reached.

TABLE IV

Data at Very Low Concentrations

Host	Impurity	Minimum concentration (ppm)	Property	Isolated impurity limit observed?	Ref.
Cu	Cr	12	ρ	Yes	[40]
	Fe	22	ρ	Yes	[59]
	Cr	15	χ	No	[98]
	Fe	54	χ	No	[49]
	Fe	81	C_v	?	[101]
	Mn	4	χ	Yes	[23]
Au	V	3300	C_v	?	[76]
	Fe	0.5	Θ	Yes	[77]
	Fe	5.6	ρ	?	[102]
Zn	Mn	7.6	ρ	Yes	[25]
Cd	Mn	17	ρ	Yes	[29]

IV. Review Articles

The situation before the discovery of the Kondo effect is reviewed by van den Berg [103]. Most of the data discussed show strong interaction effects. A review of the systematic variation of T_K with position in the periodic table is available [1]. Two long papers [88, 104] appear in *Solid State Physics*, and one [88] contains a thorough discussion of much of the NMR work in Kondo alloys (see Narath [90] for more on the resonance properties of the local moment). Another excellent experimental review has also appeared [105], as has a comparison of the details of various theoretical approaches [47]. These reviews are indispensable, as more than 800 papers on the local moment problem have appeared since 1962 [106].

References

1. M. D. Daybell and W. A. Steyert, *Rev. Mod. Phys.* **40**, 380 (1968).
2. R. Aoki and T. Ohtsuka, *J. Phys. Soc. Jap.* **26**, 651 (1969).

4. THERMAL AND TRANSPORT PROPERTIES

3. J. Kondo, *Progr. Theor. Phys.* **32**, 37 (1964).
4. A. Blandin and J. Friedel, *J. Phys. Radium* **20**, 160 (1959).
5. P. W. Anderson, *Phys. Rev.* **124**, 41 (1961).
6. J. R. Schrieffer, *J. Appl. Phys.* **38**, 1143 (1967).
7. H. Suhl and D. Wong, Physics (*Long Island City, N.Y.*) **3**, 17 (1967).
8. D. R. Hamann, *Phys. Rev.* **158**, 570 (1967); P. E. Bloomfield and D. R. Hamann, *Ibid.* **164**, 856 (1967).
9. N. Rivier and M. J. Zuckermann, *Phys. Rev. Lett.* **21**, 904 (1968).
10. G. Yuval and P. W. Anderson, *Phys. Rev. B* **1**, 1522 (1970); P. W. Anderson, G. Yuval, and D. R. Hamann, *Ibid.* **1**, 4464 (1970).
11. K. D. Schotte, *Z. Phys.* **235**, 155 (1970).
12. D. L. Mills and P. Lederer, *Phys. Rev.* **160**, 590 (1967).
13. K. C. Brog, W. H. Jones, Jr., and G. S. Knapp, *Solid State Commun.* **5**, 913 (1967).
14. G. R. Caskey and D. J. Sellmyer, *J. Appl. Phys.* **40**, 1476 (1969).
15. M. P. Sarachik and G. S. Knapp, *J. Appl. Phys.* **40**, 1105 (1969).
16. D. R. Zrudsky, W. R. Savage, and J. W. Schweitzer, *J. Appl. Phys.* **40**, 1099 (1969).
17. J. J. Hauser, *Phys. Rev. B* **5**, 110 (1972).
18. Y. Nagaoka, *Phys. Rev. A* **138**, 1112 (1965).
19. K. Fischer, *Z. Phys.* **225**, 444 (1969).
20. J. Kondo, *Phys. Rev.* **169**, 437 (1968).
21. K. Yosida, *Phys. Rev.* **147**, 223 (1966); K. Yosida and A. Yoshimori, *Progr. Theor. Phys.* **42**, 753 (1969); J. A. Appelbaum and J. Kondo, *Phys. Rev.* **170**, 542 (1968); D. R. Hamann and J. A. Appelbaum, *Ibid.* **180**, 334 (1969).
22. A. D. Caplin and C. Rizzuto, *Phys. Rev. Lett.* **21**, 746 (1968); G. Boato, M. Bugo, and C. Rizzuto, Nuovo Cimento B **45**, 226 (1966).
23. E. C. Hirschkoff, O. G. Symko, and J. C. Wheatley, *Phys. Lett. A* **33**, 19 (1970).
24. R. Tournier and A. Blandin, *Phys. Rev. Lett.* **24**, 397 (1970).
25. R. S. Newrock, B. Serin, J. Vig, and G. Boato, *J. Low Temp. Phys.* **5**, 701 (1971).
26. H. Nagasawa, S. Yoshida, and T. Sugawara, *Phys. Lett. A* **26**, 561 (1968).
27. M. P. Maley and R. D. Taylor, *Phys. Rev. B* **1**, 4213 (1970).
28. H. Nagasawa, *J. Phys. Soc. Jap.* **28**, 1171 (1970).
29. H. Alloul, R. Deltour, and R. Clad, *J. Phys. Soc. Jap.* **28**, 661 (1970).
30. T. Sugawara and H. Eguchi, *J. Phys. Soc. Jap.* **26**, 1322 (1969).
31. J. G. Booth, K. C. Brog, and W. H. Jones, *Proc. Phys. Soc. London* **92**, 1083 (1967).
32. K. Kume, *J. Phys. Soc. Jap.* **23**, 1226 (1967).
33. K. Kume and O. Kogure, *J. Phys. Soc. Jap.* **25**, 930 (1968).
34. D. K. C. MacDonald, W. B. Pearson, and I. M. Templeton, *Proc. Roy. Soc. Ser. A* **266**, 161 (1962).
35. J. W. Loram, T. E. Whall, and P. J. Ford, *Phys. Rev. B* **3**, 953 (1971).
36. P. Costa-Ribeiro, J. Souletie, and D. Thoulouze, *Phys. Rev. Lett.* **24**, 900 (1970).
37. M. B. Maple, J. G. Huber, B. R. Coles, and A. C. Lawson, *J. Low Temp. Phys.* **3**, 137 (1970).
38. P. E. Bloomfield, R. Hecht, and P. R. Sievert, *Phys. Rev. B* **2**, 3714 (1970).
39. R. B. Frankel N. A. Blum, B. B. Schwartz, and D. J. Kim, *Phys. Rev. Lett.* **18**, 1051 (1967).
40. M. D. Daybell and W. A. Steyert, *Phys. Rev. Lett.* **20**, 195 (1968).
41. J. W. Loram, A. D. C. Grassie, and G. A. Swallow, *Phys. Rev. B* **2**, 2760 (1970).
42. T. A. Kitchens, W. A. Steyert, and R. D. Taylor, *Phys. Rev.* **138**, A 467 (1965).
43. A. Narath and A. C. Gossard, *Phys. Rev.* **183**, 391 (1969).
44. D. C. Golibersuch and A. J. Heeger, *Phys. Rev.* **182**, 584 (1969).

45. B. Giovannini, R. Paulson, and J. R. Schrieffer, *Phys. Lett.* **23**, 517 (1966).
46. D. J. Scalapino, *Phys. Rev. Lett.* **16**, 937 (1966).
47. K. Fischer, in "Springer Tracts in Modern Physics" (G. Höhler, ed.), Vol. 54. Springer-Verlag, Berlin, 1970.
48. K. D. Schotte and U. Schotte, *Phys. Rev. B* **4**, 2228 (1971).
49. M. D. Daybell and W. A. Steyert, *Phys. Rev.* **167**, 536 (1968).
50. J. E. Van Dam and P. C. M. Gubbens, *Phys. Lett. A* **34**, 185 (1971).
51. J. L. Tholence and R. Tournier, *Phys. Rev. Lett.* **25**, 867 (1970).
52. L. L. Hirst, G. Williams, D. Griffiths, and B. R. Coles, *J. Appl. Phys.* **39**, 844 (1968).
53. C. M. Hurd, *J. Phys. Chem. Solids* **28**, 1345 (1967).
54. J. Bensel and J. A. Gardner, *J. Appl. Phys.* **41**, 1157 (1970).
55. T. Sugawara and S. Yoshida, *J. Low Temp. Phys.* **4**, 657 (1971).
56. B. Coqblin and J. R. Schrieffer, *Phys. Rev.* **185**, 847 (1969).
57. A. S. Edelstein, *Phys. Rev. Lett.* **20**, 1348 (1968).
58. F. T. Hedgcock and W. B. Muir, *J. Phys. Soc. Jap.* **16**, 2599 (1961); G. Boato and J. Vig, *Solid State Commun.* **5**, 649 (1967).
59. M. D. Daybell and W. A. Steyert, *Phys. Rev. Lett.* **18**, 398 (1967).
60. N. E. Alekseevskii and Iu. P. Gaidukov, *Sov. Phys. JETP* **4**, 807 (1957).
61. J. W. Loram, P. J. Ford, and T. E. Whall, *J. Phys. Chem. Solids* **31**, 763 (1970).
62. J. W. Loram, T. E. Whall, and P. J. Ford, *Phys. Rev. B* **2**, 857 (1970).
63. M. D. Daybell and Y. K. Yeo, *Bull. Amer. Phys. Soc.* **16**, 840 (1971).
64. B. Giovannini, *Phys. Rev. B* **3**, 870 (1971).
65. J. W. Garland, K. H. Benneman, A. Ron, and A. S. Edelstein, *J. Appl. Phys.* **41**, 1148 (1970).
66. K. K. Murata, Thesis, Cornell Univ., Ithaca, New York, 1971.
67. D. Gainon and A. J. Heeger, *Phys. Rev. Lett.* **22**, 1420 (1969).
68. K. Maki, *Progr. Theor. Phys.* **41**, 586 (1969).
69. K. Yosida, *Phys. Rev.* **106**, 893 (1957).
70. P. Monod, *Phys. Rev. Lett.* **19**, 1113 (1967).
71. R. J. Harrison and M. W. Klein, *Phys. Rev.* **154**, 540 (1967).
72. M. T. Beal-Monod and R. A. Weiner, *Phys. Rev.* **170**, 552 (1968).
73. C. M. Hurd and J. E. A. Alderson, *Phys. Rev. B* **4**, 1088 (1971).
74. B. B. Triplett and N. E. Phillips, *Proc. Int. Conf. Low Temp. Phys.* 12th *Kyoto*, 1970 (E. Kanda, ed.), p. 747. Academic Press of Japan, Tokyo, 1971.
75. J. Zittartz and E. Müller-Hartmann, *Z. Phys.* **212**, 380 (1968).
76. B. M. Boerstoel and W. M. Star, *Phys. Lett. A* **29**, 97 (1969).
77. G. Borelius, W. H. Keesom, C. H. Johansson, and J. O. Linde, *Leiden Comm.* **217e** (1932); *Leiden Comm. Suppl.* **70a** (1932).
78. E. Brewig, W. Kierspe, U. Schotte, and D. Wagner, *J. Phys. Chem. Solids* **30**, 483 (1969).
79. A. M. Guénault and M. Read, *Proc. Int. Conf. Low Temp. Phys.* 12th *Kyoto*, 1970 (E. Kanda, ed.), p. 759. Academic Press of Japan, Tokyo, 1971.
80. E. L. Christenson, *J. Appl. Phys.* **34**, 1485 (1963).
81. D. Gainon and J. Sierro, *Phys. Lett. A* **26**, 601 (1968).
82. H. L. Malm and S. B. Woods, *Can. J. Phys.* **44**, 2293 (1966).
83. C. van Baarle and F. W. Gorter, *Physica* (*Utrecht*) **32**, 1709 (1966).
84. M. D. Daybell, D. L. Kohlstedt, and W. A. Steyert, *Solid State Commun.* **5**, 871 (1967).
85. R. L. Powell, L. P. Caywood, Jr., and M. D. Bunch, *in* "Temperature, Its Measure-

ment and Control in Science and Industry." (C. M. Herzfeld, ed.), Vol. III, Pt. 2, p. 65. Van Nostrand-Reinhold, Princeton, New Jersey, 1962.
86. C. A. Domenicali, *Phys. Rev.* **112**, 1863 (1958).
86a. R. A. Weiner and M. T. Béal-Monod, *Phys. Rev. B* **3**, 145 (1971).
87. D. J. Huntley and C. W. E. Walker, *Can. J. Phys.* **47**, 805 (1969).
88. A. J. Heeger, *Solid State Phys.* **23**, 283 (1969).
89. P. Monod and A. Friederich, *Proc. Int. Conf. Low Temp. Phys.* 12th *Kyoto*, 1970 (E. Kanda, ed.), p. 755. Academic Press of Japan, Tokyo, 1971.
90. A. Narath, *Proc. Int. Conf. Low Temp. Phys.* 12th *Kyoto*, 1970 (E. Kanda, ed.), p. 675. Academic Press of Japan, Tokyo, 1971.
91. H. Suhl, *Solid State Commun.* **4**, 487 (1966).
92. J. Souletie and R. Tournier, *J. Low Temp. Phys.* **1**, 95 (1969).
93. M. T. Béal-Monod, *Phys. Rev.* **178**, 874 (1969).
94. K. Matho and M. T. Béal-Monod, *Phys. Rev. B* **5**, 1899 (1972).
95. M. W. Klein and R. Brout, *Phys. Rev.* **132**, 2412 (1963); W. Marshall, *Ibid.* **118**, 1519 (1960).
96. J. Kondo, *Progr. Theor. Phys.* **34**, 372 (1965).
97. E. C. Hirschkoff, O. G. Symko, and J. C. Wheatley, *J. Low Temp. Phys.* **5**, 155 (1971).
98. M. D. Daybell, W. P. Pratt, Jr., and W. A. Steyert, *Phys. Rev. Lett.* **22**, 401 (1969).
99. M. D. Daybell, W. P. Pratt, Jr., and W. A. Steyert, *Phys. Rev. Lett.* **21**, 353 (1968).
100. H. Suhl, *Phys. Rev. Lett.* **20**, 656 (1968).
101. J. C. Brock, J. C. Ho, G. P. Schwartz, and N. E. Phillips, *Solid State Commun.* **8**, 1139 (1970).
102. Van Rongen (Leiden), unpublished.
103. G. J. van den Berg, *Progr. Low Temp. Phys.* **4**, 194 (1964).
104. J. Kondo, *Solid State Phys.* **23**, 183 (1969).
105. A. Narath, *J. Appl. Phys.* **41**, 1122 (1970).
106. See e.g. "Magnetic Impurities in Metals," *Selected Papers in Phys.* Physical Society of Japan, Tokyo, 1971.

5. The s-d Model and the Kondo Effect: Magnetic Hyperfine-Interaction Studies

Albert Narath

Sandia Laboratories
Albuquerque, New Mexico

I. Introduction 149
II. Theory of Magnetic Hyperfine Interactions in Dilute Alloys 150
   1. General Considerations 150
   2. Impurity Hyperfine Effects 151
   3. Host Hyperfine Effects 154
III. Impurity Magnetization Studies 157
   1. Local Moment Formation 157
   2. Static Impurity Susceptibility 160
   3. Host Polarization 167
IV. Dynamic Response Studies 174
   1. Impurity Nuclear Spin–Lattice Relaxation 174
   2. Host Nuclear Spin–Lattice Relaxation 177
V. Concluding Remarks 179
   References 180

I. Introduction

Because of the hyperfine coupling between nuclear and electronic magnetic moments, nuclear magnetic resonance (NMR) and related techniques such as the Mössbauer effect (ME) and nuclear orientation (NO) provide powerful techniques for probing the magnetic response of dilute magnetic alloys. In contrast to bulk measurements such as the magnetic susceptibility which sense only spatial averages, hyperfine techniques offer the significant advantage of a local probe. In this chapter we are principally concerned with magnetic hyperfine interactions in alloys which approach the magnetic limit of the Anderson model [1], $|\varepsilon_d| \gg \varDelta$, $\varepsilon_d + \bar{U} \gg \varDelta$, where ε_d denotes the position of the impurity level relative to the Fermi energy (E_F), $\varDelta$ is the level width, and $\bar{U}$ is the effective intra-atomic Coulomb repulsion. In this limit the Anderson model is equivalent to the s–d exchange model [2] in which the impurity is characterized by a definite spin (S) and an antiferromagnetic conduction-electron–impurity exchange constant (J_{sd}) whose magnitude is proportional to $\varDelta$. Unfortunately, despite considerable effort during recent years, many aspects of the s–d model, especially as regards its relevance to real alloys, remain unclear. Much of the difficulty can be traced to uncertainties in the experimental situation. For example, as a result of the suppression of free spin behavior of the impurity local moment by the Kondo effect [3] it is difficult in many instances to distinguish experimentally between the magnetic and nonmagnetic (i.e., local spin fluctuation, $|\varepsilon_d| < \varDelta$) regimes of the Anderson model. Although a strongly temperature-dependent susceptibility is usually regarded as evidence of local moment behavior, ambiguities arise for alloys with high Kondo temperatures (T_K). The fact that experiments by necessity require finite impurity concentrations adds additional complications since impurity–impurity interactions, particularly at low temperatures, frequently dominate the observed behavior.

In the following we explore the contribution of nuclear hyperfine studies to the current understanding of the magnetic impurity problem. Most of the discussion applies specifically to 3d impurities in simple metals, where the antiferromagnetic s–d exchange model is of greatest relevance. In the case of rare earth impurities the direct ferromagnetic s–d exchange usually dominates the indirect antiferromagnetic admixture exchange. Notable exceptions are trivalent Ce and Yb for which instabilities in the 4f shell can lead to strong covalent mixing effects [4, 5].

II. Theory of Magnetic Hyperfine Interactions in Dilute Alloys

1. General Considerations

The principal magnetic coupling between the nuclear moment and its environment may be written as

$$\mathcal{H} = -\gamma_n \hbar \mathbf{I} \cdot (\mathbf{H} + \mathbf{h}_{\text{loc}}), \tag{2.1}$$

where $\mathbf{H}$ is the applied field, $\mathbf{h}_{\text{loc}}$ is the local field arising from the electrons, and the other symbols have their usual meaning. Since the mean time between electronic transitions (τ_e) is usually very much shorter

5. MAGNETIC HYPERFINE-INTERACTION STUDIES

than the nuclear Larmor period (τ_n) it is useful to separate $\mathbf{h}_{loc}$ into thermal-average and fluctuating components

$$\mathbf{h}_{loc} = \langle \mathbf{h}_{loc} \rangle + \delta \mathbf{h}_{loc} . \tag{2.2}$$

In this approximation the static component of $\mathbf{h}_{loc}$ produces a shift ($K \equiv \Delta H/H$) in the field for nuclear resonance

$$K = \langle h_{loc}^z \rangle / H, \tag{2.3}$$

where z defines the applied field direction, while the fluctuating components of $\mathbf{h}_{loc}$ yield longitudinal (T_1^{-1}) and transverse (T_2^{-1}) nuclear spin relaxation rates [6]

$$T_1^{-1} = \tfrac{1}{2}\gamma_n^2 \int_{-\infty}^{+\infty} \langle \{\delta h_{loc}^+(t)\, \delta h_{loc}^-(0)\} \rangle \exp(i\omega_n t)\, dt \tag{2.4}$$

(where ω_n may be set equal to zero because of the assumption $\tau_e \ll \tau_n$, provided $\hbar\omega_n \ll k_B T$) and

$$T_2^{-1} = \tfrac{1}{2} T_1^{-1} + \tfrac{1}{2}\gamma_n^2 \int_{-\infty}^{+\infty} \langle \{\delta h_{loc}^z(t)\, \delta h_{loc}^z(0)\} \rangle\, dt. \tag{2.5}$$

Here $\langle \{\ \} \rangle$ denotes an ensemble average of the symmetrized operator product. Thus, T_1^{-1} and T_2^{-1} increase with increasing electronic correlation times, making observation of the nuclear resonance increasingly difficult because of line broadening.

The ME and NO techniques generally measure only $\langle \mathbf{h}_{loc} \rangle$. The precision is poorer than can be achieved with NMR. On the other hand, the lack of sensitivity to the fluctuation amplitude $\delta \mathbf{h}_{loc}$ removes any restriction on the magnitudes of τ_e which are accessible to these techniques. Because of an inherently much greater sensitivity these techniques can be applied to extremely low impurity concentrations. It should be noted that NO experiments are sometimes performed under conditions where the requirement $\tau_e < \tau_n$ is not satisfied. Although the interpretation of the data is then more difficult, a qualitative measure of τ_e can be obtained [7]. Similar considerations apply to ME experiments [8].

2. Impurity Hyperfine Effects

In order to relate observed hyperfine effects to the magnetic properties of impurities it is essential to recognize that several distinct mechanisms may contribute simultaneously to $\mathbf{h}_{loc}$. The fact that the magnitudes of the various hyperfine coupling constants can often only be crudely estimated constitutes, of course, the major disadvantage of NMR and all other hyperfine techniques.

In the absence of significant spin–orbit coupling we may express $\mathbf{h}_{loc}$ in terms of the various hyperfine fields

$$\mathbf{h}_{loc} = \sum_i H_{hfs}^{(i)} \mathbf{m}_i / \mu_B , \qquad (2.6)$$

where (i) identifies the hyperfine mechanism and $\mathbf{m}_i$ is the associated atomic moment. The important contributions to $\mathbf{h}_{loc}$ at the impurity nucleus arise from d-spin (core polarization) and d-orbital hyperfine interactions with the spin and orbital electronic moments, respectively [9, 10]. They contribute separately to the NMR shift and relaxation rates

$$K = K_d + K_{orb} , \qquad (2.7)$$

$$T_1^{-1} = T_{1(d)}^{-1} + T_{1(orb)}^{-1} . \qquad (2.8)$$

Using (2.3) and (2.4) together with the high-temperature ($kT > \hbar\omega$) limit of the fluctuation–dissipation theorem [11]

$$\mathrm{Im}\, \chi_{(i)}^{\alpha\beta}(\omega) = (\omega/2k_B T) \int_{-\infty}^{+\infty} e^{(i\omega t)} \langle \{\delta m_{(i)}^\alpha(t)\, \delta m_{(i)}^\beta(0)\}\rangle \, dt, \qquad (2.9)$$

we can express (2.7) and (2.8) in terms of the real longitudinal and imaginary transverse components, respectively, of the local d-spin [$\chi_{(d)}(\omega)$] and d-orbital [$\chi_{orb}(\omega)$] susceptibilities

$$K_{(i)} = (\mu_B)^{-1} H_{hfs}^{(i)} \, \mathrm{Re}\, \chi_{(i)}^{zz}(0), \qquad (2.10)$$

$$T_{1(i)}^{-1} = k_B T (\mu_B)^{-2} \gamma_n^2 (H_{hfs}^{(i)})^2 \lim_{\omega\to 0}[\mathrm{Im}\, \chi_{(i)}^{+-}(\omega)/\omega]. \qquad (2.11)$$

We note that $H_{hfs}^{(d)}$ is intrinsically negative with a 3d free ion value of about -120 kOe/μ_B which is approximately independent of atomic number [12]. In contrast, $H_{hfs}^{(orb)}$ is positive and increases rapidly with atomic number, as indicated by the Hartree–Fock neutral atom values [12] shown in Fig. 1a.

It has been conjectured that s–d hybridization, resulting from the same covalent mixing which is responsible for J_{sd}, may be an important source of spin density at the impurity nucleus [13]. Such an effect would appear as a *positive* contribution to the d-spin hyperfine field. This has the serious consequence that the traditional assumption of a

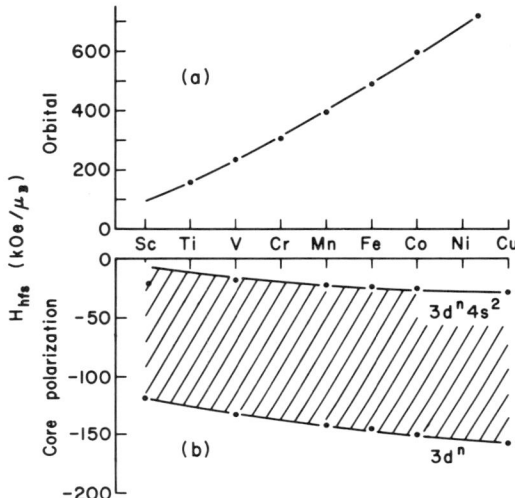

FIG. 1. Hartree–Fock hyperfine fields for $3d$ transition elements: (a) $H_{\text{hfs}}^{(\text{orb})}$ for neutral atom; (b) $H_{\text{hfs}}^{(d)}$ for neutral-atom and ionic configurations, as indicated (after Freeman and Watson [12]).

negative $H_{\text{hfs}}^{(d)}$ might not always be valid. However, for a pure d state the principal admixture effect involves the $l = 2$ conduction-electron partial wave; an $l = 0$ admixture requires an $l = 2$ potential. In a cubic crystal such a potential can only arise from nonspherical terms in the $3d$ charge distribution which are vanishingly small for orbital singlets and are probably unimportant in other cases as well [14]. A possible origin of the small d-spin hyperfine fields which are often observed is the exchange polarization of the impurity $4s$ electron. By analogy with the neutral-atom situation this can presumably lead to a greatly reduced net spin density at the nucleus. This point is illustrated in Fig. 1b which compares results of exchange-polarized Hartree–Fock calculations by Freeman and Watson [12] for $3d^n$ ions with those for the corresponding $3d^n4s^2$ atoms. The same conclusion, based on a slightly different point of view, has been reached by Hirst [15] who considered the $l = 0$ spin polarization of a plane-wave conduction band by the direct ferromagnetic s–d exchange.

Finally, we need to consider the direct s-contact interaction with the unperturbed $l = 0$ component of the conduction-electron spin density. Although $H_{\text{hfs}}^{(s)}$ is typically larger in magnitude than $H_{\text{hfs}}^{(d)}$ by a factor of about 10 [12], its contribution is nevertheless relatively unimportant for magnetic impurities for which $\chi_{(s)}/\chi_{(d)} \ll 1$.

3. Host Hyperfine Effects

A coupling between the local moment spin (**S**) and the conduction-electron spin density [$\sigma(\mathbf{r})$] of the general form

$$\mathcal{H}_{ij} = -2J(\mathbf{r})\,\mathbf{S}_i \cdot \boldsymbol{\sigma}_j(r) \tag{2.12}$$

induces a long-range oscillatory modulation of the conduction-electron spin density. The spin polarization contributes to $\mathbf{h}_{\text{loc}}$ at the host nuclei and consequently provides another means of probing the impurity magnetization. The important hyperfine mechanism in the case of simple (i.e., nearly free electron) hosts is the conduction-electron s-contact interaction; in transition metal hosts one generally expects both s- and d-electron interactions to be important, making a quantitative determination of spin polarization amplitudes from NMR data exceedingly difficult.

A formal expression for the host spin polarization is given by

$$\langle \sigma^{\alpha}(\mathbf{r},t)\rangle = (g\mu_B)^{-1} \int\!\!\int d\mathbf{r}'\,dt'\,\chi_e^{\alpha\beta}(\mathbf{r},\mathbf{r}',t-t')\,H^{\beta}(\mathbf{r}',t'), \tag{2.13}$$

where $\chi_e(\mathbf{r},\mathbf{r}',t-t')$ is the nonlocal susceptibility in the presence of the impurities and $\mathbf{H}(\mathbf{r}',t')$ is an arbitrary external magnetic field. The essential characteristics of the static polarization are easily established [16] by assuming that the s–d interaction may be represented by a weak wave-number-dependent local field $H(\mathbf{q})$ acting on an isotropic electron gas whose response is given by $\chi(\mathbf{q})$:

$$\langle \sigma^z(r)\rangle = (g\mu_B)^{-1}\sum_{\mathbf{q}} \chi(\mathbf{q})\,H^z(\mathbf{q})\,e^{i\mathbf{q}\cdot\mathbf{r}}. \tag{2.14}$$

For the special case of a free electron gas (in three dimensions) and a zero-range interaction

$$H^z(\mathbf{q}) = 2J_{sd}(g\mu_B)^{-1}\langle S^z\rangle, \tag{2.15}$$

the polarization has the well-known Ruderman–Kittel–Kasuya–Yosida (RKKY) [17–19] form

$$\langle \sigma^z(r)\rangle = \frac{9}{2}\pi(J_{sd}/E_F)\,Z^2 F(2k_F r)\langle S^z\rangle, \tag{2.16}$$

where

$$F(x) = \frac{x\cos x - \sin x}{x^4}, \tag{2.17}$$

and Z denotes the number of conduction electrons per atom. This result has been shown [20] to be valid for the s–d model in the range $T > T_K$, provided that $\langle S^z \rangle$ is assigned its actual (rather than Curie law) temperature dependence. A damped, oscillatory spatial response having a period π/k_F is a general property of an electron gas subjected to a localized perturbation and is a direct consequence of the singularity of $\chi(q)$ at $q = 2k_F$. The detailed radial dependence is, of course, dependent on the wave-number dependence of $\chi(q)$ which is generally not well known. This is particularly true for transition metal hosts where non-spherical Fermi surfaces and exchange enhancement effects can strongly modify the functional form of $\chi(q)$.

Strictly speaking, J_{sd} as used above refers to an $l = 0$ scattering potential. Actually, the matrix elements of $J(\mathbf{r})$ should be of the form $J(k, k') P_l(\cos \theta)$ [14]. In order to make the problem tractable it is customary to replace $J(k, k')$ by a constant. In discussing the magnetic properties of dilute alloys, we therefore simply replace J_{sd} by J_{eff}. The effect of the orbital degeneracy can be seen in the transport properties, which are determined by $j_{\text{eff}} = J_{\text{eff}}/(2l + 1)$, provided that the $2l + 1$ scattering channels remain independent [21]. It is often useful to relate J_{eff} to the scattering phase shifts which appear in the Friedel formulation [22] of the magnetic impurity problem. For the important case of a dominant l-resonance scattering mechanism, and in the absence of crystal-field splitting of the orbital energies,

$$J_{\text{eff}} = -\tfrac{2}{3} E_F (2l + 1)(g\mu_B/m\pi Z) \,|\sin \delta\,|, \tag{2.18}$$

where m is the impurity moment, and δ is the difference between spin-up and spin-down phase shifts at the Fermi energy. This expression, when substituted into (2.16), is valid for any arbitrary scattering potential provided $k_F r$ is sufficiently large. The behavior near the impurity depends on the detailed k dependence of the phase shifts [23]. The simple s–d model therefore yields a conduction-electron polarization which is only asymptotically correct at large distances from the impurity. The practical advantage of (2.18) lies in the fact that δ may be related to m by means of the Friedel sum rule. This yields

$$\delta = \pi(m/\mu_B)(2l + 1)^{-1}, \tag{2.19}$$

which provides a means of estimating the magnitude of J_{eff} [24]. Unfortunately, some ambiguity arises in the determination of m. In the localized spin fluctuation regime we may take $m = g\mu_B \langle S^z \rangle$ in the spirit of the Hartree–Fock approximation, and hence

$$J_{\text{eff}} = -\tfrac{4}{3}(E_F/Z). \tag{2.20}$$

Thus, in this approximation J_{eff} is independent of the nature of the impurity. In the magnetic limit (and $T > T_K$) m is traditionally identified with the saturation moment $g\mu_B S$ [25], although the validity of this approach is clearly open to question. In general, it is obvious that one needs to exercise some caution in applying expressions such as (2.16) and (2.18) to real alloys.

The static longitudinal polarization (2.16) leads to a distribution of impurity-induced host resonance shifts. In addition, components of $\sigma(\mathbf{r})$ fluctuating at a frequency ω_n will in general enhance the nuclear spin–lattice relaxation rates of the host. Unfortunately, even for simple hosts the interpretation of impurity-enhanced nuclear relaxation data is exceedingly difficult because several competing processes must be considered. Contributions arising from the RKKY coupling between nuclear and impurity spins are most conveniently discussed in terms of the nonlocal susceptibility defined by (2.13). The nuclear spin–lattice relaxation rate is given as before by (2.11). The appropriate susceptibility, Im $\chi_e(0, 0, \omega)$, has been derived by Giovannini et al. [26] within the framework of the molecular field approximation. In the low-concentration limit they obtained to lowest order in J_{eff} (for equal localized and conduction-electron g values)

$$\chi_e(0, 0, \omega) = \chi_e(0, \omega) + c(2J_{\text{eff}}/g^2\mu_B{}^2)^2 \, \chi_{(d)}(\omega) \sum_j \chi_e(\mathbf{R}_j, \omega)^2, \quad (2.21)$$

where c is the impurity concentration, $\chi_e(\mathbf{r}, \omega)$ is the conduction-electron susceptibility in the absence of the impurities, and $\chi_{(d)}(\omega)$ is the impurity susceptibility. The imaginary transverse part of (2.21) consists of three terms. For small ω [i.e., for Im $\chi_e(\mathbf{r}, \omega) \ll$ Re $\chi_e(\mathbf{r}, \omega)$] these may be conveniently written in the form

(1) Im $\chi_e^{+-}(0, \omega)$, \hfill (2.22)

(2) c Im $\chi_{(d)}^{+-}(\omega) \left[4J_{\text{eff}}^2(g\mu_B)^{-4} \int_{r_0}^{\infty} d\mathbf{r} \, \text{Re} \, \chi_e^{+-}(\mathbf{r}, \omega)^2 \right]$, \hfill (2.23)

(3) c Re $\chi_{(d)}^{+-}(\omega) \left[8J_{\text{eff}}^2(g\mu_B)^{-4} \int_{r_0}^{\infty} d\mathbf{r} \, \text{Re} \, \chi_e^{+-}(\mathbf{r}, \omega) \, \text{Im} \, \chi_e^{+-}(\mathbf{r}, \omega) \right]$, \hfill (2.24)

where r_0 is an inner cutoff radius which arises from the fact that the resonances of nuclear spins near a magnetic impurity site experience large magnetic or electric hyperfine shifts [27] and hence do not affect the macroscopic relaxation rate. It is important to recognize that the average over host sites which is implicit in (2.23) and (2.24) is only meaningful if the nuclear spin system can be characterized by a single spin temperature for $r > r_0$. This is the rapid diffusion case [28], which

5. MAGNETIC HYPERFINE-INTERACTION STUDIES

requires that r_0 be greater than the radius of the spin diffusion barrier. We also note that the radial dependence of Re $\chi_e(\mathbf{r}, \omega)$ for a free electron gas is given by the RKKY function (2.16).

The first term [Eq. (2.22)] is simply the direct Korringa process [29] for the pure host. The second term [Eq. (2.23)], which was first suggested by Benoit, de Gennes, and Silhouette (BGS) [30], represents a direct excitation of the impurity spin. The third term [Eq. (2.24)] represents an interference between the direct Korringa process and a virtual excitation of the impurity. This mechanism was first discussed by Giovanni and Heeger (GH) [31] on the basis of a perturbation treatment of the s–d model. The relative magnitudes of these three relaxation rates are determined by the temperature and magnetic field dependences of the impurity susceptibility. For sufficiently small $\chi_{(d)}^{+-}(\omega)$ (e.g., at high temperatures) the Korringa process is dominant. With increasing $\chi_{(d)}^{+-}(\omega)$ the BGS process becomes important for small H/T and the GH process for large H/T.

In addition to the RKKY mechanisms, the impurity fluctuations can couple to the host nuclei through direct dipolar (and pseudodipolar) interactions [26]. In metals the dipolar coupling constant is often small compared to the RKKY coupling constant; in that case one may neglect the dipolar mechanism in considering the effect of the *transverse* impurity fluctuations on host relaxation rates. On the other hand, because of its anisotropy the dipolar coupling provides a host relaxation mechanism which is sensitive to the *longitudinal* impurity spin and orbital fluctuations. For a host nucleus located at a position $\mathbf{r}$ relative to a single impurity the longitudinal spin fluctuation yields a relaxation rate

$$T_{1(\text{dip})}^{-1} = k_B T \gamma_n^2 [9 r^{-6} \sin^2\theta \cos^2\theta][\text{Im } \chi_{(d)}^{zz}(\omega_n)/\omega_n], \tag{2.25}$$

where θ is the angle between $\mathbf{r}$ and the external field direction. At high temperatures ($T > T_K$) the longitudinal fluctuation spectrum is centered at zero frequency, while the transverse spectrum is centered at $g\mu_B H/\hbar$. The low-frequency amplitude of the longitudinal fluctuations therefore decreases much less rapidly with increasing H/T than is the case for the corresponding transverse amplitude. For this reason the direct dipolar process may often be much more important than the GH process.

III. Impurity Magnetization Studies

1. Local Moment Formation

The occurrence of local moments in dilute alloys as well as their Kondo or spin fluctuation temperatures is strongly influenced by the local environment. Consequently, if the environment is not the same at every impurity site in a given alloy spatially inhomogeneous impurity properties may result. In such cases microscopic techniques such as NMR can play an essential role in the identification of magnetic impurity states. Attention to the importance of local environmental effects was first drawn by Jaccarino and Walker [32] who interpreted the results of ^{59}Co shift and intensity measurements in $Co_{0.01}(Rh_yPd_{1-y})_{0.99}$ alloys, and magnetic susceptibility measurements in $Fe_{0.01}(Nb_yMo_{1-y})_{0.99}$ alloys in terms of a model in which the impurities were assumed to be nonmagnetic unless a minimum of two palladium or seven molybdenum atoms, respectively, were nearest neighbors to a given impurity site. Qualitative support for this interpretation has recently been provided by the ^{57}Fe ME studies in the NbMo system by Swartzendruber [33]. Figure 2 demon-

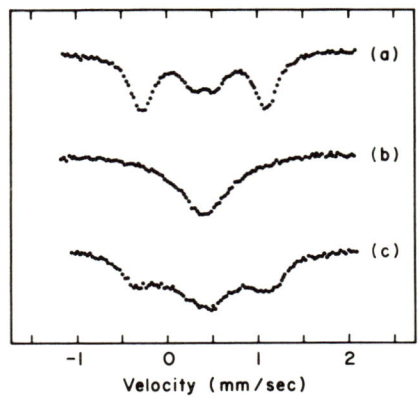

FIG. 2. Mössbauer effect absorption spectra for 1 at. % ^{57}Fe in Mo: (a) $T = 100°K$, (b) $T = 10°K$, and in $Mo_{0.8}Nb_{0.2}$: (c) $T = 10°K$. In each case a 50-kOe field was applied parallel to the direction of γ-ray propagation (after Swartzendruber [33]).

strates the coexistence of magnetic and nonmagnetic iron impurities of approximately equal abundance in a $Fe_{0.01}(Nb_{0.2}Mo_{0.8})_{0.99}$ alloy. In a pure molybdenum host the observed magnetic splitting at 100°K is dominated by the external field. With decreasing temperature the iron hyperfine field $\langle h_{loc}^z \rangle$, which is negative, increases in magnitude until it cancels the applied field near 10°K. The resulting single absorption line is to be compared with the corresponding spectrum for $Nb_{0.2}Mo_{0.8}$ which, in addition to the strong central peak, exhibits the outer lines of a Zeeman-split nonmagnetic ^{57}Fe spectrum. A similar example of discontinuous moment formation in binary alloy hosts is provided by the observation [34] of magnetic and nonmagnetic ^{59}Co NMR signals of varying intensity but essentially constant shifts in $Co_{0.01}(Ti_yMo_{1-y})_{0.99}$ alloys.

A closely related phenomenon is the discontinuous local moment behavior which often results from near-neighbor impurity–impurity interactions. For example, ^{51}V NMR experiments in **Au**V established [24] that near-neighbor V–V pairs have greatly reduced local susceptibilities compared to isolated impurities [35]. Similarly, the concentration-independent ^{59}Co NMR shifts in **Cu**Co [36], when contrasted with the rapid increase in the average cobalt susceptibility with increasing concentration [37], constitute convincing evidence that impurity clusters have greatly increased local susceptibilities in this alloy system. In this case, as in (**RhPd**)Co [32], the more magnetic species were not detected presumably because of excessively large NMR shifts and linewidths. As a final example, we consider the formation of local moments in **Mo**Co alloys. Figure 3 shows typical ^{59}Co NMR spectra [38] which indicate

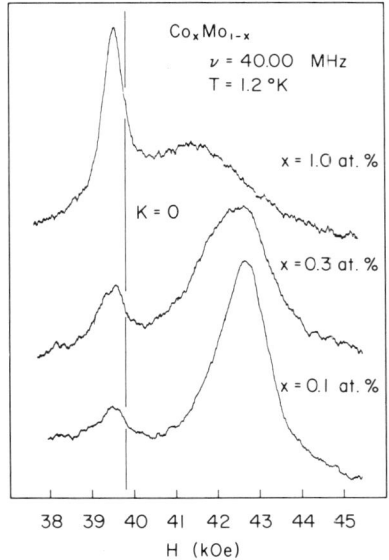

FIG. 3. Low-temperature ^{59}Co spin-echo spectra for **Mo**Co (after Narath et al. [38]).

the existence of both magnetic as well as nonmagnetic (i.e., $K \sim 0$) cobalt atoms in these alloys. The magnetic site resonance shares two characteristics with similar resonances in other dilute alloys. The first is a large resonance shift; the second is a strongly concentration-dependent linewidth which is indicative of a spatially inhomogeneous impurity polarization. The nonmagnetic site resonance has recently been associated with the intermetallic compound Mo_6Co_7 on the basis of zero-field ME studies [39]. (It is interesting to note that standard metallographic examinations had not detected the second phase.) By analogy with the **Cu**Co observations [36], it is certain that near-neighbor cobalt clusters having very low Kondo temperatures escaped detection in the NMR experiments. We may conclude from these studies that interactions between magnetic impurities result in two distinct effects. Near-neighbor interactions can produce discontinuous changes in the local impurity polarization. Longer range interactions, on the other hand, produce an essentially symmetric, continuous distribution of local polarizations. The width of this distribution can be comparable in magnitude to that of the average polarization even at relatively low impurity concentrations. It is perhaps not surprising that such large spatial inhomogeneities are occasionally accompanied by small changes in the average polarization as indicated in Fig. 3, for example, by the concentration-dependent peak positions of the ^{59}Co resonance.

2. Static Impurity Susceptibility

a. *Spin susceptibility.* The magnetic limit of the Anderson model yields a Curie law susceptibility at sufficiently high temperatures. At lower temperatures corrections arise from the antiferromagnetic covalent admixture s–d coupling. With the traditional assumptions of a zero-range s–d interaction and identical local moment and conduction-electron g values, the s–d model yields an expression for the local spin susceptibility [40]

$$\chi_{(d)} \equiv \frac{g\mu_B \langle S^z \rangle}{H},$$

$$= \frac{(g\mu_B)^2 S(S+1)}{3 k_B T} \left[1 + \frac{(2J_{eff}\rho)^2 \ln(k_B T/D)}{1 - 2J_{eff}\rho \ln(k_B T/D)}\right], \quad (3.1)$$

which results from summing the leading logarithmic terms in the perturbation expansion. Here ρ is the host density of states for one spin direction and D is an effective bandwith. In addition, the s–d coupling induces a net antiferromagnetic conduction-electron spin polarization [40]

$$\chi_e = 2J_{eff}\rho\chi_{(d)}, \quad (3.2)$$

whose spatial distribution depends on the wave-number dependence of the conduction-electron susceptibility. For the antiferromagnetic inter-

5. MAGNETIC HYPERFINE-INTERACTION STUDIES

action considered here the perturbation series diverges below the Kondo temperature

$$T_K = \frac{D}{k_B} \exp\left(\frac{-1}{2|J_{\text{eff}}|\rho}\right). \tag{3.3}$$

It is interesting to note that $\chi_{(d)}$ is proportional to χ_e. There is thus no hint of any "spin compensation" in the magnetic susceptibility. Rather, the perturbation treatment predicts a simultaneous decrease of both $\chi_{(d)}$ and χ_e with decreasing temperature as a result of the s–d coupling. It has been noted [41] that for temperatures in the range $4 \lesssim T/T_K < 100$ the total susceptibility $(\chi_{(d)} + \chi_e)$ can be approximated by a Curie–Weiss form, $\chi \propto (T + \theta_K)^{-1}$, with $\theta_K \sim 4T_K$ and a reduced Curie constant $\mu_{\text{eff}} \sim (g\mu_B)^2 S(S+1)/1.22$. Although no rigorous predictions exist for $T < T_K$ it is generally believed that the Curie–Weiss form remains approximately valid in this temperature range. Recently, it has been suggested [42, 43] that the impurity susceptibility in the limit $T \to 0$ varies as $[1 - \alpha(T/\theta_K)^2]$, where α is a constant. In contrast to the Curie–Weiss form, the quadratic form is consistent with the third law of thermodynamics which requires $(\partial \chi/\partial T)_{T=0} = 0$ [44].

An early indication of significant deviations from free spin behavior in dilute magnetic alloys appeared in ME measurements of iron impurity susceptibilities in a large number of metallic hosts [45–49]. A particularly interesting alloy, MoFe, has recently been investigated in greater detail (and for lower impurity concentration) by Maley and Taylor [50]. Mössbauer effect measurements of the ^{57}Fe hyperfine field $\langle h_{\text{loc}}^z \rangle$ as a function of temperature and magnetic field strength revealed systematic deviations from free spin behavior at low temperatures, in agreement with the earlier work, which were independent of magnetic impurity concentration and hence could be attributed to Kondo anomalies. Figure 4 summarizes the observed Curie–Weiss temperature dependence of the low-field susceptibility (i.e., $g\mu_B H/k_B T \ll 1$). These data are consistent with a Kondo temperature near $1°K$ ($\theta_K = 1.1 \pm 0.3°K$). The Kondo effect is demonstrated more convincingly by the high-field data in Fig. 5 which indicate that the susceptibility approaches a finite value below $1°K$. The solid lines represent an empirical fit of the data to a free spin Brillouin function $B_S[g\mu_B SH/k_B T]$ using an adjustable, field-dependent saturation hyperfine field. A reasonably accurate fit could also be achieved with the modified Brillouin function

$$\langle h_{\text{loc}}^z \rangle = h_{\text{loc}} B_S\left[\frac{g\mu_B SH}{k_B(T + \theta_K)}\right], \tag{3.4}$$

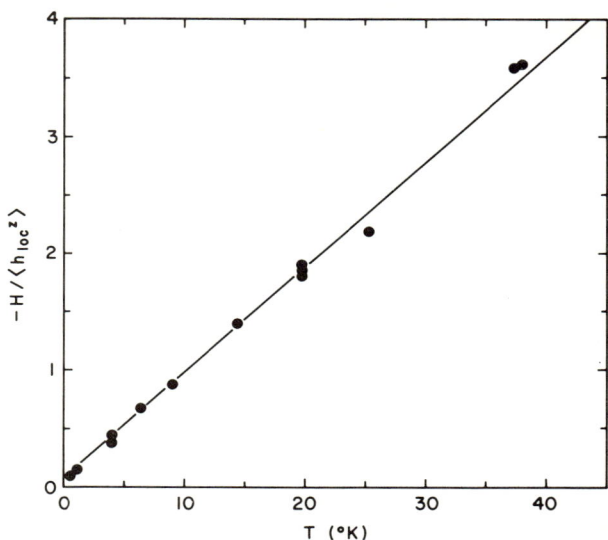

FIG. 4. Temperature dependence of the ^{57}Fe hyperfine field in **MoFe**, from ME measurements in weak applied fields, $H/T < 3$ kOe/°K (after Maley and Taylor [50]).

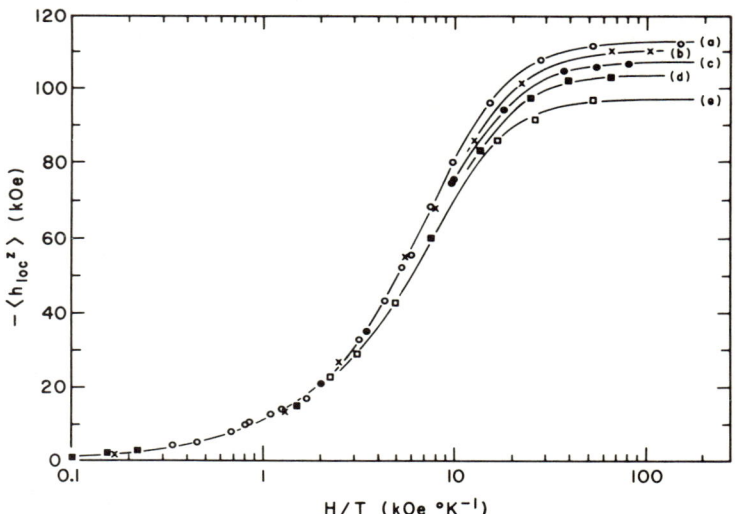

FIG. 5. Temperature dependence of the ^{57}Fe hyperfine field in **MoFe** for several applied field strengths: (a) 61.5, (b) 49.4, (c) 39.4, (d) 29.6 and (e) 19.7 kOe. The solid lines are best fits to a free spin Brillouin function with an adjustable saturation hyperfine field: (a) -112.4, (b) -109.9, (c) -107.1, (d) -102.8, and (e) -96.2 kOe (after Maley and Taylor [50]).

where h_{loc} is the saturation hyperfine field at $T = 0$. A best fit was achieved with $\theta \approx 0.8°\mathrm{K}$ although systematic deviations occurred for the lowest temperatures and highest field strengths. This is not surprising, of course, since there exists no rigorous theoretical support for (3.4). It is also noteworthy that recent bulk susceptibility measurements on **Mo**Fe by Claus [51] revealed significant deviations from Curie–Weiss behavior at high temperatures. Whereas the low-temperature data yielded $\theta_\mathrm{K} \approx 1°\mathrm{K}$, a fit to the data above about $100°\mathrm{K}$ required $\theta_\mathrm{K} > 10°\mathrm{K}$. Claus attributed these results to impurity–impurity interaction effects with Fe clusters giving rise to a much smaller value of θ_K than that of isolated impurities. However, the linear concentration dependence of the measured impurity susceptibility makes this interpretation very suspect. Moreover, the ME experiments [50] have established unequivocally that the reported low-temperature behavior is due to isolated impurities. (This is in marked contrast to alloys such as **Cu**Fe, **Cu**Co, and **Mo**Co for which hyperfine-interaction data demonstrate that the low-temperature magnetic properties are strongly influenced by impurity–impurity interactions even at relatively low impurity concentrations.) The complex behavior of the **Mo**Fe susceptibility is most likely related to the effects of orbital degeneracy which are ignored in the simple s–d model.

Another indication of a low Kondo temperature in **Mo**Fe is the observation by Maley and Taylor [50] of ^{57}Fe ME line-shape anomalies at low temperatures which were suggestive of slow electronic relaxation effects. Results of a related study have been reported by Maletta *et al.* [52].

Qualitatively similar magnetization data have been obtained by means of the NO technique for the $3d$ impurities ^{54}Mn in **Cu**Mn ($0.01°\mathrm{K}$) [53–55], **Ag**Mn ($0.04°\mathrm{K}$ [7, 53], **Au**Mn ($<0.01°\mathrm{K}$) [53, 56, 57], and **Zn**Mn ($0.3°\mathrm{K}$) [58], and ^{51}Cr in **Cu**Cr ($1.6°\mathrm{K}$) [59], **Ag**Cr ($0.03°\mathrm{K}$) [59], and **Au**Cr ($0.01°\mathrm{K}$) [59]. The temperatures in parentheses indicate estimates of θ_K based on (3.4) and the experimental field dependence of $\langle h_{\mathrm{loc}}^z \rangle$ (i.e., assuming rapid relaxation) in the range $g\mu_B H < k_B \theta_\mathrm{K}$ and $T < \theta_\mathrm{K}$. Related NO experiments have been performed for the rare earth impurity ^{137}Ce in **La**Ce [60], **Ag**Ce [60], and **Au**Ce [60, 61]. Aside from uncertainties associated with the use of (3.4) the quantitative interpretation of NO experiments is made difficult by the generally unknown magnitude of local moment relaxation rates at very low temperatures. This is illustrated in Fig. 6 which compares the **Ag**Mn data [7] with predictions based on a free spin model. Curve a assumes fast relaxation ($\tau_e < \tau_n$); in this case the γ-ray anisotropy measures the field-induced $\langle h_{\mathrm{loc}}^z \rangle$. Curve b assumes slow relaxation ($\tau_e > \tau_n$); in this case the

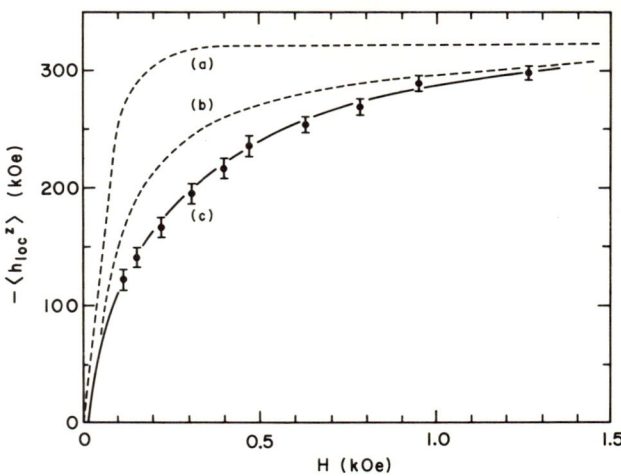

FIG. 6. Field dependence of the ^{54}Mn hyperfine field in **AgMn** from NO measurements at 0.013°K. The predicted free spin behavior for rapid electronic relaxation (a) and slow electronic relaxation (b) is compared with the experimental results (c) (after Flouquet [7]).

anisotropy must be calculated from the full hyperfine plus Zeeman Hamiltonian

$$\mathcal{H} = A\mathbf{I}\cdot\mathbf{S} - \gamma_n\hbar\mathbf{I}\cdot\mathbf{H} - \gamma_e\hbar\mathbf{S}\cdot\mathbf{H}, \tag{3.5}$$

where $A = -\gamma_n\hbar(h_{loc}/S)$. The result has been expressed in terms of the fictitious field h_{eff} in Fig. 6. The experimental data deviate from both predictions at low field strengths and thus give evidence of a reduction in $\langle S \rangle$ due to the Kondo effect. It can be argued that $\tau_e \sim \hbar/k_B T_K$ for $T \lesssim T_K$. This yields $\tau_e \approx 5 \times 10^{-9}$ sec for $T_K = 0.01°$K, which compares with a ^{54}Mn Larmor period $\tau_n \approx 4 \times 10^{-9}$ sec in a hyperfine field of 100 kOe. It would appear, therefore, that the fast-relaxation limit is not necessarily appropriate for the low-field data in Fig. 6. Since τ_n decreases with increasing local moment polarization while τ_e presumably increases, it is not surprising that the **AgMn** NO data follow the slow relaxation prediction in applied fields above about 1500 Oe [7]. In this range the Kondo state is presumably destroyed since $g\mu_B H > k_B T_K$.

b. *Orbital susceptibility.* In most discussions of 3d impurity states it is implicitly assumed that the orbital angular momentum is in some sense "quenched" and hence does not contribute significantly to the observed magnetic properties. The fact that $\langle h_{loc}\rangle$ is negative for most magnetic impurities supports this view. In the nonmagnetic (spin fluctuation) regime the spin susceptibility is usually dominant since the Hartree–Fock (HF) enhancement factor for the spin response is always greater than that of the orbital response [1, 62]. In the magnetic limit, as in the case of impurities in nonmetallic crystals, the quenching of the orbital angular momentum presumably requires crystal-field interactions of sufficiently low symmetry to remove the orbital degeneracy. Any orbital contribution to the susceptibility is then of the temperature-independent (VanVleck) form. In view of the large magnitude of $H_{\text{hfs}}^{(\text{orb})}$ relative to $H_{\text{hfs}}^{(d)}$, particularly near the end of the $3d$ period, the orbital susceptibility is expected to make a disproportionately large contribution to h_{loc}. It is obvious that orbital effects should be especially important in the hyperfine structure of nonmagnetic and high-T_{K} magnetic impurities because of their smaller spin susceptibilities. In fact, *positive* impurity NMR shifts have been observed in the nonmagnetic alloys **Cu**Ni [63], **Au**Ni [63], **Cu**Co [36], and **Au**Co [64, 65]. These have been attributed [63] to dominant orbital hyperfine effects. The large magnitude of the ^{59}Co shift in **Au**Co ($K = +29.2\%$ in the temperature range 1–4°K) [65] is particularly interesting. It suggests that the localized spin and orbital fluctuations are strongly coupled in this case, leading to large enhancements of both spin and orbital susceptibilities. The positive hyperfine field in **Pd**Co [66], on the other hand, is probably related to the strong exchange enhancement of the palladium host susceptibility which may lead to an anomalously large positive conduction-electron spin polarization at the impurity nucleus. This view is consistent with the observation [67] of a negative cobalt hyperfine field in platinum with its much weaker susceptibility enhancement.

Direct evidence for significant orbital hyperfine interactions in Kondo alloys has been obtained for **W**Co ($\theta_{\text{K}} \approx 2°$K) [38, 68], **Mo**Co ($\theta_{\text{K}} \approx 50°$K) [38, 68] and **Au**V ($\theta_{\text{K}} \approx 290°$K) [69] from temperature-dependence measurements of the NMR shifts. In each of these alloys the shifts are negative at low temperatures and positive at high temperatures, suggesting an orbital susceptibility of the VanVleck form. Moreover, the temperature-dependent terms are consistent with a Curie–Weiss form over a wide temperature range in agreement with bulk susceptibility experiments. This is demonstrated by the **Au**V results in Fig. 7. Except for significant deviations at the lowest temperatures, which agree with the T^2 variation of the susceptibility reported by Van Dam and Gubbens [70], the observed ^{51}V shifts are a linear function of $(T + \theta_{\text{K}})^{-1}$. The infinite-temperature intercept corresponds to a combined s-contact plus d-orbital shift of approximately $+1.8\%$. Estimating $K_s \approx +0.2\%$ yields $K_{(\text{orb})} \approx +1.6\%$.

In the case of **W**Co it has been possible, because of its low Kondo temperature, to obtain direct evidence for a VanVleck term in the low-temperature impurity susceptibility from measurements of the ^{59}Co NMR shift as a function of applied field strength. As indicated in Fig. 8

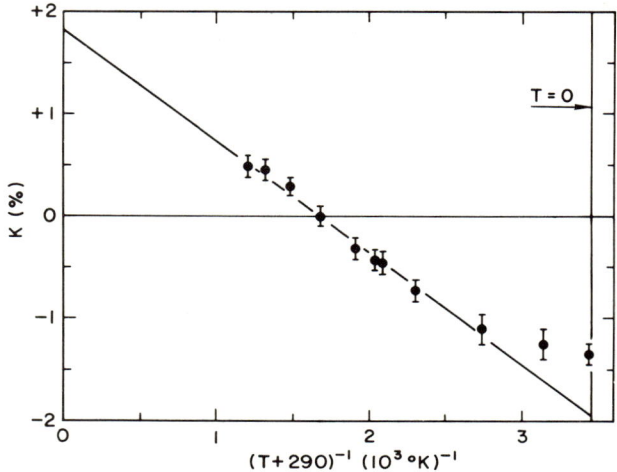

FIG. 7. Temperature dependence of the ^{51}V NMR shift in **Au**V (0.2 at. %) (after Narath [69]).

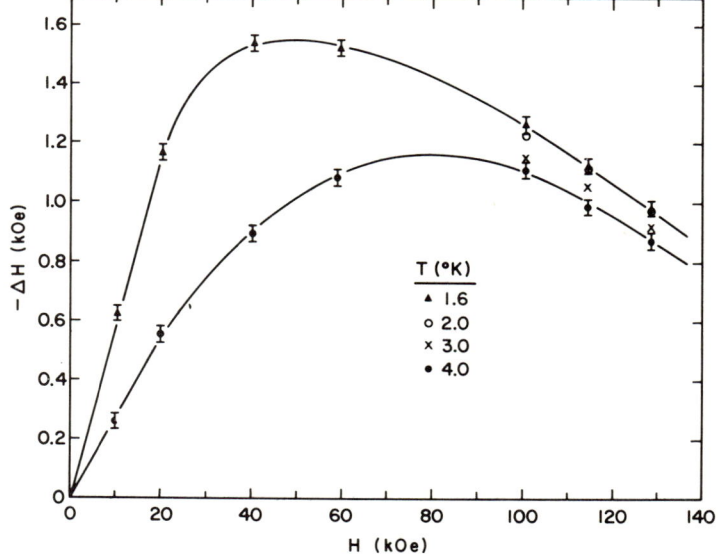

FIG. 8. Field dependence of the ^{59}Co NMR shift in **W**Co (0.1 at. %) (after Narath [68]).

the low-field slope $\Delta H/H$ is proportional to $(T+2)^{-1}$; at high fields, however, $\Delta H/H$ reverses sign and ΔH becomes essentially temperature independent. The low-field behavior is indicative of a Curie–Weiss term in the susceptibility which saturates for sufficiently large $H/(T+\theta_K)$. The remaining field dependence can be attributed to Van-Vleck paramagnetism. Two additional features of these results are noteworthy. In the first place, the ^{59}Co saturation hyperfine field is extraordinarily small; an extrapolation of the high-field shift data [68] yields a zero-field intercept $h_{loc} = -2.0$ kOe. Secondly, the high-field slope $\Delta H/H \approx +0.8\%$ is significantly smaller than the infinite-temperature intercept, $K(T \to \infty) = +1.6\%$. The small hyperfine field is undoubtedly a consequence of an appreciable orbital contribution to the Co moment, leading to near cancellation of negative spin and positive orbital hyperfine terms. Such behavior is not unexpected for a Co^{2+} ion. Crystal-field interactions cannot completely quench its orbital angular momentum, in contrast to the well-known cases of V^{2+}, Cr^{2+}, Cr^{3+}, Mn^{2+}, and Ni^{2+} in cubic environments as well as Fe^{2+} in noncubic environments. It follows that the separation of $\mathbf{h}_{loc}$ into independent spin and orbital terms, as carried out in (2.6), is probably invalid for magnetic cobalt impurities. Such a separation may also be a poor approximation in the strongly exchange-enhanced nonmagnetic regime (e.g., **Au**Co). The difference between high- and low-temperature VanVleck susceptibilities may indicate the existence of low-lying excited levels of the cobalt impurity. The importance of excited impurity states $(\Delta E \sim k_B T)$ in dilute magnetic alloys has recently been stressed by Flynn *et al.* [71], although these authors explicitly ignored orbital effects.

It is clear that the Kondo effect in **W**Co, as well as in **Mo**Co, involves both spin and orbital degrees of freedom. A detailed understanding of these alloys must necessarily await more realistic models than the simple s–d model. Finally we note that the observability of the ^{59}Co NMR in **W**Co despite its nearly free spin behavior is directly related to the anomalously small hyperfine field. A more typical 3d hyperfine field of approximately 10^5 Oe would have limited the range of observation to large values of H/T [72].

3. Host Polarization

a. High-temperature $(T > T_K)$ NMR linewidths. Because of the rapid decrease in the RKKY amplitude with increasing distance from the impurity site, the range of the average polarization $\chi(0) H^z(0) = 2g\mu_B \langle S^z \rangle J_{eff}\rho$ usually falls within the wipe-out radius r_0. The principal effect of the impurities on the host NMR in dilute alloys is therefore a symmetric broadening. Discrete satellites of magnetic origin associated with nuclei for which $r < r_0$ can sometimes be observed for weakly magnetic alloys. Examples are **AlMn** [73, 74], **YCe** [75], and **RhCo** [76]. The satellite positions in the aluminum alloy are in fair agreement with the direct NMR determination of the average polarization in liquid **AlMn** by Flynn *et al.* [23].

The magnitude of the symmetric magnetic hyperfine broadening may be estimated from (2.16). The full width, $W = 4H_{\text{hfs}} |\langle \sigma^z \rangle|_{\text{av}}$, is found to be

$$(W/H)_{\text{host}} = 18\pi(\chi_{(d)}/g\mu_B) H_{\text{hfs}}^{(s)} Z^2 (|J_{\text{eff}}|/E_F)[F(2k_F r)]_{\text{av}}, \qquad (3.6)$$

where $[F(2k_F r)]_{\text{av}}$ denotes the magnitude of $F(2k_F r)$ (due to all impurities) averaged over all lattice sites for which $r > r_0$. (The average is usually taken to be the half-amplitude point of the distribution.) Thus for $T > T_K$ the inhomogeneous NMR linewidth is proportional to the impurity susceptibility, provided other sources of broadening (e.g., nuclear spin–spin interactions) can be neglected [77]. Since the observed broadening represents a spatial average, measurements of the temperature and field dependences of host linewidths in dilute alloys are essentially equivalent to bulk susceptibility measurements. Both techniques suffer from a lack of discrimination in the presence of more than one impurity type. In such cases the NMR technique suffers from the additional complication that both $[F(2k_F r)]_{\text{av}}$ as well as J_{eff} will, in general, be different for different impurities. The impurity contribution to the NMR lindewidth is therefore not simply related to the total sample magnetization unless only a single-impurity type is present.

Most magnetic hyperfine-linewidth studies have been aimed at the determination of J_{eff}. The earliest effort of this kind was that of Behringer [78] who analyzed the ^{63}Cu linewidths in **CuMn** [79] on the basis of a free electron histogram of $F(2k_F r)$ for a single impurity and obtained $|J_{\text{eff}}| = 1.3$ eV. A more detailed study of **CuMn** by Chapman and Seymour [80], in which the $(2k_F r)$ histogram was calculated for the more realistic case of a random impurity distribution, gave $|J_{\text{eff}}| = 2.6$ eV. Another example is Mizuno's [81] study of **AgMn** which yielded $|J_{\text{eff}}| = 1.1$ eV. It is noteworthy that s–d exchange constants inferred from NMR data are typically much larger than those derived from transport measurements. This undoubtedly reflects the difference between J_{eff} and j_{eff} discussed earlier.

It is obvious that the broadening mechanism discussed above also applies to the impurity. However, in the impurity case the direct broadening mechanism becomes unimportant for sufficiently large χ_d relative to an indirect mechanism in which the conduction-electron polarization induces (via the s–d interaction) a nonuniform impurity polarization [24]. This polarization is sensed by the nuclear spins through the d-spin hyperfine coupling. The linewidth corresponding to the indirect mechanism follows directly from the RKKY coupling energy between two impurities. (Equivalently, it may be estimated from Caroli's [82] d–d double-resonance theory of impurity–impurity interactions.) Treating this interaction in the molecular field approximation leads to a linewidth

$$(W/H)_{\text{impurity}} = 18\pi(\chi_{(d)}/g\mu_B)^2 H_{\text{hfs}}^{(d)} Z^2 (J_{\text{eff}}^2/E_F\mu_B)[F(2k_F r)]_{\text{av}} . \quad (3.7)$$

We note that the linewidth is proportional to $\chi_{(d)}^2$. A much stronger temperature dependence is thus predicted for the impurity linewidth than for the host linewidth in qualitative agreement with observations in **MoCo** and **WCo** [38].

b. **Low-temperature ($T < T_K$) NMR linewidths.** With decreasing temperature, anomalies in the impurity susceptibility are reflected in the host NMR linewidth. Whether the width remains proportional to $\chi_{(d)}$, however, is still a matter of considerable speculation. For example, the possible existence below T_K of nonperturbative positive-definitive contributions to the host susceptibility (i.e., a negative-definite spin polarization) has been the subject of considerable interest [10, 41, 83] since it would provide an important distinction between the low-temperature magnetic properties of spin fluctuation and Kondo alloys. The available experimental as well as theoretical evidence suggests, however, that such an effect does not exist. On the other hand, there still remains the possibility of enhanced RKKY amplitudes below T_K [84, 85]. In this regard it is important to recognize that the field-induced polarization is related to $\langle (\mathbf{S} + \boldsymbol{\sigma}(\mathbf{r})) \cdot \boldsymbol{\sigma}(\mathbf{r}) \rangle$, which is not necessarily equivalent to the ground-state correlation function $\langle \mathbf{S} \cdot \boldsymbol{\sigma}(\mathbf{r}) \rangle$.

The original arguments in support of nonperturbative spin polarization "clouds" around impurities for $T < T_K$ were based on ^{57}Fe ME and ^{63}Cu NMR linewidth data in **CuFe** ($\theta_K \approx 29°K$) [86, 87]. From an analysis of the temperature dependences of the local susceptibility (as measured by the ^{57}Fe hyperfine field) and the bulk susceptibility it was concluded that the former accounts for only about one-half of the total low-temperature impurity susceptibility. Moreover, the observation of a nonlinear field dependence of the ^{63}Cu linewidth above about 10 kOe was taken as evidence that the excess susceptibility can be destroyed in fields far weaker than $k_B T_K/g\mu_B$. Because of the absence of measurable shifts in the ^{63}Cu NMR it was proposed that the range of the enhanced polarization was less than the wipe-out radius ($\sim$ 9 Å; i.e., about 12 near-neighbor shells). In order to rationalize the linewidth data it was conjectured that a self-consistent treatment of the conduction-electron scattering from the relatively localized "giant" moment might lead to an enhanced RKKY amplitude. Subsequently, it was shown [83, 88] that the low-temperature susceptibility of **Cu**Fe is strongly influenced by impurity–impurity interactions which are important even at very low concentrations. Superparamagnetic clusters and strongly magnetic ($\theta_K \approx 0$) Fe–Fe pairs have been identified. (Evidence for such interactions can also be seen in the specific heat [89].) By separating the susceptibility into its various contributions it was established that the temperature dependences of the local and total susceptibilities of an isolated iron impurity in **Cu**Fe are indistinguishable. Recently, Welsh and Potts [84, 85] have carefully reexamined the concentration and field dependences of the magnetic hyperfine contribution to the ^{63}Cu linewidth. Their low-temperature results are summarized in Fig. 9. The

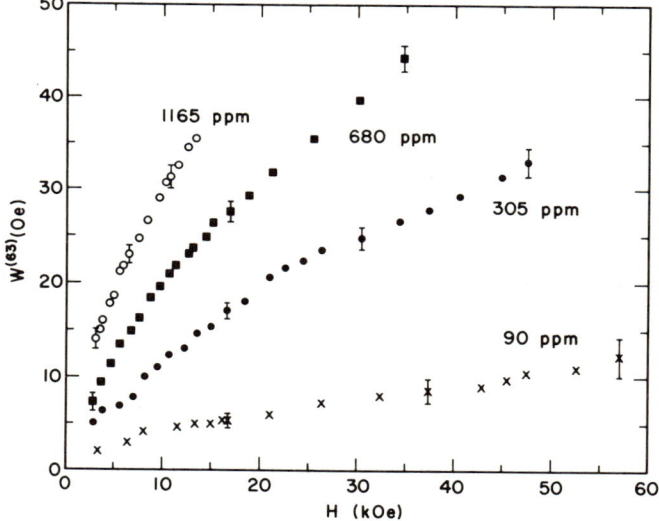

FIG. 9. Field dependence of the ^{63}Cu NMR linewidth in **Cu**Fe for several impurity concentrations at 1.65°K (after Potts and Welsh [85]).

pronounced curvature in the data is in general agreement with the earlier work. The high-field slope $(dW/dH)_H$ has the same $(T + 29)^{-1}$ temperature dependence as has been observed for the iron hyperfine field as well as for the isolated impurity contribution to the bulk susceptibility. The low-field slope $(dW/dH)_L$, however, exhibits an anomalous increase at low temperatures, as illustrated in Fig. 10, which Welsh and

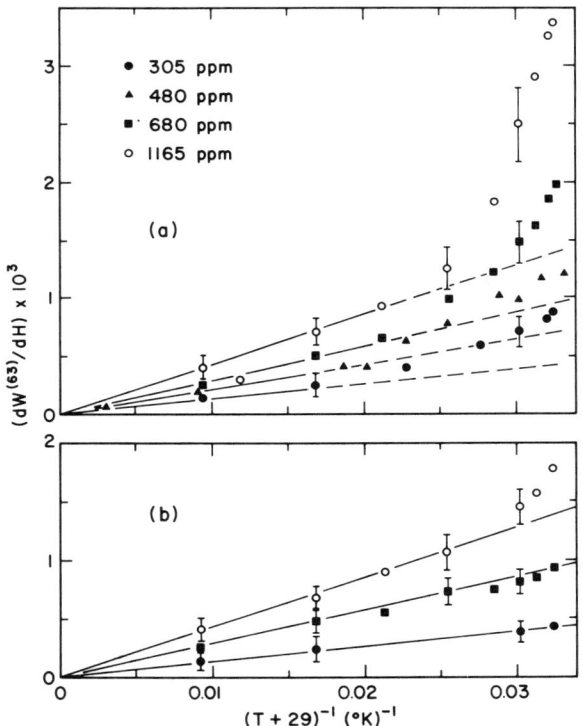

FIG. 10. Temperature dependence of the field derivatives of the ^{63}Cu NMR linewidth in **Cu**Fe: (a) low-field slope, (b) high-field slope (after Potts and Welsh [85]).

Potts [84, 85] have attributed to many-body correlation effects. Since both $(dW/dH)_L$ and $(dW/dH)_H$ are approximately linear functions of concentration, the experimental linewidths, in contrast to the bulk susceptibility, seem to be relatively insensitive to iron pairs. (This disparity can be understood by supposing that r_0 is much greater for clusters than for isolated iron impurities.) The transition between low-field and high-field regimes, however, occurs at field strengths which are inversely proportional to the impurity concentration. This has led Welsh and Potts [84, 85] to conclude that the excess polarization saturates when the effective field acting on a given iron impurity reaches some critical magnitude. In this regard it should be remarked that the local iron susceptibility (Fig. 11) also exhibits an anomalous field dependence at low temperatures for field strengths far below $k_B\theta_K/g\mu_B$ [46, 48]. It would appear, therefore, that the low-temperature properties of **CuFe** warrant additional attention before one can conclude with any certainty that the ^{63}Cu linewidths reflect an intrinsic property of the Kondo state.

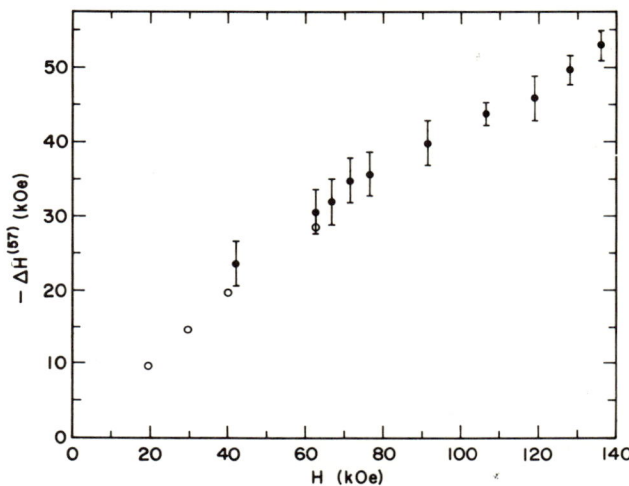

FIG. 11. Field dependence of the ^{57}Fe hyperfine field in **CuFe** at low temperatures ($T \ll \theta_K$) (after Kitchens et al. [46]—open circles, and Frankel et al. [48]—closed circles).

A related study has been carried out on **AuV** ($\theta_K \approx 290°$K) [24]. Measurements of the ^{109}Ag resonance shifts in **Au(Ag)V** [where ^{109}Ag($I = \frac{1}{2}$) was used as a convenient probe of the host response] again demonstrated the absence of any longe-range negative-definite spin polarization. This is illustrated in Fig. 12 which compares the observed dependence of the ^{109}Ag shift on vanadium concentration with the expected behavior if half of the experimental impurity susceptibility were uniformly distributed over the host. Because of its smaller susceptibility compared to **CuFe**, the wipe-out radius in **Au(Ag)V** corresponds to only one or two near-neighbor shells. Thus any anomalous conduction-electron polarization must be highly localized.

In contrast to the absence of any observable impurity-induced shift, the ^{109}Ag NMR in **Au(Ag)V** broadens rapidly with increasing vanadium concentration. As expected from (3.6) and (3.7), the inhomogeneous broadening is even greater for the impurity NMR. Because of its high T_K **AuV** affords an interesting comparison of NMR linewidths with the

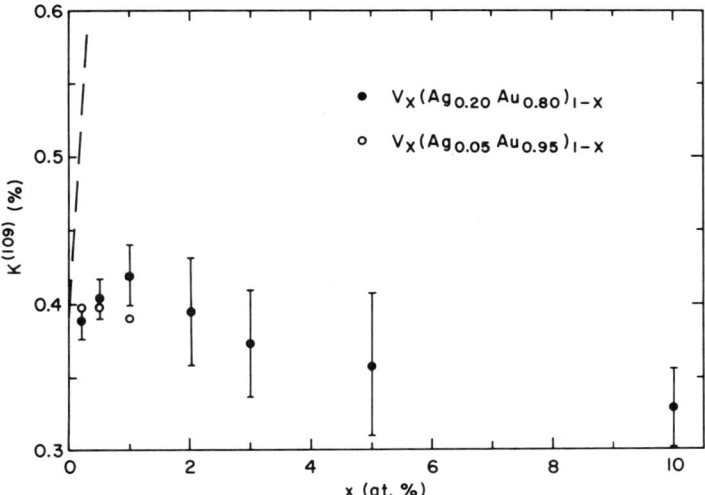

FIG. 12. Variation of the ^{10a}Ag NMR shift with vanadium concentration in Au(Ag)V. The dashed line indicates the predicted behavior if half of the measured impurity susceptibility were uniformly distributed over the host sites (after Narath and Gossard [24]).

spin fluctuation alloy AlMn for which similar data have been reported [74]. In order to obtain an estimate for J_{eff} we base our comparison on the HF treatment of RKKY broadening. [According to (2.20) this implies $J_{\text{eff}} = 7.3$ eV for AuV and 5.2 eV for AlMn.] Combining (3.6) and (3.7) with (2.20) gives

$$(W/H)_{\text{host}} = 24\pi(\chi_{(d)}/g\mu_B) H_{\text{hfs}}^{(s)} Z[F(2k_F r)]_{\text{av}}, \tag{3.8}$$

$$(W/H)_{\text{impurity}} = 32\pi(\chi_{(d)}/g\mu_B)^2 H_{\text{hfs}}^{(d)}(E_F/\mu_B)[F(2k_F r)]_{\text{av}}. \tag{3.9}$$

The quantities appearing in these expressions can be estimated with reasonable accuracy. The resulting predictions for Au(Ag)V and AlMn (Table I) are seen to be in excellent agreement with experiment [24, 74]. The fact that (3.8) and (3.9) account satisfactorily for the observed ratios of host to impurity linewidths supports the indirect broadening mechanism postulated for the impurity. [Additional support for this mechanism is provided by the observation [69] that the ^{51}V linewidth in AuV is approximately proportional to $(T + \theta_K)^{-2}$ in the temperature range 4–500°K.] Of greater significance is the surprising result that the HF theory appears to be equally valid for the localized spin fluctuation as well as Kondo regimes.

TABLE I

COMPARISON OF OBSERVED IMPURITY-INDUCED LINEWIDTHS[a]

	Impurity[b]	Host[b]
Au(Ag)V	5.0(5.9) %	0.5 (0.48) %
AlMn	2.3(2.1) %	0.08(0.09) %

[a] Per atomic percent.
[b] The numbers in parentheses are predictions based on the HF theory.

IV. Dynamic Response Studies

1. Impurity Nuclear Spin–Lattice Relaxation

As mentioned earlier, qualitative information concerning local moment fluctuation rates may sometimes be inferred from ME and NO studies. A more quantitative approach to the study of dynamic properties of impurities is provided by nuclear spin–lattice relaxation rates which, according to (2.11), are determined by the low-frequency transverse spin and orbital fluctuation amplitudes of the local moment. It is frequently useful to eliminate the hyperfine fields between (2.10) and (2.11) which yields for each of the hyperfine mechanisms

$$K_{(i)}^2 T_{1(i)} T = (k_B \gamma_n^2)^{-1} C_{(i)}, \tag{4.1}$$

$$C_{(i)} \equiv \chi_{(i)}^{zz}(0)^2 \lim_{\omega \to 0} \left[\frac{\omega}{\operatorname{Im} \chi_{(i)}^{+-}(\omega)} \right], \tag{4.2}$$

where the quantity $K^2 T_1 T$ is often referred to as the Korringa product. Since only two mechanisms (d spin and d orbital) have to be considered for the impurity case, a measurement of K and $T_1 T$ is sufficient to determine the individual magnitudes of the spin and orbital contributions—provided that the $C_{(i)}$ are known.

In order to establish contact between theory and experiment it is instructive to consider first the nonmagnetic regime where the susceptibilities are isotropic [i.e., $\chi_{(i)}^{+-}(\omega) = 2\chi_{(i)}^{zz}(\omega)$] and can be calculated in the random phase approximation (RPA). Assuming full fivefold orbital degeneracy (i.e., the weak crystal-field limit) one obtains [63, 90]

$$\chi_{(d)}(0) = 2\rho_d \mu_B{}^2 [1 - \tfrac{1}{5}(\bar{U} + 4\bar{J}) \rho_d]^{-1}, \tag{4.3}$$

$$\chi_{(\text{orb})}(0) = 4\rho_d \mu_B{}^2 [1 - \tfrac{1}{5}(\bar{U} - \bar{J}) \rho_d]^{-1}, \tag{4.4}$$

5. MAGNETIC HYPERFINE-INTERACTION STUDIES

and

$$\lim_{\omega \to 0} \operatorname{Im} \chi_{(d)}(\omega)/\omega = \tfrac{2}{5} \pi \hbar \rho_d{}^2 \mu_B{}^2 [1 - \tfrac{1}{5}(\bar{U} + 4\bar{J})\rho_d]^{-2}, \qquad (4.5)$$

$$\lim_{\omega \to 0} \operatorname{Im} \chi_{(\text{orb})}(\omega)/\omega = \tfrac{4}{5} \pi \hbar \rho_d{}^2 \mu_B{}^2 [1 - \tfrac{1}{5}(\bar{U} - \bar{J})\rho_d]^{-2}, \qquad (4.6)$$

where $\bar{J}$ is the effective intra-atomic d–d (Hund's rule) exchange parameter. Using Eqs. (4.3) to (4.6), the Korringa products may be written

$$K_{(d)}^2 T_{1(d)} T = 5\mathscr{S}, \qquad K_{(\text{orb})}^2 T_{1(\text{orb})} T = 10\mathscr{S}, \qquad (4.7)$$

where

$$\mathscr{S} \equiv \hbar(\gamma_e/\gamma_n)^2 (4\pi k_B)^{-1}. \qquad (4.8)$$

As a consequence of the quadratic dependence of $T_{1(i)}^{-1}$ on $K_{(i)}$ the relaxation rate is usually dominated by either the d-spin or d-orbital hyperfine mechanism. The application of the RPA expressions is best illustrated by considering low-temperature NMR data for **AlV**, **AlCr**, and **AlMn** [73, 91, 92]. The susceptibility enhancement in these alloys increases strongly in the sequence V, Cr, Mn. The relevant NMR results together with the d-spin and d-orbital shifts which are obtained by means of (4.7) (with the assumption $K_d < 0$) are listed in Table II. The variation in $K_{(d)}$ parallels closely the observed trend in the impurity susceptibilities [93]. It is interesting to note that $\chi_{(d)}$ increases approximately ninefold between **AlV** and **AlMn**. In contrast, $\chi_{(\text{orb})}$ increases only slightly as can be seen by adjusting the orbital shifts in Table II according to the orbital fields in Fig. 1a. From (2.10) we estimate $\chi_{(\text{orb})} \approx 2 \times 10^{-4}$ emu/g-atom for **AlMn**. This compares with the experimental impurity susceptibility $\chi \approx 14 \times 10^{-4}$ emu/g-atom, indicating that the enhancement as given by the appropriate denominator

TABLE II

Summary of Low-Temperature (1–4°K) Impurity NMR Data for Dilute, Nonmagnetic Aluminum Alloys[a]

	c (at. %)	K (%)	$\gamma_n{}^2 T_1 T$ (10^6 sec^{-1}°K Oe^{-2})	$K^2 T_1 T/5\mathscr{S}$	K_d (%)	K_{orb} (%)
Al^{51}V	0.1	$+0.30(3)$	28. (2)	0.21	-0.4	$+0.7$
Al^{53}Cr	0.5	$-0.38(3)$	3.5(7)	0.05	-1.4	$+1.0$
Al^{55}Mn	0.04	$-2.01(5)$	0.8(1)	0.35	-3.2	$+1.2$

[a] The first column gives the lowest impurity concentration for which data were obtained. The last two columns list estimates of the d-spin and d-orbital shift contributions (assuming $K_s = 0$) as discussed in the text.

in (4.3) and (4.4) is approximately 12 times greater for $\chi_{(d)}$ than that of $\chi_{(\text{orb})}$. Thus **AlMn** is an example of a dilute alloy which lies close to the spin, *but not orbital*, HF instability boundary.

Impurity nuclear spin–lattice relaxation data have been reported for three alloys which exhibit Kondo anomalies (**AuV** [24, 94], **PtCr** [10], **MoCo** [38]). In the temperature range investigated (1–4°K), the data are characterized by $T_1T = $ constant and values of $K^2T_1T/\mathscr{S}$ which are similar in magnitude to that of the corresponding **AlMn** ratio. The similarity between **AlMn** and **AuV** (Table III) is particularly remarkable

TABLE III

SUMMARY OF LOW-TEMPERATURE (1–4°K) IMPURITY NMR DATA FOR DILUTE MAGNETIC ALLOYS[a]

	c (at. %)	K (%)	$\gamma_n{}^2 T_1 T$ (10^6 sec^{-1}°K Oe^{-2})	$K^2T_1T/5\mathscr{S}$	K_d (%)	K_{orb} (%)
Au^{51}**V**	0.1	−1.5 (1)	0.8 (1)	0.20	−3.1	+1.6
					(−3.0)	(+1.5)
Pt^{53}**Cr**	0.5	−0.82(2)	2.1 (2)	0.15	−2.0	+1.2
Mo^{59}**Co**·	0.1	−6.8 (2)	0.10(1)	0.51	−9.5	+2.7

[a] The *d*-spin and *d*-orbital shift estimates are based on the HF/RPA analysis except for the parenthetical **AuV** values which are based on Dworin's atomic model.

in view of the fact that the former has been identified as a nonmagnetic spin fluctuation alloy, whereas the latter is quite certainly a Kondo alloy [95]. This leads to the speculation that the general form of the spin susceptibility which is valid in the nonmagnetic regime [41]

$$\chi_{(d)}(\omega) = C_{(d)}(\omega_0 + i\omega)^{-1}, \tag{4.9}$$

where ω_0 is some characteristic fluctuation frequency, also describes the low-frequency response in the magnetic case provided that $g\mu_B H \ll k_B T_K$ and $T \ll T_K$. Recently, Dworin [95, 96] has developed a theory for the magnetic limit of the Anderson model (for the case of a crystal-field-quenched orbital angular momentum) which is consistent with this view. His low-temperature results

$$\chi_{(d)}(0) = (g\mu_B)^2 S(S+1)(3k_B \theta_K)^{-1}, \tag{4.10}$$

$$\lim_{\omega \to 0} \operatorname{Im} \chi_{(d)}(\omega)/\omega = \hbar(g\mu_B)^2\, 2S(S+1)^2\, (3k_B \theta_K)^{-2}, \tag{4.11}$$

give

$$K_{(d)}^2 T_{1(d)} T = \pi S \mathscr{S}, \tag{4.12}$$

which is (somewhat fortuitously) close to the HF prediction for $S = \frac{3}{2}$. For the assumed orbitally nondegenerate ground state, appropriate for a ^4F ion (e.g., V^{2+}) in an octahedral crystal field, Dworin [95] estimated an orbital Korringa product equal to approximately $6\mathscr{S}$. Using this estimate together with (4.12) to partition the **AuV** NMR shift gives the results listed in Table III. Also listed for purposes of comparison are the HF/RPA results for **AuV** as well as for **PrCr** and **MoCo**. (The **MoCo** shifts differ from those given by Narath et al. [38] in that they were computed using the more recent [97] reference ratio $\nu/H = 1.005$ kHz/Oe.)

The good agreement for **AuV** between $K_{(\text{orb})}$ estimated from the low-temperature relaxation data $(+1.5\%)$ and from the temperature dependence of the resonance shift $(+1.8\%)$ supports the assumption of a temperature-independent orbital susceptibility. (Both of these estimates should be reduced by $\sim 0.2\%$ to correct for a small s-contact contribution, as noted in Section III2b.) Thus the Kondo effect in **AuV** involves only the spin but not the orbital moment. The orbital shift corresponds to $\chi_{(\text{orb})} \approx 4 \times 10^{-4}$ emu/g-atom; subtracting this value from the measured susceptibility [35, 70, 98] gives $\chi_{(d)} \approx 41 \times 10^{-4}$ emu/g-atom for $T/\theta_K \ll 1$. As noted by Dworin [95] a small orbital susceptibility implies a large crystal-field splitting in his model. The estimate of $\chi_{(\text{orb})}$ given here corresponds to a separation of about 0.3 eV between the $^4A_2$ ground state and the first excited state $(^4T_2)$.

2. Host Nuclear Spin–Lattice Relaxation

Although host nuclear spin–lattice relaxation rates offer a potentially powerful probe of low-frequency impurity spin dynamics below T_K, only limited progress in that direction has been made to date. Low-temperature measurements have been reported for **CuFe** [85, 99–102] and **CuCr** [101, 103] without, however, yielding any convincing conclusions. This is not surprising in view of the fact that even the interpretation of high-temperature $(T > T_K)$ data has generated considerable controversy. In particular, it has been difficult to establish the relative strengths of the GH and longitudinal dipolar mechanisms in the high-field regime despite the availability of rather detailed data for **CuMn** [104–106], **CdMn** [107], and **ScGd** [108]. Several complications are responsible for this situation. The multiplicity of relaxation mechanisms [Eqs. (2.22)–(2.25)] has already been mentioned. Additional factors are the uncertain functional forms of the low-frequency impurity susceptibility and RKKY range function. Of perhaps even greater importance is the strong dependence of the experimental relaxation rates on poorly defined quantities [28] such as the wipe-out radius, diffusion-barrier radius, and diffusion constant, which are themselves functions of H and T.

Recently, the importance of the longitudinal dipolar relaxation process has been unambiguously demonstrated by McHenry et al. [109] who studied the ^{27}Al spin–lattice relaxation rates in Gd-doped LaAl$_2$. The interpretation of these experiments was greatly simplified by two factors. In the first place the ^{27}Al spins are subjected to large electric quadrupole interactions because of the low symmetry of the aluminum sites in this intermetallic compound. The resulting decoupling of the ^{27}Al nuclear spins allowed the relaxation rates to be studied in the absence of nuclear spin diffusion, thereby eliminating the dependence of T_1 on any diffusion parameters. The impurity contribution to the relaxation of the longitudinal nuclear magnetization is characterized in this case by the time-dependence $\exp[-(t/\tau)^{1/2}]$ for large t (in contrast to the exponential time dependence in the rapid diffusion case) with an effective time constant τ which is independent of r_0. A second important factor was the choice of sufficiently high impurity concentrations to make the exchange coupling between impurities the dominant local moment relaxation mechanism. The Gd impurity spin correlations can therefore be described by a spatial distribution of Gaussian fluctuation spectra. In a dilute alloy this yields an average correlation function which can be approximated by an exponential time dependence whose characteristic time τ_{Gd} is essentially independent of H/T. The relationship between τ and τ_{Gd} for the longitudinal dipolar mechanism, Eq. (2.25), can then be expressed by the angular average

$$\tau^{-1} = \left(\tfrac{16}{9}\right) \pi^3 (N_0 c)^2 \gamma_n^2 (g\mu_B S)^2 \frac{\partial B_S(X)}{\partial X} \left[\frac{\tau_{Gd}}{1+(\omega_n \tau_{Gd})^2}\right], \quad (4.13)$$

where $N_0 c$ denotes the number of impurities per unit volume. The dependence of τ^{-1} on the derivative of the Brillouin function arises from the relationship

$$\langle (\delta S^z)^2 \rangle = \langle (S^z)^2 \rangle - \langle S^z \rangle^2 \equiv \partial \langle S^z \rangle / \partial X, \quad (4.14)$$

where $X = g\mu_B SH/k_B T$, and $\langle (\delta S_z)^2 \rangle$ is equal to the area under the longitudinal fluctuation spectrum. [The relationship between the spectral density function and Im $\chi_{(d)}^{zz}(\omega)/\omega$ is given by Eq. (2.9) as before.] The GH rate in the diffusionless regime, on the other hand, depends on Re $\chi_{(d)}^{+-}(\omega)/\omega$ and thus on $B_S(X)/X$. The data of McHenry et al. [109] (Fig. 13) are described by (4.13) over a wide range of H and T with $\omega_n \tau_{Gd} \ll 1$ and $\tau_{Gd}^{-1} \propto c$ as expected for impurity–impurity exchange. The GH mechanism is therefore ineffective in this system. This lends

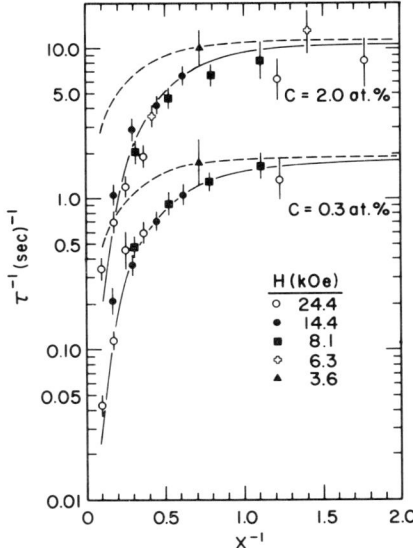

FIG. 13. Comparison of ^{27}Al spin–lattice relaxation rates in $La_{1-c}Gd_cAl_2$ with predictions based on the GH [dashed lines $= B_S(X)/X$] and longitudinal dipolar [solid lines $= \partial B_x(X)/\partial X$] models, where $X = g\mu_B SH/k_B T$ (after McHenry et al. [109]).

support to speculations [26, 107, 110] that the dipolar mechanism is also responsible for the observed high-field host relaxation rates in other dilute alloys. Recent work by Alloul and Bernier [106] on **CuMn**, in which account was taken of the H/T dependences of r_0 and the diffusion-barrier radius, has led to the same conclusion.

The dominance of the longitudinal dipolar mechanism probably does not extend into the Kondo regime, $T < T_K$, $g\mu_B H < k_B T_K$. Here the fluctuation spectrum is presumably isotropic and it is likely therefore that the low-temperature host nuclear relaxation rates are determined by transverse fluctuations via the RKKY interaction.

V. Concluding Remarks

Although considerable progress has been achieved toward a fundamental understanding of the magnetic impurity problem it is clear that many uncertainties remain. One of these concerns the validity of the simple s–d model. For example, the large values of J_{eff} (i.e., $2J_{\text{eff}}\rho \sim 1$) which are required in many alloys to account for the observed conduction-electron polarization amplitudes argue that the requisite weak coupling between the impurity d states and conduction electrons may be an exceptional case. Furthermore, the neglect of any orbital degeneracy in the s–d model severely limits its usefulness even in the weak coupling regime. It has already been shown [4, 111] that the coupling has a much more complex form in the presence of orbital degeneracy. Another area of concern is the relationship between the localized spin fluctuation and s–d models. The available nuclear resonance data suggest that it may be possible to order all dilute alloys according to a scale which defines a gradual, continuous transition between nonmagnetic and magnetic regimes in terms of a single parameter. In view of the many independent parameters which appear in the Anderson model, however, it is unfortunately not obvious that such a simplification, despite its intuitive appeal, is justified. One may hope that future hyperfine-interaction studies will help clarify these and related problems.

References

1. P. W. Anderson, *Phys. Rev.* **124**, 41 (1961).
2. J. R. Schrieffer and P. A. Wolff, *Phys. Rev.* **149**, 491 (1966).
3. J. Kondo, *Progr. Theor. Phys.* **32**, 37 (1964).
4. B. Coqblin and J. R. Schrieffer, *Phys. Rev.* **185**, 847 (1969).
5. L. J. Tao, D. Davidov, R. Orbach, and E. P. Chock, *Phys. Rev. B* **4**, 5 (1971).
6. T. Moriya, *Progr. Theor. Phys.* **16**, 23, 641 (1956).
7. J. Flouquet, *Phys. Rev. Lett.* **25**, 288 (1970).
8. L.L. Hirst, E. R. Seidel, and R. L. Mössbauer, *Phys. Lett. A* **29**, 673 (1969).
9. A. Narath, *J. Appl. Phys.* **41**, 1122 (1970).
10. A. Narath, *Proc. Int. Conf. Low Temp. Phys. Kyoto*, 1970 (E. Kanda, ed.), p. 675. Academic Press of Japan, Tokyo, 1971.
11. R. Kubo, *J. Phys. Soc. Jap.* **12**, 570 (1957).
12. A. J. Freeman and R. E. Watson *in* "Magnetism" (G. T. Rado and H. Suhl, eds.), Vol. IIA. Academic Press, New York, 1965.
13. I. A. Campbell, *Solid State Commun* **9**, 301 (1971).
14. R. E. Watson, A. J. Freeman, and S. Koide, *Phys. Rev.* **186**, 625 (1969).
15. L. L. Hirst, *Z. Phys.* **245**, 378 (1971).
16. C. Kittel, *Solid State Phys.* **22**, 1 (1968).
17. M. A. Ruderman and C. Kittel, *Phys. Rev.* **96**, 99 (1954).
18. T. Kasuya, *Progr. Theor. Phys.* **16**, 45 (1956).
19. K. Yosida, *Phys. Rev.* **106**, 893 (1957).
20. H. V. Everts and B. N. Ganguly, *Phys. Rev.* **174**, 594 (1968).
21. A. Blandin, *J. Appl. Phys.* **39**, 1285 (1968).
22. J. Friedel, *Advan. Phys.* **3**, 446 (1954); *Nuovo Cimento* **7**, 287 (1958).
23. C. P. Flynn, D. A. Rigney, and J. A. Gardner, *Phil. Mag.* **15**, 1255 (1967).
24. A. Narath and A. C. Gossard, *Phys. Rev.* **183**, 391 (1969).
25. A. Blandin, *Proc. Int. School of Phys.*, "Enrico Fermi", Course XXXVII (W. Marshall, ed.). Academic Press, New York, 1967.
26. B. Giovannini, P. Pincus, G. Gladstone, and A. J. Heeger, *J. Phys.* **32**, (*Paris*) C1-163 (1971).

27. G. Grüner, E. Kovàcs-Csetényi, K. Tompa, and C. R. Vassel, *Phys. Status Solidi* **45**, 663 (1971).
28. I. J. Lowe and D. Tse, *Phys. Rev.* **166**, 279 (1968).
29. J. Korringa, *Physica (Utrecht)* **16**, 601 (1950).
30. H. Benoit, P. G. de Gennes, and D. Silhouette, *C.R. Acad. Sci.* **256**, 3841 (1963).
31. B. Giovannini and A. J. Heeger, *Solid State Commun.* **7**, 287 (1969).
32. V. Jaccarino and L. R. Walker, *Phys. Rev. Lett.* **15**, 258 (1965).
33. L. J. Swartzendruber, *Int. J. Magn.* **2**, 129 (1972); see also H. Nagasawa and N. Sakai, *J. Phys. Soc. Jap.* **27**, 1150 (1969).
34. K. C. Brog and W. H. Jones, *Phys. Rev. Lett.* **24**, 58 (1970).
35. L. Creveling, Jr., and H. L. Luo, *Phys. Rev.* **176**, 614 (1968).
36. S. Wada and K. Asayama, *J. Phys. Soc. Jap.* **30**, 1337 (1971).
37. R. Tournier and A. Blandin, *Phys. Rev. Lett.* **24**, 397 (1970).
38. A. Narath, K. C. Brog, and W. H. Jones, Jr., *Phys. Rev. B* **2**, 2618 (1970).
39. R. C. Knauer, unpublished data, 1972.
40. K. Yosida and A. Okiji, *Progr. Theor. Phys.* **34**, 505 (1964).
41. A. J. Heeger, *Solid State Phys.* **23**, 283 (1969).
42. P. W. Anderson, G. Yuval, and D. R. Hamann, *Phys. Rev. B* **1**, 4464 (1970).
43. K. D. Schotte and U. Schotte, *Phys. Rev. B* **4**, 2228 (1971).
44. B. B. Triplett and N. E. Phillips, *Phys. Rev. Lett.* **27**, 1001 (1971).
45. R. D. Taylor, T. A. Kitchens, D. E. Nagle, W. A. Steyert, and W. E. Millet, *Solid State Commun.* **2**, 209 (1964).
46. T. A. Kitchens, W. A. Steyert, and R. D. Taylor, *Phys. Rev. A* **238**, 467 (1965).
47. N. A. Blum, A. J. Freeman, and L. Grodzins, *Rev. Mod. Phys.* **36**, 406 (1964).
48. R. B. Frankel, N. A. Blum, B. B. Schwartz, and D. J. Kim, *Phys. Rev. Lett.* **18**, 1051 (1967).
49. B. B. Schwartz, D. J. Kim, R. B. Frankel, and N. A. Blum, *J. Appl. Phys.* **39**, 698 (1968).
50. M. P. Maley and R. D. Taylor, *Phys. Rev. B* **1**, 4213 (1970).
51. H. Claus, *Phys. Rev. B* **5**, 1134 (1972).
52. H. Maletta, K. R. P. Rao, and I. Nowik, *Z. Phys.* **249**, 189 (1972).
53. J. A. Cameron, I. A. Campbell, J. P. Compton, R. A. G. Lines, and G. V. H. Wilson, *Phys. Lett* **20**, 569 (1966).
54. I. A. Campbell, J. P. Compton, I. R. Williams, and G. V. H. Wilson, *Phys. Rev. Lett.* **19**, 1319 (1967).
55. J. Flouquet, *J. Phys. F* **1**, 87 (1971).
56. J. P. Compton, I. R. Williams, and G. V. H. Wilson, in "Hyperfine Structure and Nuclear Radiations" (E. Matthias and D. A. Shirley, eds.), p. 793. North-Holland Publ., Amsterdam, 1968.
57. E. Lagendijk, L. Niesen, and W. J. Huiskamp, *Phys. Lett. A* **30**, 326 (1969).
58. J. D. Marsh, *Phys. Lett. A* **33**, 207 (1970).
59. I. R. Williams, I. A. Campbell, C. J. Sanctuary, and G. V. H. Wilson, *Solid State Commun.* **8**, 125 (1969).
60. J. Flouquet, *Phys. Rev. Lett.* **27**, 515 (1971).
61. J. Flouquet and D. Marsh, *Phys. Lett. A* **32**, 501 (1970).
62. K. Yosida, A. Okiji, and S. Chikazumi, *Progr. Theor. Phys.* **33**, 559 (1965).
63. L. Dworin and A. Narath, *Phys. Rev. Lett.* **25**, 1278 (1970).
64. R. J. Holliday and W. Weyhmann, *Phys. Rev. Lett.* **25**, 243 (1970).
65. A. Narath and D. C. Barham, *Phys. Rev. B* **7**, 2195 (1973).
66. T. Ericsson, M. T. Hirvonen, T. E. Katila, and P. Reivari, *Proc. Int. Conf. Low*

Temp. Phys. 12th, Kyoto, 1970 (E. Kanda, ed.), p. 765. Academic Press of Japan, Tokyo, 1971.
67. T. Ericsson, M. T. Kirvonen, T. E. Katila, and V. K. Typpi, *Solid State Commun.* **8**, 765 (1970).
68. A. Narath, *Phys. Rev.* to be published.
69. A. Narath, *Solid State Commun.* **10**, 521 (1972).
70. J. E. Van Dam and P. C. M. Gubbens, *Phys. Lett. A* **34**, 185 (1971).
71. C. P. Flynn, J. J. Peters, and C. A. Wert, *Phys. Lett. A* **35**, 157 (1971).
72. R. E. Walstedt and A. Narath, *Phys. Rev. B* **6**, 4118 (1972).
73. H. Launois and H. Alloul, *Solid State Commun.* **7**, 535 (1969); H. Alloul and H. Launois, *J. Appl. Phys.* **41**, 923 (1970).
74. H. Alloul, P. Bernier, H. Launois, and J. P. Pouget, *J. Phys. Soc. Jap.* **30**, 101 (1971).
75. D. Silhouette, *J. Phys. (Paris)* **32**, C1-216 (1971).
76. R. E. Walstedt and J. H. Wernick, *Phys. Rev. Lett.* **20**, 856 (1968).
77. T. Sugawara, *J. Phys. Soc. Jap.* **14**, 643 (1959).
78. R. E. Behringer *J. Phys. Chem. Solids* **2**, 209 (1957).
79. J. Owen, M. E. Brown, W. D. Knight, and C. Kittel, *Phys. Rev.* **102**, 1501 (1956).
80. A. C. Chapman and E. F. W. Seymour, *Proc. Phys. Soc. London* **72**, 797 (1958).
81. K. Mizuno, *J. Phys. Soc. Jap.* **30**, 742 (1971).
82. B. Caroli, *J. Phys. Chem. Solids* **28**, 1427 (1967).
83. J. L. Tholence and R. Tournier, *Phys. Rev. Lett.* **25**, 867 (1970).
84. L. B. Welsh and J. E. Potts, *Phys. Rev. Lett.* **26**, 1320 (1971).
85. J. E. Potts and L. B. Welsh, *Phys. Rev. B* **5**, 3421 (1972).
86. A. J. Heeger, L. B. Welsh, M. A. Jensen, and G. Gladstone, *Phys. Rev.* **172**, 302 (1968).
87. D. C. Golibersuch and A. J. Heeger, *Phys. Rev.* **182**, 584)1969); *Solid State Commun.* **8**, 17 (1970).
88. H. E. Ekström and H. P. Myers, *Phys. Kondens. Materie* **14**, 265 (1972).
89. J. C. F. Brock, J. C. Ho, G. P. Schwartz, and N. E. Phillips, *Solid State Commun.* **8**, 1139 (1970).
90. B. Caroli, P. Lederer, and D. Saint-James, *Phys. Rev. Lett.* **23**, 700 (1969).
91. A. Narath and H. T. Weaver, *Phys. Rev. Lett.* **23**, 233 (1969).
92. Y. Oda, H. Yamagata, and K. Asayama, *J. Phys. Soc. Jap.* **25**, 629 (1968).
93. R. Aoki and T. Ohtsuka, *J. Phys. Soc. Jap.* **26**, 651 (1969).
94. K. Kume, K. Mizuno, S. Kazama, and Y. Nakamura, *J. Phys. Soc. Jap.* **27**, 508 (1969).
95. L. Dworin, *Phys. Rev. Lett.* **26**, 1372 (1971).
96. L. Dworin, to be published.
97. R. E. Walstedt, J. H. Wernick, and V. Jaccarino, *Phys. Rev.* **162**, 301 (1967).
98. K. Kume, *J. Phys. Soc. Jap.* **23**, 1226 (1967).
99. L. B. Welsh, A. J. Heeger, and M. A. Jensen, *J. Appl. Phys.* **39**, 696 (1968).
100. M. Hanabusa, *J. Phys. Soc. Jap.* **30**, 1756 (1971).
101. M. Hanabusa and T. Kushida, *Phys. Rev. B* **5**, 3751 (1972).
102. J. E. Potts and L. B. Welsh, *Phys. Lett. A* **34**, 397 (1971).
103. G. Gladstone, *J. Appl. Phys.* **41**, 1150 (1970).
104. O. J. Lumpkin, *Phys. Rev.* **164**, 324 (1967).
105. R. Levine, *Phys. Lett. A* **28**, 504 (1969).
106. H. Alloul and P. Bernier, to be published.
107. P. Bernier, H. Launois, and H. Alloul, *J. Phys. (Paris)* **32**, C1-513 (1971).
108. F. Y. Fradin, *Phys. Rev. Lett.* **26**, 1033 (1971); *Phys. Rev. B* **5**, 1119 (1972).

109. M. R. McHenry, B. G. Silbernagel, and J. H. Wernick, *Phys. Rev. Lett.* **27**, 426 (1971); *Phys. Rev. B* **5**, 2958 (1972).
110. B. G. Silbernagel, private communication, 1972.
111. L. L. Hirst, *Z. Phys.* **241**, 9 (1970); **244**, 230 (1971).

6. Perturbative, Scattering, and Green's Function Theories of the s–d Model

W. Brenig

Max Planck Institut für Festkörperforschung
Stuttgart, Germany, and
Physik Department der TU München
Germany

J. Zittartz

Institut für Theoretische
Physik der Universität Köln
Germany

I. Introduction	185
II. Hamiltonian and Green's Functions	186
1. The Model	186
2. Green's Function at $T = 0$	187
3. Green's Functions at $T \neq 0$	188
III. Dispersion Theory	189
1. Derivation of Dispersion Relation	189
2. Solution of Dispersion Equations	190
IV. Equation of Motion Method	192
1. Decoupling Procedure	192
2. Solution of the Integral Equation	195
V. Properties of the Solution	198
VI. Thermal Properties	200
1. Entropy and Specific Heat	201
2. Ground-State Energy	203
VII. Electrical Conductivity	205
VIII. Magnetic Field Effects	205
1. Magnetoresistance	206
2. Magnetic Susceptibility	207
Appendix A. Diagrammatic Methods	210
Appendix B. Electronic Susceptibility	213
References	214

I. Introduction

Research in the general area of local moments in dilute magnetic alloys was stimulated a great deal by a discovery of Kondo in 1964 [1].

He noticed that the spin-flip amplitude for the scattering of conduction electrons by local moments exhibits a divergence which occurs in second order of perturbation theory at low temperatures and low energies. The origin of this divergence is a threshold anomaly at the Fermi surface. For non-spin-flip scattering the corresponding anomaly for single-particle intermediate states is completely canceled by the corresponding one for single-hole intermediate states. However, due to the quantum nature of the spin, the cancellation is only incomplete for spin-flip scattering.

Two improved approximations have been successfully applied which are free of divergences: Suhl's dispersion theory [2] and Nagaoka's equation of motion method [3]. In the dispersion approach intermediate states with more than one particle or one hole are neglected. The equation of motion for higher order Green's functions is decoupled, neglecting correlations between the conduction electrons. In spite of these rather drastic simplifications the overall behavior of the solutions is quite reasonable. Both approximations turn out to be essentially equivalent and yield identical results for the single-particle scattering amplitudes.

The perturbative theoretical expansions of these amplitudes agree with the exact expansions to all leading orders in $\ln(T_K/T)$, T_K being the so-called Kondo temperature. The solutions are free of divergences at all temperatures and energies. They allow the correlation of various anomalies in transport phenomena, thermal properties, and effects in magnetic fields in a qualitative, sometimes even quantitative way by means of one simple underlying physical concept: a resonance in the spin-flip scattering at the Fermi level. Only at very low temperatures $(T \ll T_K)$ can new effects associated with the emission and reabsorption of particle–hole excitations become relevant, such as those treated by Anderson and Yuval in Chapter 7.

II. Hamiltonian and Green's Functions

1. The Model

We consider the s–d model with a short-range interaction between the conduction (s) electrons and a localized (d) electron

$$\psi_0^+(V - J_{\text{ex}} \mathbf{s} \cdot \mathbf{S}) \psi_0 , \qquad (2.1)$$

where $\psi_0^+(\psi_0)$ creates (annihilates) an s electron in a Wannier state, conveniently placed at the origin, $\mathbf{s}$ is the conduction-electron spin with matrix elements $\mathbf{s}_{\mu\nu}(\mu, \nu = \pm\frac{1}{2})$, and $\mathbf{S}$ is the localized d spin with matrix elements $S_{\kappa\lambda}(\kappa, \lambda = -S,..., S)$. The normal potential V can be eliminated by a suitable transformation [4, 5]. We therefore consider only the exchange potential J_{ex}. The complete Hamiltonian then reads†

$$H = \sum_{\mathbf{k}} \varepsilon_{\mathbf{k}\mu} c^+_{\mathbf{k}\mu} c_{\mathbf{k}\mu} - \frac{J_{ex}}{N} \sum_{\mathbf{k},\mathbf{k}'} c^+_{\mathbf{k}\mu} \mathbf{S}_{\mu\mu'} \cdot \mathbf{S} c_{\mathbf{k}'\mu'} + \omega_d S^z. \quad (2.2)$$

The $\mathbf{k}$ sums run over the Brillouin zone of the s electrons. Here $\varepsilon_{\mathbf{k}\mu}$ is the band energy measured from the chemical potential, including a possible Zeeman term $\omega_s \mu = \mu_B g_s B \mu$, $\omega_d = \mu_B g_d B$, the Zeeman energy of the local moment, and N is the number of atomic cells.

The density of states will be denoted as $n_0 \rho(\varepsilon)$ where n_0 is the density of states at the Fermi energy $\varepsilon = 0$, so that $\rho(0) = 1$. Then

$$\frac{1}{N} \sum_{\mathbf{k}} \varphi(\varepsilon_{\mathbf{k}}) = n_0 \int d\varepsilon\, \rho(\varepsilon)\, \varphi(\varepsilon) \quad (2.3)$$

for arbitrary φ.

We shall neglect all details of band structure and often use as prototype functions

$$\rho(\varepsilon) = \left(1 - \left(\frac{\varepsilon}{D}\right)^2\right)^{1/2} \quad (2.4a)$$

$$\rho(\varepsilon) = \frac{D^2}{\varepsilon^2 + D^2}. \quad (2.4b)$$

2. Green's Function at $T = 0$

We start from the ground state of the noninteracting system with $B = 0$. Assuming that the Fermi sea of the s electrons has spin zero, this state will be $(2S + 1)$-foldly degenerate due to the impurity spin. We assume that each of the $2S + 1$ states after switching on the interaction J_{ex} adiabatically transforms into an eigenstate $|\kappa\rangle$ of H and the z component of the *total* spin. In an external B field the degeneracy of the ground-state energy E_0 is lifted and $|\kappa\rangle$ obeys the Schrödinger equation

$$H|\kappa\rangle = (E_0 + \omega_\kappa)|\kappa\rangle. \quad (2.5)$$

Here in the limit $J_{ex} \to 0$, ω_κ is the Zeeman energy $\omega_d \kappa$. At $T = 0$ the eigenstates $|\kappa\rangle$ can be used to define Green's functions

$$\langle A; B\rangle_{\kappa\kappa'} = \langle\kappa|A(z - H + E_0)^{-1} B|\kappa'\rangle$$
$$+ \langle\kappa|B(z - \omega_\kappa - \omega_{\kappa'} + H - E_0)^{-1} A|\kappa'\rangle, \quad (2.6)$$

† In dilute magnetic alloys corrections to macroscopic quantities A are expected to be of the form: $\Delta A = ca + O(c^2)$, where c is the concentration. The one-impurity Hamiltonian (2.2) is then sufficient to determine a.

which obey the equations of motion

and
$$(z - \omega_\kappa)\langle A; B\rangle_{\kappa\kappa'} - \langle [A, H]; B\rangle_{\kappa\kappa'} = \langle \kappa | [A, B]_+ | \kappa'\rangle$$
$$(z - \omega_{\kappa'})\langle A; B\rangle_{\kappa\kappa'} - \langle A; [H, B]\rangle_{\kappa\kappa'} = \langle \kappa | [A, B]_+ | \kappa'\rangle. \quad (2.7)$$

In particular one has

$$(z - \varepsilon_{\mathbf{k}\mu\kappa})\langle c_{\mathbf{k}\mu}; c^+_{\mathbf{k}'\mu'}\rangle_{\kappa\kappa'} - \langle j_\mu; c^+_{\mathbf{k}'\mu'}\rangle_{\kappa\kappa'} = \delta_{\mathbf{k}\mu\kappa,\mathbf{k}'\mu'\kappa'}, \quad (2.8)$$

where $\varepsilon_{\mathbf{k}\mu\kappa} = \varepsilon_{\mathbf{k}\mu} + \omega_\kappa$ and the "current operator" (independent of $\mathbf{k}$)

$$j_\mu = [c_{\mathbf{k}\mu}, H_{\text{int}}] = -\frac{J_{\text{ex}}}{N} \sum_{\mathbf{k},\nu} \mathbf{S} \cdot \mathbf{s}_{\mu\nu} c_{\mathbf{k}\nu}. \quad (2.9)$$

Applying the equation of motion once more to (2.8) one is led to a Green's function completely independent of $\mathbf{k}$ and $\mathbf{k}'$

$$\langle \mu\kappa | T(z) | \mu'\kappa'\rangle = \langle \kappa | [j_\mu, c^+_{\mathbf{k}'\mu'}]_+ | \kappa'\rangle + \langle j_\mu; j^+_{\mu'}\rangle_{\kappa\kappa'}$$
$$= (z - \varepsilon_{\mathbf{k}\mu\kappa})\langle c_{\mathbf{k}\mu}; j^+_{\mu'}\rangle_{\kappa\kappa'}$$
$$= \langle j_\mu; c^+_{\mathbf{k}'\mu'}\rangle_{\kappa\kappa'} (z - \varepsilon_{\mathbf{k}'\mu'\kappa'}). \quad (2.10)$$

This function is closely related to the amplitudes for the scattering of single particles and holes by the impurity (Section III).

3. Green's Functions at $T \neq 0$

At nonzero temperatures one no longer specifies the quantum numbers κ but considers thermodynamic Green's functions $\langle A; B\rangle$. They are defined as the analytical continuation of the Fourier coefficients $G_{AB}(z_n)[z_n = (2n + 1) \pi i/\beta]$ of the thermodynamic averages $\langle TA(t)B(0)\rangle$ ($0 \leq t \leq \beta$) into the complex z plane [6].

The equations of motion for these functions are completely equivalent to (2.7). One only has to omit the ω_κ and replace the ground-state averages by thermodynamic averages. In particular

$$(z - \varepsilon_{\mathbf{k}\mu})\langle c_{\mathbf{k}\mu}; c^+_{\mathbf{k}'\mu}\rangle - \langle j_\mu; c^+_{\mathbf{k}'\mu}\rangle = \delta_{\mathbf{k},\mathbf{k}'}. \quad (2.11)$$

Applying the equation of motion once more to $\langle j_\mu; c^+_{\mathbf{k}'\mu}\rangle$ one is, on the one hand, led to a "scattering amplitude" T_μ by

$$T_\mu(z) = -\frac{J_{\text{ex}}}{N} \langle S^z\rangle \mu + \langle j_\mu; j_\mu^+\rangle = (z - \varepsilon_{\mathbf{k}\mu})\langle c_{\mathbf{k}\mu}; j_\mu^+\rangle$$
$$= \langle j_\mu; c^+_{\mathbf{k}'\mu}\rangle(z - \varepsilon_{\mathbf{k}'\mu}). \quad (2.12)$$

6. PERTURBATIVE, SCATTERING, AND GREEN'S FUNCTION THEORIES

On the other hand, one can express the current j_μ as in (2.9) and use

$$(z - \varepsilon_{\mathbf{k}\nu})\langle S c_{\mathbf{k}\nu} ; c_{\mathbf{k}'\mu}^+\rangle - \langle[S c_{\mathbf{k}\nu}, H_{\text{int}}]; c_{\mathbf{k}'\mu}^+\rangle = \langle S \rangle \delta_{\mathbf{k}\nu, \mathbf{k}'\mu}. \quad (2.13)$$

The commutator in this equation then yields a higher order Green's function containing four electron field operators besides S.

Equation (2.10) forms the basis of Suhl's dispersion approach [7–9], and (2.11) and (2.13) that of Nagaoka's equation of motion approach [3]. Both methods introduce some generalized "single-particle approximations," neglecting higher order correlations, and yield identical results [4, 12] for the scattering amplitudes.

III. Dispersion Theory

1. Derivation of Dispersion Relation

The S-matrix theory of the Kondo effect has been developed by Suhl in a series of papers [7–9]. For a detailed account of the method we refer to Suhl's Varenna summer school lecture [9] (see also papers by Maleev [10]). Here we give a brief derivation of Suhl's dispersion relations at zero temperature. We start out from an expression for $T(z)$ on the "energy shell" $z = \varepsilon_{\mathbf{k}\mu\kappa}$ in terms of scattering states $|\mathbf{k}\mu\kappa\pm\rangle$

$$\langle\mu\kappa | T(\varepsilon_{\mathbf{k}\mu\kappa} \pm i\delta) | \mu'\kappa'\rangle = \langle\pm\mathbf{k}\mu\kappa | j_{\mu'}^+ | \kappa'\rangle \theta(\varepsilon_{\mathbf{k}\mu}) + \langle\kappa | j_{\mu'}^+ | \mathbf{k}\mu\kappa'\rangle \theta(-\varepsilon_{\mathbf{k}\mu}), \quad (3.1)$$

where $\theta(\varepsilon)$ is the usual step function (1 for positive, 0 for negative ε). Equation (3.1) is a simple consequence of (2.10) and the asymptotic condition [11]

$$\frac{\pm i\delta}{\epsilon_{\mathbf{k}\mu} + \omega_\kappa + E_0 - H \pm i\delta} c_{\mathbf{k}\mu}^+ | \kappa\rangle \Rightarrow |\mathbf{k}\mu\kappa\pm\rangle \theta(\varepsilon_{\mathbf{k}\mu}), \quad \delta \to +0 \quad (3.2)$$

for single-particle scattering and

$$\frac{\mp i\delta}{\epsilon_{\mathbf{k}\mu} - \omega_\kappa - E_0 + H \mp i\delta} c_{\mathbf{k}\mu} | \kappa\rangle \Rightarrow |\mathbf{k}\mu\kappa\pm\rangle \theta(-\varepsilon_{\mathbf{k}\mu}), \quad \delta \to +0 \quad (3.3)$$

for single-hole scattering.

The Green's function $\langle j_\mu ; j_{\mu'}^+\rangle_{\kappa\kappa'}$ occurring in $T(z)$ according to (2.10) may be evaluated by introducing a complete set of eigenfunctions $|n\rangle$ of H as intermediate states. Then $T(z)$, according to (2.10), is expressed in terms of matrix elements of the form $\langle\kappa | j_\mu | n\rangle$, and so on. The essential step of Suhl's theory now is to keep only single-particle or single-hole intermediate states, neglecting particle–hole excitations in the scattering process. Using (3.1) the corresponding matrix elements of j_μ then can be expressed in terms of $T(z)$ itself. Thus one obtains a closed set of equations.

For vanishing external field ($B = 0$) these equations become particularly simple. Because of rotational invariance T can be decomposed as

$$\langle \mu\kappa \mid T(z) \mid \mu'\kappa' \rangle = \frac{J_{\text{ex}}}{N} \{t(z) \, \delta_{\mu\mu'}\delta_{\kappa\kappa'} + \tau(z) \, \mathbf{s}_{\mu\mu'} \cdot \mathbf{S}_{\kappa\kappa'}\}, \quad (3.4)$$

that is, into a non-spin-flip amplitude t and a spin-flip amplitude τ. If this is used together with (3.1) and Suhl's approximation to evaluate (2.10), one finds [putting $t(\varepsilon \pm i\delta) = t_\pm$ and $J = n_0 J_{\text{ex}}$]

$$t(z) = J \int d\varepsilon \, \frac{\rho(\varepsilon)}{z - \varepsilon} [t_+ t_- + \tau_+ \tau_- S(S+1)/4] \quad (3.5a)$$

$$\tau(z) = -1 + J \int d\varepsilon \, \frac{\rho(\varepsilon)}{z - \varepsilon} [t_+ \tau_- + t_- \tau_+ - \tau_+ \tau_- (\tfrac{1}{2} - f(\varepsilon))]. \quad (3.5b)$$

In deriving (3.5), we have used (2.9) to evaluate the "Born term" of $T(z)$ and in

$$\langle \kappa \mid [j_\mu, c^+_{\mathbf{k}'\mu'}]_+ \mid \kappa' \rangle = -\frac{J_{\text{ex}}}{N} \langle \kappa \mid \mathbf{S} \mid \kappa' \rangle \cdot \mathbf{s}_{\mu\mu'} \quad (3.6)$$

$\langle \kappa \mid \mathbf{S} \mid \kappa' \rangle$ is approximately replaced by its first-order value $\mathbf{S}_{\kappa\kappa'}$. In our derivation of (3.5b) for $T = 0$, $f(\varepsilon)$ is nothing but the step function $\theta(-\varepsilon)$. Suhl [7, 9] has tried to generalize the scattering theory to nonzero temperatures. He finds (3.5a) unchanged and in (3.5b) the Fermi function $2f(\varepsilon) = 1 - \tanh(\beta\varepsilon/2)$ instead of the step function. A derivation of this result using diagrammatic methods is given in Appendix A.

Taking the discontinuity of t and τ across the real axis one finds the generalized "unitarity relations"

$$t_+ - t_- = -2\pi i J\rho(\varepsilon)[t_+ t_- + \tau_+ \tau_- S(S+1)/4] \quad (3.7a)$$

$$\tau_+ - \tau_- = -2\pi i J\rho(\varepsilon)[t_+ \tau_- + t_- \tau_+ - \tau_+ \tau_- (\tanh(\beta\varepsilon/2))/2]. \quad (3.7b)$$

2. Solution of Dispersion Equations[†]

To solve (3.5) for t and τ it is useful to introduce analytic continuations $\rho(z)$ of the density of states functions. For (2.4a) we use simply $\rho(z) = [1 - (z/D)^2]^{1/2}$. This function is holomorphic except for a cut

[†] See [4, 8, 10, 12–14].

6. PERTURBATIVE, SCATTERING, AND GREEN'S FUNCTION THEORIES

which we choose from $-D$ to $+D$ such that $\rho_\pm(\varepsilon) = \pm \rho(\varepsilon)$. For (2.4b) the same can be achieved by choosing

$$\rho(z) = \text{sgn}(\text{Im } z) \frac{D^2}{z^2 + D^2}. \tag{3.8}$$

This function is not sectionally holomorphic but has poles at $\pm iD$. Now an auxiliary function $X(z)$ is introduced by

$$1 - 2\pi i J \rho(z) t(z) = -\tau(z) X(z). \tag{3.9}$$

The discontinuity of X is determined by (3.7b)

$$X_+ - X_- = i\pi J \rho(\varepsilon) \tanh(\beta\varepsilon/2), \tag{3.10}$$

which suggests the ansatz

$$X(z) = R(z) + 1 + \chi(z) \tag{3.11}$$

with

$$R(z) = -\frac{J}{2} \int d\varepsilon \, \rho(\varepsilon) \frac{\tanh(\beta\varepsilon/2)}{z - \varepsilon}. \tag{3.12}$$

Then $\chi(z)$ is continuous across the real axis and behaves asymptotically for large z as

$$\chi \Rightarrow -\frac{2\pi J}{D} \lim_{z\to\infty} zt(z), \quad \text{square root} \tag{3.13a}$$

$$\chi \Rightarrow 0, \quad \text{Lorentzian}. \tag{3.13b}$$

According to (2.4a) and (3.13a), in the first case, χ is a bounded entire function and thus a constant. This constant can be determined from the asymptotic behavior of τ by a transcendental equation. For the Lorentzian density of states $\rho(z)$, and thus $X(z)$, is not holomorphic but determined by the poles of $\rho(z)$ far away from the real axis via a nonsingular integral equation. For both density of states functions it turns out that χ is small of order J^2. It can therefore be neglected for most practical purposes when compared to $R(z) + 1$ along the real axis. (For general density of states functions see Zittartz [12].) Having determined $X(z)$ one can proceed further, using (3.7a), and obtain $\tau(z)$ from

$$[\tau_+ \tau_-]^{-1} = |X_+(\varepsilon)|^2 + (\pi J \rho(\varepsilon))^2 S(S+1) \equiv K(\varepsilon). \tag{3.14}$$

Since both density of states functions are even in ε it follows from (3.5) that $\tau_+(\varepsilon) = \tau_+^*(-\varepsilon)$. In particular $\tau_+(0)$ is real and thus completely determined by (3.14).

One finds

$$Jt(0) = \frac{1}{2\pi i}(1 + \tau(0) X(0)) \tag{3.15a}$$

$$J\tau(0) = -\frac{1}{[K(0)]^{1/2}}. \tag{3.15b}$$

A complete determination of the phase of τ can be obtained from (3.14) by taking the logarithm and assuming that $\tau(z)$ is free of zeros.[†] Then after multiplying (3.14) by the appropriate discontinuous functions one obtains

$$\tau(z) = -\exp\left\{\frac{1}{2\pi i}\int d\epsilon A \frac{\ln K(\varepsilon)}{z - \varepsilon}\right\}. \tag{3.16}$$

Here the quantity A is somewhat different for the two density of states functions. One has

$$A = \rho(z)/\rho_+(\epsilon), \quad \text{square root} \tag{3.17a}$$

$$A = \text{sgn}(\text{Im } z), \quad \text{Lorentzian.} \tag{3.17b}$$

IV. Equation of Motion Method

The equation of motion approach to the Kondo problem is due to Nagaoka [3] who first introduced a suitable decoupling procedure. The solution of the resulting Nagaoka equations proceeds in two steps. First the equations are reduced algebraically to one singular integral equation for the amplitude [15, 16]. In the second step the integral equation is solved exactly [17, 18] by using Mushkelishvili methods for singular integral equations [19].

1. Decoupling Procedure

If one considers again only the case of vanishing external field ($B = 0$) and introduces the notations

$$G_{\mathbf{k}\mathbf{k'}} = \tfrac{1}{2}\sum_\mu \langle c_{\mathbf{k}\mu} ; c^+_{\mathbf{k'}\mu}\rangle \tag{4.1}$$

[†] Dispersion relations usually do not determine the solution uniquely [7, 8]. The solution with $\tau(z) \neq 0$ is the analytical continuation in J of the high-temperature perturbation expansion [13] and it agrees with the unique solution of the equation of motion approach (Section IV).

6. PERTURBATIVE, SCATTERING, AND GREEN'S FUNCTION THEORIES 193

and

$$\Gamma_{\mathbf{k}\mathbf{k}'} = \tfrac{1}{2} \sum_{\mu,\nu} \langle \mathbf{S} c_{\mathbf{k}\nu} ; c^+_{\mathbf{k}'\mu} \rangle \cdot \mathbf{s}_{\mu\nu} , \qquad (4.2)$$

then (2.11) takes the form

$$(z - \varepsilon_{\mathbf{k}}) G_{\mathbf{k}\mathbf{k}'} + \frac{J_{\text{ex}}}{N} \sum_l \Gamma_{l\mathbf{k}'} = \delta_{\mathbf{k}\mathbf{k}'} , \qquad (4.3)$$

and (2.12) becomes, with $\tfrac{1}{2} \sum_\mu T_\mu(z) = (J_{\text{ex}}/N) t(z)$,

$$t(z) = -(z - \varepsilon_{\mathbf{k}'}) \sum_l \Gamma_{l\mathbf{k}'}(z). \qquad (4.4)$$

The equation of motion (2.13) for Γ contains higher order Green's functions with four electron field operators. In order to get a closed set of equations one uses a standard decoupling by "linearizing" the product of electron operators in a generalized Hartree–Fock treatment [3]

$$c^+ cc \sim \langle c^+ c \rangle c + \text{all other combinations.} \qquad (4.5)$$

This approximation, which neglects electronic correlations, is in the same spirit as the restriction to one-particle intermediate states in the scattering approach (Section III). Therefore it is not too surprising that both methods finally yield identical scattering amplitudes [4, 12]. The result of this approximation is

$$(z - \varepsilon_{\mathbf{k}}) \Gamma_{\mathbf{k}\mathbf{k}'} = \frac{J_{\text{ex}}}{N} \left[\left(m_{\mathbf{k}} - \frac{S(S+1)}{4} \right) \sum_l G_{l\mathbf{k}'} - (n_{\mathbf{k}} - \tfrac{1}{2}) \sum_l \Gamma_{l\mathbf{k}'} \right], \qquad (4.6)$$

where

$$n_{\mathbf{k}} - \tfrac{1}{2} = \tfrac{1}{2} \left[\sum_l \langle c^+_{l\mu} c_{\mathbf{k}\mu} \rangle - 1 \right] = \sum_l \frac{P}{\beta} \sum_n G_{\mathbf{k}l}(z_n) \qquad (4.7a)$$

$$m_{\mathbf{k}} = \tfrac{1}{2} \sum_l \langle c^+_{l\mu} \mathbf{s}_{\mu\nu} c_{\mathbf{k}\nu} \cdot \mathbf{S} \rangle = \sum_l \frac{P}{\beta} \sum_n \Gamma_{\mathbf{k}l}(z_n). \qquad (4.7b)$$

Here P denotes principal part summation. The set of Eqs. (4.3), (4.6), and (4.7) has been derived by Nagaoka [3]. Now we proceed by reducing this set of equations to one integral equation for the quantity $t(z)$ [15].

Using Eqs. (4.3) and (4.4), and $m_{\mathbf{k}} = m_{\mathbf{k}}^*$ we obtain

$$n_{\mathbf{k}} - \tfrac{1}{2} = \frac{P}{\beta} \sum_n \frac{1 + F(z_n) t(z_n)}{z_n - \varepsilon_{\mathbf{k}}} \qquad (4.8a)$$

and

$$m_k = -\frac{P}{\beta}\sum_n \frac{t(z_n)}{z_n - \varepsilon_k} \qquad (4.8b)$$

with

$$F(z) = \frac{J_{ex}}{N}\sum_k \frac{1}{z - \varepsilon_k} = J\int_{-\infty}^{\infty} d\varepsilon \frac{\rho(\varepsilon)}{z - \varepsilon}. \qquad (4.8c)$$

Thus m_k and n_k are linear functionals of t. Summing Eq. (4.6) over $\mathbf{k}$ and regrouping terms finally leads to the integral equation

$$\phi(z)\,t(z) = N(z), \qquad (4.9)$$

where

$$N(z) = -\frac{J_{ex}}{N}\sum_k \frac{m_k - S(S+1)/4}{z - \varepsilon_k}$$

$$\phi(z) = 1 - F(z)N(z) + \frac{J_{ex}}{N}\sum_k \frac{n_k - \frac{1}{2}}{z - \varepsilon_k}. \qquad (4.10)$$

The two functions N and ϕ are obviously sectionally holomorphic. They can be expressed in a more suitable form by using (4.8) and the relation

$$\frac{J_{ex}}{N}\sum_k \frac{1}{(z - \varepsilon_k)(z_n - \varepsilon_k)} = \frac{F(z_n) - F(z)}{z - z_n}.$$

We obtain the following representations [20]

$$N(z) = L_1(z) - F(z)[L_0(z) - S(S+1)/4], \qquad (4.11)$$

$$\phi(z) = 1 + R_1(z) - F(z)R_0(z) + L_2(z) - 2F(z)L_1(z)$$
$$+ F^2(z)[L_0(z) - S(S+1)/4], \qquad (4.12)$$

where the functions R_ν and L_ν are defined by

$$R_\nu = \frac{P}{\beta}\sum_n \frac{F^\nu(z_n)}{z - z_n}, \qquad R = R_1 - FR_0. \qquad (4.13)$$

$$L_\nu = \frac{P}{\beta}\sum_n \frac{F^\nu(z_n)\,t(z_n)}{z - z_n}. \qquad (4.14)$$

Note that the L_ν and R_ν are meromorphic functions with poles at the discrete set of points z_n on the imaginary axis. However, in the combinations (4.11), (4.12) the residues at these poles just compensate such that N and ϕ are sectionally holomorphic as noticed before.

2. Solution of the Integral Equation

It is clear from (4.10) that a perturbational expansion in terms of J should be possible if z stays away from ε_k on the real axis. In this case we obtain $\phi = 1 + O(J)$, $N = O(J)$, and $t = O(J)$. On the other hand, both functions ϕ and N blow up if we consider z along the real axis, and we have to use more refined methods to solve the problem. A more detailed analysis may be found in the work of Zittartz and Müller-Hartmann [18, 20].

The following conditions for the density of states function $\rho(\varepsilon)$ will be required: (a) ρ falls off at infinity rapidly enough such that $\int \rho \, d\varepsilon$ is finite; (b) either ρ is analytic near the whole real axis (such as the Lorentzian in (2.4b) with singularities a distance of the order D away from the origin or $\rho(z)$ is sectionally holomorphic itself such as the square root function in (2.4a).

We then define $\bar{F}(z)$ as the analytical continuation of $F(z)$ [Eq. (4.8c)] through the cut. Explicitly we get from $\bar{F}(\varepsilon \pm i\delta) = F(\varepsilon \mp i\delta)$ and the discontinuity relation

$$F(\varepsilon + i\delta) - F(\varepsilon - i\delta) = -2\pi i J\rho(\varepsilon) \tag{4.15}$$

the representation

$$\bar{F}(z) = F(z) + 2\pi i J\rho(z). \tag{4.16}$$

Next consider the auxiliary function

$$X(z) = \phi(z) + (F - \bar{F}) N(z) = 1 + R(z) + \chi(z), \tag{4.17}$$

where

$$\chi(z) = L_2 - (F + \bar{F}) L_1 + F\bar{F}[L_0 - S(S+1)/4]. \tag{4.18}$$

It is easy to check that χ has no poles at the set $z = z_n$, as again the residues compensate in the expression. Furthermore χ has no cut along the real axis as the functions $F + \bar{F}$ and $F\bar{F}$ are analytic there ($\bar{F}$ is the analytic continuation of F, and vice versa). Thus χ is determined by the singularities of ρ (via $\bar{F}$) far away from the origin and by its behavior at infinity. Then χ is bounded on the real axis, and applying perturbation theory we get

$$|\chi(\varepsilon)| \leq AJ^2, \tag{4.19}$$

where A is some number. Thus for all practical purposes χ may be neglected in (4.17) when compared with unity as long as ε is real and $|J| \ll 1$.

The method of solution will proceed by expressing every function of interest in terms of X which apart from the unimportant χ is known explicitly. From (4.17) we have N in terms of ϕ and X. To determine ϕ we consider the "unphysical" functions $\bar\phi$ and $\bar X$ which one obtains from ϕ and X by replacing F by $\bar F$, and vice versa.

By explicit calculation we get from (4.12) and (4.17):

$$\phi\bar\phi - X\bar X = (F - \bar F)^2 H, \tag{4.20}$$

$$H(z) = [1 + R_1 + L_2]\left[L_0 - \frac{S(S+1)}{4}\right] - L_1(L_1 + R_0), \tag{4.21}$$

where H obviously is a meromorphic function without a cut along the real axis as F and $\bar F$ do not occur in (4.21). As the L_ν and R_ν appear quadratically in (4.21), one would expect double poles at the discrete set $z = z_n$ along the imaginary axis. However, it is easy to show that

$$\lim_{z \to z_n}(z - z_n)^2 H(z) = 0, \tag{4.22a}$$

$$\lim_{z \to z_n}(z - z_n) H(z) = \lim_{z \to z_n}(z - z_n)\frac{1}{\beta}(t\phi - N) = 0, \tag{4.22b}$$

where the integral equation (4.9) is used in the last step. Therefore H is free of poles and, as no other singularities occur, H is analytic in the whole z plane and is determined by its behavior at infinity. From (4.21) and (4.13), (4.14) it follows that

$$H(z) \equiv -S(S+1)/4, \tag{4.23}$$

which relation is the key step toward the final solution. Along the cut we now get from (4.20), (4.23) the relation

$$\phi(\varepsilon + i\delta)\phi(\varepsilon - i\delta) = K(\varepsilon) \equiv |X_+(\varepsilon)|^2 + (\pi J\rho(\varepsilon))^2 S(S+1) \tag{4.24}$$

[we also have used $X_+^*(\varepsilon) = X_-(\varepsilon)$ and (4.15)]. Next one shows that $\phi(z) \neq 0$. First it is clear from (4.24) that no zeros occur along the real axis as $K > 0$. If $\phi = 0$ at some point off the axis, then the requirement that t is sectionally holomorphic would imply from (4.9) that $N = 0$ at the same point. However, $\phi = 0$ and $N = 0$ are incompatible with (4.23). Therefore $\phi \neq 0$ and the function ϕ^{-1} also is sectionally holomorphic. If we define

$$\psi = \begin{cases} \phi; & \text{Im } z > 0 \\ \phi^{-1}; & \text{Im } z < 0, \end{cases}$$

6. PERTURBATIVE, SCATTERING, AND GREEN'S FUNCTION THEORIES

then the analog of (4.24) for ψ is

$$\psi_+(\varepsilon) = K(\varepsilon) \psi_-(\varepsilon), \tag{4.25}$$

which is the standard homogeneous Hilbert problem [19]. With the ansatz[†]

$$\psi(z) = \alpha(z) \exp\left\{-\frac{1}{2\pi i}\int \frac{d\varepsilon}{z-\varepsilon} \ln K(\varepsilon)\right\} \tag{4.26}$$

one gets

$$\alpha_+(\varepsilon) = \alpha_-(\varepsilon). \tag{4.27}$$

Thus α is analytic everywhere and the behavior of ϕ and ψ at infinity implies $\alpha \equiv 1$. The final solution for ϕ follows from (4.26)

$$\phi(z) = \exp\left\{-\frac{\text{sgn}(\text{Im } z)}{2\pi i}\int_{-\infty}^{\infty} \frac{d\varepsilon}{z-\varepsilon} \ln K(\varepsilon)\right\}; \tag{4.28}$$

ϕ is therefore expressed as a functional of K and X via (4.24). Strictly speaking we are still left with a formidable regular integral equation involving the function χ. However, as mentioned before, χ is bounded and of order J^2 along the real axis and will henceforth be neglected.

On the real axis we write

$$\phi_+(\varepsilon) = [|X_+(\varepsilon)|^2 + (\pi J\rho)^2 S(S+1)]^{1/2} e^{i\varphi(\varepsilon)}, \tag{4.29}$$

where the phase function is

$$\varphi(\varepsilon) = \frac{P}{2\pi}\int \frac{d\varepsilon'}{\varepsilon-\varepsilon'} \ln K(\varepsilon') \tag{4.30}$$

and P denotes the principal part. Finally we obtain the amplitude for non-spin-flip scattering t from (4.9) and (4.17)

$$Jt_+(\varepsilon) = \frac{1}{2\pi i\rho(\varepsilon)}\left\{1 - \frac{X_+(\varepsilon)}{\phi_+(\varepsilon)}\right\}. \tag{4.31}$$

Finally we note that the equivalence of the two methods in Sections III and IV may be seen from a comparison of (3.16) and (4.28) with the proper identification: $\tau(z) = -\phi^{-1}(z)$.

[†] For a finite cut density of states function such as the square root in (2.4a), the integral in (4.26) is modified to

$$\int_{-D}^{D} \frac{\rho(z) \ln K(\varepsilon)}{\rho(\varepsilon)}\frac{d\varepsilon}{z-\varepsilon}.$$

V. Properties of the Solution

Having established the solutions for the transition amplitudes we might start to calculate various thermodynamic properties. There exists a vast literature on perturbative calculations of such quantities starting with the famous paper by Kondo in 1964 on the resistance minimum [1]. These perturbational calculations have been reviewed extensively [5, 21], and we refrain from repeating them here.

The main result of perturbation theory is the well-known low-temperature logarithmic divergence of the transition amplitudes at the Fermi surface, which behavior implies Kondo anomalies. As these Kondo anomalies are typical Fermi surface effects and thus do not depend on the actual band structure, we shall first simplify the solution of the last sections by going to the weak coupling limit. In this limit the typical Kondo behavior will become more transparent. In the weak coupling limit the coupling constant J_{ex} is considered to be very small compared to the bandwidth D. In fact we shall let $D \to \infty$ wherever this is possible. Thus we replace

$$\rho(z) \to 1. \tag{5.1}$$

The mathematical structure of the solution in this limit has been investigated thoroughly by Müller-Hartmann [22]. Quite generally the function X can be written as

$$X_\pm(\varepsilon) = 1 + J \ln(\alpha_0 \beta W) - J g_\pm(\beta \varepsilon/2\pi), \tag{5.2}$$

where W is a characteristic band energy of the order of the bandwidth D

$$W = \begin{cases} 2De^{-1}, & \text{square root} \\ D, & \text{Lorentzian} \end{cases} \tag{5.3}$$

and $\alpha_0 = (2/\pi) e^{\mathscr{E}} = 1.134$ ($\mathscr{E}$ is Euler's constant). The functions $g_\pm(x)$ are

$$g_\pm(x) = \psi(\tfrac{1}{2} \mp ix) - \psi(\tfrac{1}{2}), \tag{5.4}$$

where ψ is the Digamma function $[g_\pm(x) \to \ln |x|, |x| \to \infty]$. Using (5.1) and (5.2) one obtains for the solution:

$$K(\varepsilon) \Rightarrow J^2 K_0(\varepsilon); \quad K_0(\varepsilon) = |J^{-1}X_+(\varepsilon)|^2 + \pi^2 S(S+1) \tag{5.5a}$$

$$\varphi(\varepsilon) = \frac{P}{2\pi} \varepsilon \int \frac{d\varepsilon'}{\varepsilon^2 - \varepsilon'^2} \ln K_0(\varepsilon') \tag{5.5b}$$

$$J\tau_+(\varepsilon) = -(\text{sgn } J)[|J^{-1}X_+(\varepsilon)|^2 + \pi^2 S(S+1)]^{-1/2} e^{-i\varphi(\varepsilon)} \tag{5.5c}$$

$$2\pi i J t_+(\varepsilon) = 1 + X_+(\varepsilon) \tau_+(\varepsilon). \tag{5.5d}$$

In these formulas the band structure has been eliminated.

6. PERTURBATIVE, SCATTERING, AND GREEN'S FUNCTION THEORIES

The function X (5.2) depends characteristically on the parameter

$$\lambda = J^{-1} + \ln(\alpha_0 \beta W), \tag{5.6}$$

which for ferromagnetic coupling ($J > 0$) is a large positive quantity. For antiferromagnetic coupling we define the Kondo temperature T_K by setting

$$\lambda = \ln\left(\frac{T_K}{T}\right), \qquad T_K = \frac{\alpha_0 W}{K_B} e^{-1/|J|} \tag{5.7}$$

[more generally $X(\varepsilon = 0, T = T_K) = 0$]. For dilute magnetic alloys T_K is mainly in the range of 0.1 to 10°K. In the case of $J > 0$ we may introduce the renormalized coupling constant

$$\bar{J}(T) = \lambda^{-1}, \qquad 0 < \bar{J} < J, \tag{5.8}$$

which is a small quantity for all temperatures. In terms of this coupling constant $\bar{J}$ the transition amplitudes (5.5) at the Fermi surface ($\varepsilon = 0$) are given as

$$J_\tau(0) = -\bar{J}[1 + \pi^2 S(S+1)\,\bar{J}^2]^{-1/2} = O(\bar{J}), \tag{5.9a}$$

$$2\pi i\, Jt(0) = 1 - [1 + \pi^2 S(S+1)\,\bar{J}^2]^{-1/2} = O(\bar{J}^2). \tag{5.9b}$$

Therefore the transition amplitudes are always small and no anomalies will occur in macroscopic quantities for $J > 0$.

In the case of antiferromagnetic coupling ($J < 0$) we obtain at the Fermi surface

$$J_\tau(0) = [\lambda^2 + \pi^2 S(S+1)]^{-1/2}, \tag{5.10a}$$

$$2\pi i\, Jt(0) = 1 + \lambda[\lambda^2 + \pi^2 S(S+1)]^{-1/2}. \tag{5.10b}$$

Thus we see that the spin-flip amplitude goes to zero both for high and low temperatures, $\lambda \to \pm\infty$. It reaches a maximum at T_K where $\lambda = 0$. The non-spin-flip amplitude increases monotonically as $T \to 0$, reaching the unitarity limit

$$Jt(0) = 1/i\pi \qquad \text{for} \quad T = 0, \tag{5.11}$$

which corresponds to resonance scattering (maximal possible scattering). The resonance is a real many-body resonance, due to the sharpness of the Fermi surface. The meaning of T_K is that it is the characteristic temperature scale for Kondo anomalies in contrast to the Fermi temperature $T_F = D/K_B$ which is the characteristic scale in normal electronic properties.

VI. Thermal Properties

We now investigate the thermal properties of the s–d model on the basis of the solutions for the transition amplitudes. All quantities of interest can be derived by differentiation from the free energy which, neglecting potential scattering again, is given by the well-known formula

$$F(T) - F_0(T) \equiv \Delta F(T) = \int_0^J \frac{dJ'}{J'} \langle H' \rangle, \qquad (6.1)$$

where H' is the interaction part of the Hamiltonian with coupling constant J'. Using (4.2), (4.4), and (4.8c) we can express the average as

$$\langle H' \rangle = \frac{2}{\beta} \sum_n F(z_n) t(z_n), \qquad (6.2)$$

that is, a frequency sum involving the t matrix which also may be written as a contour integral along the real axis. One important conclusion may be drawn from this representation. Though the perturbational t matrix contains logarithmic divergences in every order at $T = 0$ for $\varepsilon \to 0$, the perturbation theoretical expansion for $\langle H' \rangle$, and thus the free energy, exists at least in the asymptotic sense. This is because the $\ln \varepsilon$ terms in the t-matrix expansion are perfectly integrable. Kondo anomalies only appear as small corrections beyond the power series expansions in J or in temperature-dependent quantities. The general structure of $\Delta F(T)$ is expected to be of the form

$$\Delta F(T) = f_1(J) D \left[1 + O\left(\left(\frac{T}{T_F}\right)^2\right) \right] + K_B T_K f_2(J) \left[1 + O\left(\frac{T}{T_K}\right) \right]. \qquad (6.3)$$

The first part is the "normal" part. The bandwidth $D \approx K_B T_F$ is the characteristic energy, temperature corrections are measured on the scale T_F, and $f_1(J)$ is a function of the coupling constant for which an asymptotic expansion exists starting with a J^2 term, the coefficients depending on the particular band structure. The second part, proportional to $K_B T_K \approx D e^{-1/|J|}$ ($J < 0$), is the "anomalous" part which does not show up in perturbation theory and which is small when compared with the first part. However, the temperature scale for the second part is T_K ($\ll T_F$), and therefore we expect large Kondo anomalies in derivatives with respect to temperature such as the entropy and specific heat.

Instead of using (6.2) for our calculations we shall start with a formula derived by Brenig and Götze [13]:

$$\langle H' \rangle = 2 \lim_{z \to \infty} z \left[t(z) - \frac{S(S+1)}{4} F(z) \right]. \qquad (6.4)$$

6. PERTURBATIVE, SCATTERING, AND GREEN'S FUNCTION THEORIES   201

This relation can be proved by using (2.12), which is equivalent to rewriting $t(z)$ in terms of the "current–current" Green's function

$$\frac{J_{ex}}{N} \lim zt(z) = \lim \frac{z}{2} \sum_\mu \langle j_\mu ; j_\mu{}^+\rangle = \frac{1}{2} \sum_\mu \langle [j_\mu, j_\mu{}^+]_+\rangle, \tag{6.5}$$

and explicit evaluation via (2.9). Within our approximation the equivalence of (6.2) and (6.4) also follows from the explicit asymptotic behavior ($z \to \infty$) of the integral equation (4.9): $\lim_{z \to \infty} zt(z) = \lim_{z \to \infty} zN(z)$ [20].

1. Entropy and Specific Heat

As Kondo anomalies will appear only for antiferromagnetic coupling, we confine ourselves to the case $J < 0$. According to (6.3) the additional entropy, $\Delta S = -\partial \Delta F / \partial T$, contains a "normal part" which vanishes in the weak coupling limit ($D \to \infty$, $J \to 0$, $T_K = $ constant) and an "anomalous" part which we want to calculate. Explicitly we have

$$\Delta S(T) = -2 \int_0^J \frac{dJ'}{J'} \lim_{\varepsilon \to \infty} \varepsilon \frac{\partial}{\partial T} t_+(\varepsilon). \tag{6.6}$$

In the weak coupling limit we use the expressions (5.5) and we obtain

$$\lim_{\varepsilon \to \infty} \varepsilon \frac{\partial}{\partial T} t_+(\varepsilon) = -\frac{1}{2\pi i J} \lim_{\varepsilon \to \infty} \varepsilon \frac{\partial}{\partial T} e^{-i\varphi(\varepsilon)}$$

$$= \frac{1}{4\pi^2 J} \int d\varepsilon' \frac{\partial}{\partial T} \ln K_0(\varepsilon'). \tag{6.7}$$

Next we use (5.2), take the temperature derivative, and change the integration variable to $x = \beta\varepsilon/2\pi$. This leads to

$$\frac{\Delta S(T)}{K_B} = \frac{2}{\pi} \int_0^J \frac{dJ'}{J'^2} \int_{-\infty}^{\infty} \frac{[\lambda' - g_-(x)][1 - xg_+'(x)]}{|\lambda' - g_+(x)|^2 + \pi^2 S(S+1)} dx. \tag{6.8}$$

Then the J' integration is transformed into one over λ'. With $d\lambda' = -dJ'/J'^2$, $0 \geq J' \geq J \to -\infty < \lambda' \leq \lambda = \ln(T_K/T)$, we obtain [18]

$$\frac{\Delta S(T)}{K_B} = -\frac{2}{\pi} \int_{-\infty}^{\lambda} d\lambda' \int_{-\infty}^{\infty} \frac{[\lambda' - g_-(x)][1 - xg_+'(x)]}{|\lambda' - g_+(x)|^2 + \pi^2 S(S+1)} dx. \tag{6.9}$$

The universal function on the right-hand side still may be simplified by performing the λ' integral first where some care must be taken in the limit $\lambda' \to -\infty$.

The final result is

$$\frac{\Delta S}{K_B} = -\frac{2}{\pi} \int_{-\infty}^{\infty} dx [1 - xg_+'(x)] \left\{ \frac{1}{2} \ln \left[\frac{|\lambda - g_+|^2 + \pi^2 S(S+1)}{\lambda^2 + \pi^2 S(S+1)} \right] \right.$$

$$+ \frac{1}{4} \frac{\tanh(\pi x)}{[S(S+1) + \frac{1}{4}\tanh^2(\pi x)]^{1/2}}$$

$$\left. \times \left[\ln \frac{\lambda - \frac{1}{2}(g_+ + g_-) + i\pi \, (S(S+1) + \frac{1}{4}\tanh^2(\pi x))^{1/2}}{\lambda - \frac{1}{2}(g_+ + g_-) - i\pi \, (S(S+1) + \frac{1}{4}\tanh^2(\pi x))^{1/2}} - 2\pi i \right] \right\},$$

(6.10)

where $g_- - g_+ = i\pi \tanh(\pi x)$ has been used.

The integral in (6.10) expresses the additional entropy as a universal function of the variable $\lambda = \ln(T_K/T)$. For finite temperatures this function must be calculated numerically. However, it is more instructive to calculate the entropy change at $T = 0$ which may be done analytically [18]. One gets ($\lambda \to \infty$)

$$-\frac{\Delta S(0)}{K_B} = (2S + 2)\ln(2S + 2) + 2S \ln(2S) - 2(2S + 1)\ln(2S + 1).$$

(6.11)

This is a somewhat surprising result, as it seems to contradict a theorem by Mattis [23] according to which the impurity spin S should be reduced by a cloud of conduction electrons to $S - \frac{1}{2}$. This would lead to

$$\frac{\Delta S(0)}{K_B} = \ln \frac{2S}{2S+1}, \qquad \frac{S(0)}{K_B} = \ln(2S),$$

corresponding to a 2S-fold ground-state degeneracy [$S(0) = K_B \ln G$, $G = $ degeneracy]. The discrepancy does not necessarily indicate that (6.11) is a poor approximation. One has to exercise care in the way the limits $T \to 0$, volume $\Omega \to \infty$, are taken. Mattis' proof at $T = 0$ holds for finite volume, whereas in our thermodynamic calculation we use $\Delta S = \lim_{T \to 0} \lim_{\Omega \to \infty} \Delta S(T, \Omega)$. In the latter case, which is realistic, low-lying excitations presumably do not separate from the true ground state and thus may contribute as well. In fact, the result (6.11) indicates that the thermodynamic limiting state is a mixture of a singlet, a doublet, and a triplet for $S = \frac{1}{2}$.

The additional specific heat [18], $\Delta C = T \partial \Delta S/\partial T$, immediately follows from (6.9):

$$\frac{\Delta C}{K_B} = \frac{2}{\pi} \int_{-\infty}^{\infty} dx \, \frac{[\lambda - g_-(x)][1 - xg_+'(x)]}{|\lambda - g_+(x)|^2 + \pi^2 S(S+1)}.$$

(6.12)

6. PERTURBATIVE, SCATTERING, AND GREEN'S FUNCTION THEORIES

Again we get a universal function which has been calculated numerically by Müller-Hartmann [22]. The result is shown in Fig. 1. We have a typical low-temperature Schottky-type anomaly with a maximum near $\frac{1}{3}T_K$ and independent of the coupling constant. In the limit $[\lambda^2 + \pi^2 S(S+1)] \gg 1$, that is, for $T \to 0, \infty, S \to \infty$, the integral can be expanded and one gets the analytical result [18]

$$\frac{\Delta C}{K_B} \approx \frac{\pi^2 S(S+1)}{[\lambda^2 + \pi^2 S(S+1)]^2}. \tag{6.13}$$

In particular, $\Delta C \propto \ln^{-4} T$, $T \to 0$, which is a rather unusual result for a specific heat.

2. Ground-State Energy

The ground-state energy is not a quantity of particular interest from the experimental point of view. However, since many other theories essentially deal with ground-state properties, we shall briefly discuss the result within the framework of the theory of this paper. The calculation of Müller-Hartmann and Zittartz [24] is somewhat involved and will not be repeated here. We just note that in order to get a convergent result for the "normal" part [compare Eq. (6.3)] which is proportional to the bandwidth D one cannot work in the weak coupling limit as before; instead one must use a particular density of states function and the full t-matrix solution of Section IV. Starting out with formula (6.4) and exploiting the analytical properties of the solutions, one gets for integral spin S:

$$E(J) - E(0) = De(J) - (-1)^S K_B T_K (1 + O(J)) \theta(-J), \tag{6.14}$$

where

$$\theta(x) = \begin{cases} 1, & x > 0, \\ 0, & x < 0. \end{cases}$$

Thus for ferromagnetic coupling, $J > 0$, we have only a "normal" part $De(J)$. It has been shown [24] that the function $e(J)$ is analytic in J, which means that the perturbational power series expansion converges. For antiferromagnetic coupling, $J < 0$, one has in addition to the same analytical "normal" part an "anomalous" contribution which represents an isolated essential singularity at $J = 0$, $K_B T_K \approx De^{-1/|J|}$. The fact that this term oscillates in sign with S is rather amusing and presumably reflects no particular physical property. However, this behavior definitely rules out the interpretation that the "anomalous" term is the binding

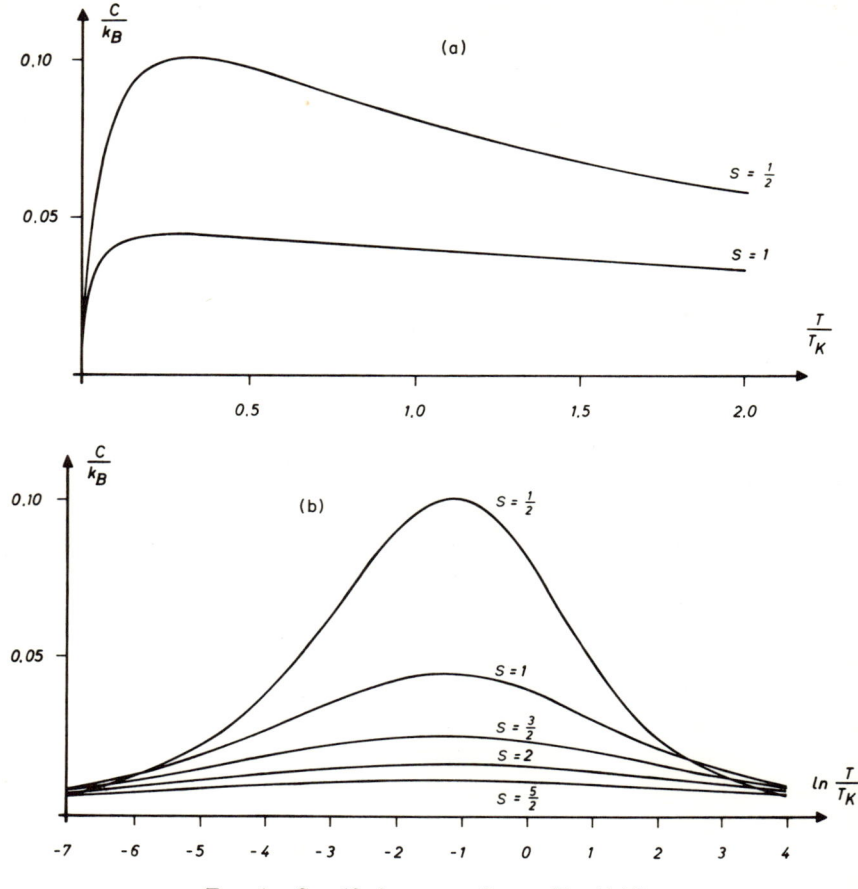

FIG. 1. Specific heat according to Eq. (6.12).

energy associated with an electron bound to the impurity spin as suggested in the Yosida–Yoshimori theory [25] and others.

For half-integral spin $S(S = \frac{1}{2}, \frac{3}{2},...)$ the expression for the ground-state energy actually can no longer be separated into a "normal" part and an "anomalous" part. In fact, if one writes

$$E(J) - E(0) = Dg(J), \qquad (6.15)$$

one proves [24] that the function $g(J)$ has a cut extending from $J = 0$ to infinity. This means that the function $g(J)$ has no convergent power series expansion, but diverges like $\sum_n n! J^n$. For details and discussion we refer to Müller-Hartmann and Zittartz [24].

VII. Electrical Conductivity

The dc conductivity can be related rather straightforwardly to the total scattering cross section. Since the scattering potential (2.1) is momentum independent, there are no vertex corrections to the current–current correlation function and the transport lifetime is equal to the single-particle lifetime. Thus

$$\sigma = \frac{ne^2}{m}\tau \tag{7.1}$$

with

$$\tau = \frac{1}{c}\int_{-\infty}^{\infty} d\varepsilon \left\{-\frac{\partial f}{\partial \varepsilon}\right\} \{-2\,\text{Im}\,Jt_+(\varepsilon)\}^{-1}. \tag{7.2}$$

Here we have expressed the total scattering cross section by the imaginary part of the forward scattering amplitude, using the optical theorem (3.7a). In general the integral (7.2) can only be done numerically. However, one obtains a reasonable estimate by taking the scattering amplitude $t(\varepsilon)$ at the Fermi surface $\varepsilon = 0$. Then the conductivity obeys, according to (5.10b),

$$\frac{\sigma(0)}{\sigma(T)} = \frac{1}{2}\left\{1 + \frac{\lambda}{[\lambda^2 + \pi^2 S(S+1)]^{1/2}}\right\}. \tag{7.3}$$

Another interesting quantity is the thermopower. It turns out that in our oversimplified model with a symmetric density of states functions $\rho(\varepsilon)$ and vanishing normal potential V in (2.1) the thermopower vanishes. However, if $V \neq 0$, one finds also a nonvanishing thermopower with an anomalously large value ("giant thermopower") at $T \approx T_K$ [5].

VIII. Magnetic Field Effects

The inclusion of an external magnetic field increases the mathematical complexity of the problem considerably. The results of the theory are therefore in most cases less satisfactory than those discussed in Sections V–VII. The calculations are in general more involved. Sometimes only numerical calculations exist. In this section we shall therefore confine ourselves mostly to the presentation of results and refer to the original literature concerning the details of the calculations. We also consider $J < 0$ only.

Two quantities have been studied rather extensively: (a) the magnetic field dependence of the resistance, and (b) the susceptibility.

1. Magnetoresistance

In order to study the magnetoresistance in Suhl's or Nagaoka's approximation one has to start from Eqs. (2.10), (2.12), and (2.13). The invariant decomposition of the scattering matrix $T(z)$ then becomes more involved. Instead of two invariant components t and τ as in (3.4) one now has six invariant components [26]. The Nagaoka decoupling procedure also becomes more complicated [27]. An analytical solution so far has been impossible in both cases. Only the low-field, low-temperature case is simple. In this case the magnetoresistance can be expressed in terms of t and τ alone. The analytical calculation leads to [28]

$$\rho_B(T=0) = \frac{1}{\sigma_B} \propto \left\{ |t(\omega_L)|^2 + |\tau(\omega_L)|^2 \frac{S(S+1)}{4} \right\} (1 + O(B^2)). \quad (8.1)$$

Here t and τ are the scattering amplitudes in an external field at the energy $\varepsilon = \omega_L = \omega_s = \omega_d$ (equal g factors being assumed). For general temperature and field numerical solutions of the dispersion equations [26] as well as the Nagaoka equation [27] exist. Some results are shown in Fig. 2. The physical origin of the *negative* magnetoresistance can be traced back to the combined effect of the Zeeman splitting of the Fermi level and the "quenching" of the Kondo resonance in the spin-flip scattering.

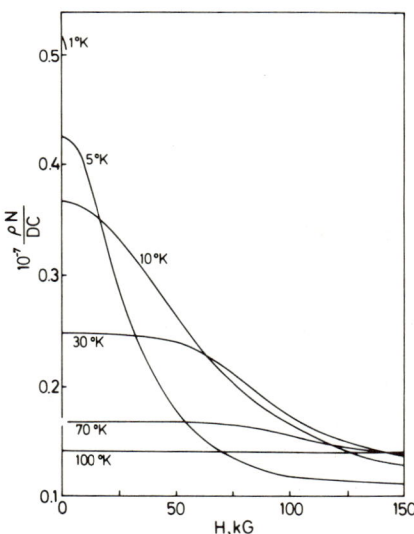

FIG. 2. Magnetoresistance according to Bloomfield *et al.* [27].

2. Magnetic Susceptibility

The linear response of the magnetization

$$M(\mathbf{r}) = \mu_B(g_s s^z(\mathbf{r}) + g_d S^z \delta(\mathbf{r})) \qquad (8.2)$$

to a magnetic field $B \propto \exp(i\mathbf{k} \cdot \mathbf{r} - \omega t)$ is described by the wave number and frequency-dependent susceptibility

$$\chi(\mathbf{k}, z) = \langle M_\mathbf{k} ; M_\mathbf{k}^+ \rangle_z , \qquad (8.3)$$

which is the analytical continuation of the Fourier transform of the retarded commutator

$$i\langle [M(\mathbf{r}, t), M(0, 0)] \rangle \theta(t) \qquad (8.4)$$

to complex frequencies z. The s–d interaction leads to Kondo anomalies in $\chi(\mathbf{k}, z)$ as a function of $\mathbf{k}$, z, and temperature T.

First we consider the static susceptibility in zero magnetic field, $\chi(0, 0)|_{B=0} \equiv \chi$

$$\chi = \tfrac{1}{3}\mu_B^2 \langle g_s \mathbf{S}_{\text{el}} + g_d \mathbf{S}; g_s \mathbf{S}_{\text{el}} + g_d \mathbf{S} \rangle_0 , \qquad (8.5)$$

which is generally given by

$$\chi = i \int_0^{-i\beta} \langle M_0(t) M_0(0) \rangle \, dt. \qquad (8.6)$$

Inserting (8.2) into (8.6) one may obviously decompose the magnetization correlation function into four spin–spin correlation functions

$$K_{ij} = \frac{i}{\beta} \int_0^{-i\beta} \langle \mathbf{S}_i(t) \cdot \mathbf{S}_j(0) \rangle \, dt$$

such that

$$\chi = \chi_d + \chi_s = \frac{\beta \mu_B^2}{3} \{g_d^2 K_{dd} + g_s g_d (K_{sd} + K_{ds}) + g_s^2 K_{ss}\}. \qquad (8.7)$$

Since the total spin $\mathbf{S}_{\text{el}} + \mathbf{S}$ is a conserved quantity in the s–d model we also have

$$K_{sd} + K_{dd} = K_{sd}^0 + K_{dd}^0 \qquad (8.8)$$
$$K_{ds} + K_{ss} = K_{ds}^0 + K_{ss}^0 ,$$

where $K_{ij}^0 = \langle \mathbf{S}_i \cdot \mathbf{S}_j \rangle$ are the equal time spin–spin correlations. It is instructive to consider the perturbation theoretical values of the correlation functions. They are given up to second order in Table I [29]. Here $K_P(K_C)$ is the zero-order Pauli (Curie) paramagnetic value for the $s(d)$ electrons. From the table one can read off the result for the static susceptibilities to second order

$$\chi_s - \chi_p = \beta O(T/D), \qquad \text{negligible},\tag{8.9}$$

$$\chi_d = \frac{\beta}{3}(\mu_B g_d)^2 [S(S+1) + \langle \mathbf{S} \cdot \mathbf{S}_{\text{el}} \rangle] = \frac{\beta}{3}(\mu_B g_d)^2 \left[K_{dd} + O\left(\frac{T}{D}\right) \right].\tag{8.10}$$

TABLE I

PERTURBATION THEORETICAL VALUES OF THE CORRELATION FUNCTIONS

	K	K^0
$K_{ss} - K_P$	$O(T/D)$	$S(S+1)J^2 \ln(D/T)$
$K_{sd} = K_{ds}$	$O(T/D)$	$-S(S+1)J^2 \ln(D/T)$
$K_{dd} - K_C$	$-S(S+1)J^2 \ln(D/T)$	0

A general nonperturbative derivation of (8.9) and (8.10) is given in Appendix B. The two different relations (8.10) for χ_d suggest two different interpretations of the logarithmic corrections to the Curie "constant," either as being due to an antiparallel s–d spin correlation at equal times or due to a rapid decrease of a parallel d–d spin correlation in time.

The first relation (8.10) suggests an evaluation of the susceptibility by using the Γ Green's function (4.2). Writing

$$\langle \mathbf{S} \cdot \mathbf{S}_{\text{el}} \rangle = \frac{2}{\beta} \sum_{\mathbf{k},n} \Gamma_{\mathbf{kk}}(z_n),\tag{8.11}$$

one may use (4.6) and the t-matrix solutions (5.5). The calculation is straightforward, though somewhat involved, and leads to the Curie "constant" [30, 31]

$$C = S(S+1) + \frac{1}{\pi^3}\int_{-\infty}^{\infty} dx$$

$$\times \frac{[g_1^2 + \pi^2 S(S+1)]\, g_1 g_0' + [g_0 - \lambda][\frac{1}{2}g_0'^2 - g_1'g_1^2] + \frac{1}{2}g_0'g_1'g_1}{(\lambda - g_0)^2 + g_1^2 + \pi^2 S(S+1)},$$
$$\tag{8.12a}$$

where $g_0(x) = \operatorname{Re} g_+(x)$, $g_1(x) = \operatorname{Im} g_+(x)$. This universal function of

$\lambda = \ln(T_K/T)$ may be calculated numerically. For $|\lambda| \to \infty$ one gets the asymptotic expansion

$$C = S(S+1) - \theta(\lambda)(S+\tfrac{1}{2}) + \frac{S(S+1)}{\lambda} + O(\lambda^{-2}). \tag{8.12b}$$

In the theories presented in Sections III and IV no quantity occurs from which one could calculate $\langle \mathbf{S}(t) \cdot \mathbf{S}(0) \rangle$ and thus could use the second relation (8.10). Extending the diagrammatic approach one may, however, introduce some further approximations. For instance, one can replace bare vertices in the second-order calculation of K_{dd} by renormalized vertices which contain both elastic and inelastic parts. As the inelastic vertices, or transition amplitudes, are beyond the scope of Suhl's approach, one may approximate them by the elastic ("on shell") values. Such a calculation then leads to [29, 32]

$$C = S(S+1)\left\{1 - \frac{1}{S+\tfrac{1}{2}}\left[\frac{1}{2} + \frac{1}{\pi}\arctan\frac{\lambda}{\pi(S+\tfrac{1}{2})}\right]\right\}. \tag{8.13}$$

At high temperatures, $\lambda \to -\infty$, both (8.12) and (8.13) agree with the perturbation expansion of χ up to third order in J.

As the two results (8.12) and (8.13) do not agree with one another, the problem of determining the susceptibility is not yet settled. In principle one could also determine the susceptibility thermodynamically by taking the second derivative of the free energy with respect to the magnetic field B and using formulas (6.1) and (6.4) generalized to $B \neq 0$. Since there exists no analytical solution of the problem and since the numerical solutions are not accurate enough, this procedure cannot be applied so far. If in Suhl's treatment all of the six (for $S = \tfrac{1}{2}$) invariant amplitudes are neglected except t and τ, one can find an analytical solution. For the magnetoresistance this t–τ approximation is reasonable, for the susceptibility χ it agrees with (8.13), the prefactor $(S + \tfrac{1}{2})^{-1}$ changed into $(S + \tfrac{1}{2})/[S(S+1)] - \ln(1 + 1/S)$ [28]. In the case of $S = \tfrac{1}{2}$, (8.12) yields more than complete screening at $T = 0$, (8.13) shows complete screening, and the t–τ approximation in the free energy gives incomplete screening.

The spatial dependence of the equal time correlation $\langle \mathbf{s}(\mathbf{r}) \cdot \mathbf{S} \rangle$ has also been investigated [31, 33, 34]. It exhibits a long-range nonoscillatory term $(1/J + \ln(k_f r))^{-2} r^{-3}$ besides the usual Rudermann–Kittel oscillations. This is in fact the origin of the $\ln T$ terms in the integrated function K^0_{sd}. Note that the corresponding static (zero frequency) correlation K_{sd} contains no such Kondo anomalies [29].

The frequency dependence of $\chi(\mathbf{k}, z)$ has been investigated in a series of papers [35–37]. The EPR linewidth can be described in terms of two width parameters

$$\gamma_s = 2\pi n_0 c \frac{S(S+1)}{3} |\tau(0)|^2, \qquad (8.14)$$

$$\gamma_d = \pi n_0^2 T |\tau(0)|^2 \qquad (8.15)$$

besides the spin–lattice relaxation times. The Kondo anomalies inherent in (8.14) and (8.15) due to the spin–flip amplitude $\tau(0)$ are masked in many cases by the spin–lattice relaxation "bottle neck." The bottleneck situation can, however, be broken by sufficiently large spin–lattice relaxation, large g-factor differences, or large $\mathbf{k}$ values [36]. At larger $\mathbf{k}$ values also the spin diffusion exhibits Kondo anomalies [37] due to the occurrence of the width parameter

$$\gamma = 2\pi n_0 c \left\{ |t(0)|^2 + |\tau(0)|^2 \frac{S(S+1)}{4} \right\}, \qquad (8.16)$$

which is closely related to the electrical resistance (7.2).

Further anomalies occur in the line shift and line shape which have been studied to second order in J [38, 39].

Appendix A. Diagrammatic Methods

The dispersion relations in Section III.1 have been derived only for $T = 0$. At nonzero temperatures one can give a derivation using infinite partial summations of the perturbation theoretical series for the scattering matrix. A straightforward application of diagrammatical methods is not possible. The occurrence of impurity spin operators besides the Fermion operators for the s electrons invalidates the simple form of Wick's theorem. Two methods have been successfully applied to the derivation of Suhl's diagrammatical equation.

1. A generalization of Wick's theorem and the linked diagram expansion theorem including spin operators [40].
2. The introduction of so-called pseudo-Fermions [42] for the impurities, carrying the spin.

Although, of course, the d-electron wave function actually not only depends on spin but also on spatial coordinates, the introduction of creation and annihilation operators presents some problems. The d states are strongly localized, and one does not want to treat this problem of localization explicitly in the s–d model. If one simply introduces Fermion operators for strongly localized states, one introduces spurious states with occupation numbers different from 1 (which are the only ones allowed in the s–d model). The effect of spurious states can be eliminated by considering an ensemble of impurities with a concentration $c \ll 1$ [13]. After averaging over the positions of the impurities the ensemble becomes translationally invariant so that a momentum $\mathbf{q}$ can be ascribed to the impurity intermediate states in spite of their localization. The only effect of localization is to make the impurity energy independent of momentum and the s–d vertex likewise (apart from a delta function describing conservation of total momentum)

$$\frac{J_{ex}}{N} \delta_{\mathbf{k}+\mathbf{q},\mathbf{k}'+\mathbf{q}'} \mathbf{S}_{\kappa\kappa'} \cdot \mathbf{s}_{\mu\mu'}. \qquad (A.1)$$

Then the standard finite-temperature Feynman diagrams can be used for perturbation theory. Each nth-order Feynman diagram can (again according to standard rules [41]) be decomposed into $n!$ time-labeled (Goldstone) diagrams.

The scattering matrix $T(z)$ can then be defined as the sum of all connected Goldstone diagrams with one ingoing and one outgoing s-electron line as well as one "upward-running" impurity line. The restriction to only upward-running lines corresponds to the neglect of corrections of order c due to impurity–impurity interactions.

A dispersion relation can then be derived diagrammatically by considering the discontinuity across the real energy axis of each Goldstone diagram. Since an nth-order diagram contains a product of $(n-1)$ energy denominators its discontinuity is a sum of $(n-1)$ terms. Each of these terms contains a delta function describing energy conservation at the intermediate lines. It can be represented by a "cut" crossing those lines. The cuts can be classified according to the number of intermediate lines being intersected. There are two possible types of two-line cuts (Fig. 3). It is easy to see [13] that after insertion of such contributions into a dispersion integral they produce a Kondo singular function $R(z)$ [Eq. (3.12)]. Higher cuts with more intermediate lines are less singular. Thus one generates a sum of diagrams containing all the leading powers of $R(z)$ (in nth-order R^{n-1}) if one iterates the diagrams of Fig. 3 in the following way. The nth-order contribution $T^{(n)}$ to $T(z)$ is found by inserting the quantities $T^{(i)}$ and $T^{(j)}$ with $i+j=n$ into the circles of Fig. 3 and summing over i from 1 to $n-1$. After summing over n and adding the Born term $T^{(1)}$ from (A.1) one arrives at Suhl's

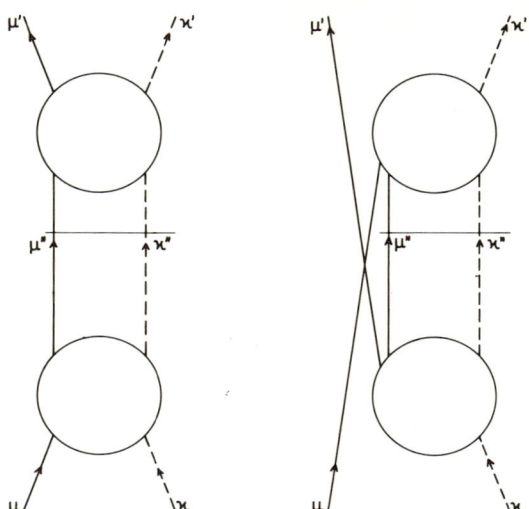

FIG. 3. The two single-particle cuts for T-matrix diagrams.

equations (3.5a), (3.5b) with $f(\varepsilon)$ being the Fermi function at temperature T.

Notice that Suhl's equations sum up neither an infinite partial series of Feynman diagrams nor of Goldstone diagrams. Up to third order in J all Goldstone diagrams are included. From fourth order on, however not only are some of the Goldstone diagrams omitted but some are taken into account only partially according to the "cut prescription" above.

At this point some remarks might be appropriate concerning the comparison of the pseudo-Fermions of Brenig and Götze [13] and Abrikosov [42]. Abrikosov introduces creation and annihilation operators for localized states directly. The effect of the spurious states is being eliminated by introducing something like a chemical potential (λ in his notation). The effect of the spurious states then can be "frozen out" by picking up the leading terms of order $\exp(-\lambda/T)$ for $\lambda \to \infty$. In doing so, however, one has to expand the numerator and denominator of the Green's functions separately, and it turns out that this procedure invalidates the straightforward use of Wick's theorem [40] (see also Keiter [43] and Perrier [44]. In the work of Brenig and Götze [13] the spurious states are eliminated by picking up the leading terms of order c for $c \to 0$. The chemical potential μ^i of the pseudo-Fermions in this method is not a formal quantity but has a physical meaning. It has to be determined so as to yield the correct number cN of pseudo-Fermions

6. PERTURBATIVE, SCATTERING, AND GREEN'S FUNCTION THEORIES

and thus is subject to renormalizations depending on the interaction. This is the only price one has to pay for being able to use Wick's theorem leading to standard Feynman and Goldstone diagrams. A simple reinterpretation of Abrikosov's λ as a true chemical potential, on the other hand, is not possible because his states belong to a single center. The particle number at such a center is a quantity of order unity undergoing large fluctuations.

Fortunately the above-mentioned details of differences between the two approaches do not show up in the derivation of Suhl's equations. They do become relevant, however, in higher order terms considered by Mattuck [45] and Abrikosov [46].

Appendix B. Electronic Susceptibility[†]

Let $t_{\uparrow,\downarrow}(z, B)$ denote non-spin-flip amplitudes for $B \neq 0$. Then using (4.3) and (4.4) one can write

$$\langle S_{\mathrm{el}}^z \rangle - \langle S_{\mathrm{el}}^z \rangle_0 = -\frac{1}{2\beta} \sum_{n,\mu} F'\left(z_n + \frac{\mu}{2} g_s \mu_B B\right) t_\mu(z_n, B), \quad (B.1)$$

where $\langle S_{\mathrm{el}}^z \rangle_0$ is the Pauli contribution. Linearizing in B the part $F''(z_n) t(z_n)$ in (B.1) may be dropped as the logarithmic terms of t at the Fermi surface are perfectly integrable and thus lead to corrections of order J^2. Then

$$\langle S_{\mathrm{el}}^z \rangle - \langle S_{\mathrm{el}}^z \rangle_0 = -B \frac{1}{\beta} \sum_n F'(z_n) \frac{\partial}{\partial B} [t_\uparrow(\infty, B) + (t_\uparrow(z_n, B) - t_\uparrow(\infty, B))]. \quad (B.2)$$

The contribution of $t_\uparrow(\infty, B) = -\langle S^z \rangle/2$ [see (2.12)] gives just the Knight shift $J\langle S^z \rangle/2$, which again is perturbationally small. To estimate the second integral we set $F'(z) \approx F(z)/D$ as both functions vary on the scale D. The remaining n sum converges uniformly and is bounded of order 1 such that the whole contribution is of order B/D. This proves (8.9). As the result holds independently of the values of the g factors g_s and g_d, we have separately

$$K_{sd} = K_{ds} = O(T/D); \quad K_{ss} - K_P = O(T/D). \quad (B.3)$$

As K_{sd} is negligible we have at once the second relation (8.10). The first relation (8.10) follows when we replace the negligible term g_s by g_d and use (8.8).

[†] See [30].

References

1. J. Kondo, *Progr. Theor. Phys.* **32**, 37 (1964).
2. H. Suhl, *Phys. Rev. A* **138**, 515 (1965).
3. Y. Nagaoka, *Phys. Rev. A* **138**, 1112 (1965); *Progr. Theor. Phys.* **37**, 13 (1967).
4. D. Schotte Z. *Phys.* **212**, 467 (1968).
5. K. Fischer, *Springer Tracts Mod. Phys.* **54**, 1 (1970).
6. A. A. Abrikosov, L. P. Gorkov, and I. E. Dzyaloshinski, "Methods of Quantum Field Theory in Statistical Physics." Prentice-Hall, Englewood Cliffs, New Jersey, 1963.
7. H. Suhl, *Physics (Long Island City, N.Y.)* **2**, 39 (1965); *Phys. Rev.* **141**, 483 (1966).
8. H. Suhl and D. Wong, *Physics (Long Island City, N.Y.)* **3**, 17 (1967).
9. H. Suhl "Theory of Magnetism in Transition Metals" *in*(W. Marshall, ed.), p. 116. Academic Press, New York, 1967.
10. S. V. Maleev, *Zh. Eksp. Reor. Fiz.* **24**, 1300 (1967); *Sov. Phys. JETP* **26**, 620 (1968).
11. M. H. Ross, "Quantum Scattering Theory." Indiana Univ. Press, Bloomington, 1963.
12. J. Zittartz, *Z. Phys.* **217**, 43 (1968).
13. W. Brenig and W. Götze, *Phys. Lett. A* **27**, 276 (1968); *Z. Phys.* **217**, 188 (1968).
14. J. Kondo, *Progr. Phys.* **40**, 659 (1968).
15. D. R. Hamann, *Phys. Rev.* **158**, 570 (1967).
16. D. S. Falk and M. Fowler, *Phys. Rev.* **158**, 567 (1967).
17. P. E. Bloomfield and D. R. Hamann, *Phys. Rev.* **164**, 856 (1967).
18. J. Zittartz and E. Müller-Hartmann, *Phys. Lett. A* **26**, 284 (1968); *Z. Phys.* **212**, 380 (1968).
19. N. I. Muskhelishvili, "Singular Integral Equations." P. Nordhoff, Groningen, Holland, 1953.
20. J. Zittartz, Lectures. *Finish Summer School*, 5th *Liperi, Finland*, 1969. to be published.
21. J. Kondo, *Solid State Phys.* **23**, 00 (1969).
22. E. Müller-Hartmann, Thesis, Univ. of Cologne, Cologne, Germany, unpublished, 1968; *Z. Phys.* **223**, 267 (1969).
23. D. C. Mattis, *Phys. Rev. Lett.* **19**, 1478 (1967).
24. E. Müller-Hartmann and J. Zittartz, *Z. Phys.* **225**, 455 (1969).
25. K. Yosida, *Phys. Rev.* **147**, 223 (1966); A. Yoshimori and K. Yosida, *Progr. Theor. Phys.* **39**, 1413 (1968).
26. R. Moore and H. Suhl, *Phys. Rev. Lett.* **20**, 500 (1968).
27. P. E. Bloomfield, R. Hecht, and P. R. Sievert, *Phys. Rev.* **2**, B 3714 (1970).
28. J. A. Gonzalez, Thesis, T. U. Munich, unpublished.
29. W. Brenig, J. A. Gonzalez, W. Götze, and P. Wölfe, *Z. Phys.* **235**. 52 (1970).
30. J. Zittartz, *Z. Phys.* **217**, 155 (1968).
31. E. Müller-Hartmann, *Z. Phys.* **223**, 277 (1969).
32. C. S. Ting, *Phys. Chem. Solids* **32**, 395 (1971).
33. H. U. Everts and B. N. Ganguly, *Phys. Rev.* **174**, 595 (1968).
34. H. Keiter, *Z. Phys.* **223**, 289 (1969).
35. W. Brenig, W. Götze, and P. Wölfle, *Phys. Lett. A* **30**, 448 (1969); *Z. Phys.* **235**, 59 (1969).
36. W. Brenig, W. Götze, and P. Wölfle, *Phys. Rev. B* **2**, 4533 (1970).
37. P. Wölfe, *Z. Phys.* **242**, 262 (1971).
38. W. Götze and P. Wölfle, *Z. Phys.* to be published.
39. H. J. Spencer and S. Doniach, *Phys. Rev. Lett.* **18**, 994 (1967).
40. H. Keiter, *Z. Phys.* **213**, 466 (1968); **214**, 22 (1968).

6. PERTURBATIVE, SCATTERING, AND GREEN'S FUNCTION THEORIES

41. G. Baym and A. M. Sessler, *Phys. Rev.* **131**, 2345 (1963) C. Bloch, *in* "Studies in Statistical Mechanics" (D. De Boer and D. Uhlenbeck, eds.), Vol. III. Amsterdam, 1964.
42. A. A. Abrikosov, *Physics (Long Island City, N.Y.)* **2**, 5 (1965).
43. H. Keiter, *Phys. Lett.* **36**A, 257 (1971).
44. J. Perrier, Thesis, Université de Paris, Orsay (1971), (unpublished).
45. C. Y. Cheung and R. D. Mattuck, *Phys. Rev.* **2B**, 2735 (1970).
46. A. A. Abrikosov and A. A. Migdal, *J. Low Temp. Phys.* **3**, 519 (1970).

7. Asymptotically Exact Methods in the Kondo Problem

P. W. Anderson

Bell Telephone Laboratories, Inc.
Murray Hill, New Jersey and
Cambridge University
Cambridge, England

G. Yuval[†]

Joseph Henry Laboratories
Princeton University
Princeton, New Jersey

I. Introduction 217
II. The Kondo Problem: A Discrete Path-Integral Approach 219
III. Eliminating the Fermi Gas 221
IV. The Method of Schotte and Schotte 224
V. Summing Up over the Paths 225
VI. Finite Temperatures 226
VII. Numerical Results 227
VIII. The Scaling Method 230
IX. The Kondo Temperature 232
X. Physical Implications 233
XI. "Renormalization Group" Methods in the Kondo Problem 233
References 235

I. Introduction

In 1964, Kondo [1] discussed the Born series for the scattering of the electrons in a Fermi gas by a single localized spin. At absolute zero the second-order term turned out to be infinite; at a finite temperature T, this term behaves like $\log T$. Thus, below a certain temperature—the Kondo temperature—the second-order term is larger than the first-order one, and the convergence of the Born series is in doubt. The behavior of the system when the interaction J approaches zero is thus

[†] Present address: Racah Institute of Physics, The Hebrew University, Jerusalem, Israel.

not a small variation of the behavior when $J = 0$. The trouble seemed to be that the system's behavior was not *understood*, and not that any great complication appeared in the mathematics.

One might have thought that these difficulties arose because the local spin was just postulated to exist, and its origin in the repulsion between electrons in a tunneling resonance [2] was ignored. However, the convergence difficulties due to Kondo appear also in the Anderson model, which is not open to that criticism. The approach to that problem by methods similar to those we discuss here is given in Chapter 8 by Hamann, and leads to essentially the same basic problem.

A great many *approximate* methods have been applied to the Kondo problem, several of which are described in this volume. While many of these give an adequate numerical account of the behavior in the region near and above the Kondo temperature, none seems to have led to an adequate qualitative understanding of the problem, specifically on two points:

1. Many authors (not all) agree that the state at $T = 0$ involves a bound compensating spin in the antiferromagnetic case, but the nature of this state and its connection to the finite T behavior were unclear; specifically whether $T = 0$ was a singular line was unknown.

2. Mathematically, $J \to 0^+$ is certainly a singular point, as can be seen most simply by noting that $J < 0$ (antiferromagnetic) and $J > 0$ (ferromagnetic) behavior are qualitatively different; the heart of the "Kondo problem" is the nature of this singularity (almost all methods specifically neglect terms of relative order $J\rho$ compared to the leading ones).

The methods discussed in this chapter have in common that, although they are only beginning to produce numerical results for comparison with experiment, the qualitative statements they make are not approximate in any real sense. That is, these methods do not modify the nature of the mathematical singularities, but do deal approximately to some extent with their coefficients. They are, then, to a real extent asymptotically exact formalisms in the $J\rho \to 0$ limit.

The most important and most powerful of these methods are the path-integral ones, which are the subject of the first ten sections of the chapter. These lead first to exact equivalences between the Kondo problem and certain classical one-dimensional statistical ones. These methods have been exploited to furnish computer results for the susceptibility in good agreement with experimental behavior. Secondly, the

† See Section II for definitions.

classical problem can be approached in certain regions to give a precise answer for the ferromagnetic case, and to demonstrate nonsingularity of the $T \to 0$ limit, as well as elucidating the real nature of the Kondo problem. Some other results, such as the mathematical nature of the ground-state energy, are available.

In Section XI we discuss the set of methods using the idea of the "renormalization group." Precise equivalences between different sets of parameters of the Kondo problem are established which can reproduce some of the results of the path-integral method.

II. The Kondo Problem: A Discrete Path-Integral Approach[†]

With the wisdom of hindsight, we can say that the Kondo problem [1] is a textbook exercise in Feynman's path-integral methods.[‡]

In the Hamiltonian of this problem, we have a noninteracting Fermi gas

$$H_0 = \sum_{k\sigma} \varepsilon_k a^+_{k\sigma} a_{k\sigma}, \qquad (2.1)$$

a localized spin $\mathbf{S}[S^2 = \frac{1}{2}(\frac{1}{2} + 1) = \frac{3}{4}]$, and an interaction between them

$$H_1 = J_z S_z \cdot s_z(0) = (J_z/2) S_z \cdot \sum_{k,k',\sigma} \langle \sigma \mid S_z \mid \sigma \rangle C^*_{k\sigma} C_{k'\sigma} \qquad (2.2)$$

$$H_2 = (J_\pm/2)[S_+ \cdot s_-(0) + S_- \cdot s_+(0)]$$

$$= (J_\pm/2) \left[S_+ \sum_{k\sigma; k'\sigma'}{}' \langle \sigma \mid S_- \mid \sigma' \rangle C^+_{k\sigma} C_{k'\sigma'} + hc \right]. \qquad (2.3)$$

The total Hamiltonian is $H_0 + H_1 + H_2$. For an isotropic system (to which we shall keep until Section V), $J_z = J_\pm$; we shall then call this number J. If it were not for H_2, the problem would be trivial—the spin-up electron would be scattered by a *scalar* potential due to S_z, the sign

---

[†] The Kondo problem's chief attraction for theorists has been the difficulty in understanding it, that is, in describing the system's behavior by a simple picture. Therefore we shall concentrate here on the physical ideas we introduce, referring the reader to the original articles for details of the mathematics.

[‡] Most of the path-integral techniques used for the Kondo problem have been developed from very similar, but simpler, methods for Mahan's x-ray edge problem [3]. In this problem, we have an infinitely heavy hole at the origin (with creation and annihilation operators b^+, b) and $H = E_h b^+ b + \sum \varepsilon_k a_k^+ a_k + b^+ b \cdot n(0)$, where $n(0)$ is the number of Fermions at the origin. Spin is ignored, and E_h is assumed to be very large.

Whenever a method is introduced for the Kondo problem, we shall, in a footnote, refer to the similar method for the x-ray problem (XRP).

depending on that of S_z, and so would the spin-down electron, but the sign would be opposite. The same result holds, of course, if H_1 were equal to $JS_x \cdot s_x(0)$ or $JS_y \cdot s_y(0)$. The sum of these last two terms is H_2 and we see that any attempt to solve the problem in a manifestly spherically symmetric manner will make the $H_0 + H_1$ problem look as difficult as the $H_0 + H_1 + H_2$ problem—the Kondo problem—which it is not. For this reason, keeping H_1 and H_2 separate seems preferable.

In classical mechanics, it is possible to find the behavior of a system by considering all its possible paths $X_i = X_i(t)$ (where X_i are the system's coordinates) between an initial configuration and a final one, calculating for each path a functional—the action—and choosing the path for which this functional is a minimum.

The quantum mechanical analog of this is Feynman's path-integral approach: We take all possible histories (paths) of the system, and give each an amplitude (which depends on the action integral along it). Instead of a wave function depending on the particles' position (and spins) at all times, integrals over the paths give us various properties of the system in a way similar to that in which integrals over the coordinates do this in ordinary quantum mechanics.

If we perform this path-integral calculation for the Kondo Hamiltonian, we can then consider all the paths of the system in which the paths $S_z = S_z(t)$ of the localized spin are identical, whatever the history of anything else may be. This partial set of paths describes the motion of the noninteracting electrons under the influence of a time-dependent single-particle Hamiltonian, due to their scattering by exchange interactions with the time-dependent spin $S_z(t)$. For this new problem, we no longer have to use path integrals; if we want to sum over the Fermi gas paths,[†] we only have to take its ground state, let the time-dependent scatterer due to S_z act on it, and project back to the ground state. The amplitude $\langle 0 \mid \exp[i \int H(t)\, dt] \mid 0 \rangle$ of this projection is then the amplitude for the relevant path $S_z = S_z(t)$ of the localized spin.[‡] Since all states of a spin are linear combinations of its spin-up ($S_z = \frac{1}{2}$) and spin-down ($S_z = -\frac{1}{2}$) states, the only possible paths for the local spin consist of alternate flips of S_z up and down at times t_i (where i has one parity when S_z flips up, and another when S_z flips down). This leaves us with two problems: (a) how to calculate the amplitude for each path of S_z; and (b) what to do with these amplitudes, once we get them.

Feynman [5] has shown how to eliminate harmonic oscillators (i.e., Bosons) from a problem, leaving only an effective interaction within the

---

[†] We assume that, in the initial and final configuration, $H_0 + H_1$ is in its ground state $\mid 0 \rangle$ with $S_z = +\frac{1}{2}$.

[‡] See Noziéres and de Dominicis [4] for this approach in the XRP.

7. ASYMPTOTICALLY EXACT METHODS IN THE KONDO PROBLEM

remainder of the physical system[†]—in the Kondo problem, this would mean a (noninstantaneous) self-interaction of S_z with itself. If the resulting self-interaction of S_z were instantaneous, the problem would be trivial. It is the interaction's noninstantaneous nature that forces us to work with path integrals rather than wave functions.

III. Eliminating the Fermi Gas[‡]

Because of the conservation of spin, an electron in the Fermi sea must flip its spin up whenever the local spin flips down, and vice versa. Thus, if the $S_z s_z$ interaction term $H_1 = J_z S_z \sum_{k\sigma}$ were not there, we would only have to follow the behavior of the two Fermi gases—one for $s_z = \frac{1}{2}$ and one for $s_z = -\frac{1}{2}$, when creation and annihilation operators[§] at the origin are applied to them whenever S_z flips. If we start with the ground state, and project upon the ground state at the end, we get the product of two many-electron Green's functions; this product must be multiplied by $(J_{\pm/2})^{2n}$ (for $2n$ spin flips) because each S term in H_2 has the coefficient $(J_{\pm/2})$ in front of it.

That is, the amplitude for this path is

$$\langle 0 | H_2(t_1) H_2(t_2) H_2(t_3) \cdots H_2(t_m) | 0 \rangle$$
$$= (J_{\pm/2})^{2n} \langle 0 | \psi_\uparrow^+(t_1, r=0) \psi_\uparrow(t_2, r=0) \psi_\uparrow^+(t_3, r=0) \cdots | 0 \rangle$$
$$\times \langle 0 | \psi_\downarrow(t_1, 0) \psi_\downarrow^+(t_2, 0) \cdots | 0 \rangle. \tag{3.1}$$

Since the particles are free, each of these Green's functions reduces (for a path with $2n$ spin flips) to a sum of $n!$ products of one-electron Green's functions

$$\langle 0 | \psi_\sigma^+(t', 0) \psi_\sigma(t'', 0) | 0 \rangle.$$

This sum of $n!$ products of n Green's function turns out [7] to be a determinant.

[†] The most familiar example is eliminating the photons from quantum electrodynamics, leaving a delayed (or advanced) interaction between the charges. This method is also used for the polaron problem.

[‡] This approach was used in the XRP by Noziéres and de Dominicis [4]. It was the first [6] path-integral method applied to the Kondo problem. After seeing these results, Hamann and Schrieffer and Evenson applied similar methods to the Anderson model; these are the subject of Chapter 8 of this volume.

[§] Flipping an electron's spin up is equivalent to creating a spin-up electron and annihilating a spin-down one.

In order to proceed, we need the form of the one-electron Green's function. As we only need this function when the creation and annihilation operators in it are at the origin, we shall ignore its behavior in the rest of (three-dimensional) space. Thus $\mathbf{r} = 0$ will be assumed from now on. The real physical exchange interaction has a form factor, but we shall limit the bandwidth rather than introduce a $J(r)$.

If we assume an infinite structureless band, the behavior of $G_0(E, r = 0)$ is clear:

$$\text{Im } G = \int \delta(E - E_k) \cdot \text{sgn}(E_k - E_F) \cdot n(E_k) \, dE_k \tag{3.2}$$

and hence it has one value for $E > E_F$, and the opposite value for $E < E_F$. If we put $E_F = 0$, we have Im $G_0(E) \pm \text{sgn}(E)$. Re G is indeterminate but may be set equal to zero. Fourier transforming, we find up to a scale factor

$$G_0(t) = 1/it. \tag{3.3}$$

In order to keep expressions real, we shall here and henceforward use imaginary times, $t = -i\tau$. Then

$$G_0(\tau) = 1/\tau,$$

and the determinant of n Green's functions is

$$\left| \frac{1}{\tau_{2i} - \tau_{2j-1}} \right|.$$

For a finite band, G_0 will have a short-time cutoff due to the bandwidth. For instance, for a rectangular band of width $2/t_0$, symmetric about the Fermi surface, we find

$$G_0(\tau) = \frac{1 - \exp(-|\tau/t_0|)}{\tau}.$$

This short-time cutoff is essential to keep the expression finite, and plays an important role in later developments. The determinant of free Green's functions equals [8]:

$$D = \frac{\prod_{i<j} (\tau_{2i} - \tau_{2j})(\tau_{2i-1} - \tau_{2j-1})}{\prod_{i,j} (\tau_{2i} - \tau_{2j-1})}. \tag{3.4}$$

The effect of a short-time cutoff in G on D is to impose such a cutoff on D too† (ref. [7]). There are two such determinants: one for the spin-up electrons, and one for the spin-down ones. Then, if it were not for the time-dependent potential scattering due to S_z, D^2 would be the amplitude of the path in which S_z flips at the times t_i.

Because of the potential scattering due to S_z, the Green's function in the electron gas differs from the free electron function. However, it is still possible to find a closed-form expression for $G(\tau, \tau')$.

The Green's function obeys Dyson's equation

$$G(\tau, \tau') = G_0(\tau - \tau') + J \int_{-\infty}^{\infty} G_0(\tau - \tau'') S_z(\tau') G(\tau'', \tau') d\tau''$$

$$\cong \frac{1}{\tau - \tau'} + J \int_{-\infty}^{\infty} \frac{1}{\tau - \tau''} S_z(\tau'') G(\tau'', \tau') d\tau''. \quad (3.5)$$

This is a singular integral equation of the type studied by Muskhelishvili [9]; for the case when S_z flips between two values, we find [7]

$$G(\tau, \tau') = \frac{1}{\tau - \tau'} \prod_i \left(\frac{\tau_i - \tau}{\tau_i - \tau'} \right)^{\pm \delta/\pi}, \quad (3.6)$$

where the $\pm$ sign depends on whether S_z flips up or down at time t, and δ is the scattering phase shift due to J_z.

Therefore, the effect of the time-dependent $S_z(t)$ term H_2 is to multiply each of the $n!$ products of n Green's functions by

$$\left[\frac{\prod_{i<j} (\tau_{2i} - \tau_{2j})(\tau_{2i-1} - \tau_{2j-1})}{\prod_{i,j} (\tau_{2i} - \tau_{2j-1})} \right]^{2(\delta/\pi)} = D^{2(\delta/\pi)}. \quad (3.7)$$

Thus D is multiplied by $D^{2(\delta/\pi)}$.

There is one other effect due to S_z It perturbs the ground-state Fermi sea. Indeed, the two many-body ground states of $H_0 + H_1$ with $S_z = +\tfrac{1}{2}$ and with $S_z = -\tfrac{1}{2}$ are orthogonal [10] for any value of J; thus a sufficiently long interval in which S_z is reversed will cause the overlap

$$\langle 0 | \exp \left[i \int H(t) dt \right] | 0 \rangle$$

to be very small.

---

† It is shown in ref. [7] that, although the cutoff modifies the Cauchy nature of the determinant when times are too close together, (3.4) holds for all widely separated times; it is asymptotically exact in the sense of the introduction. That is, as $J_\pm \to 0$, the important paths have fewer and fewer t's, and thus $t_i - t_{i-1} \to \infty$. But it will turn out that nearest neighbor times are not very important anyhow.

One should note that this effect appears without any reference to the creation and annihilation operators. This effect can be included [4], [7] by summing up over the closed-loop diagrams (Fig. 1). This

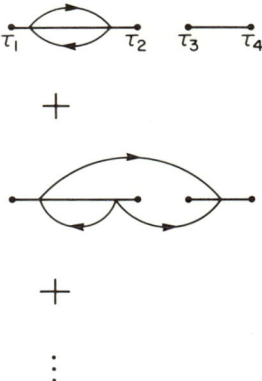

FIG. 1. The closed-loop diagrams.

sum can be obtained from $G(\tau, \tau)$, as done by Yuval and Anderson [7], and the result is to multiply D further by $D^{(\delta/\pi)^2}$. Therefore, the amplitude corresponding to the path is

$$D^{2[(\pi+\delta)/\pi]^2} = \left[\frac{\prod_{i<j}(\tau_{2i} - \tau_{2j})(\tau_{2i-1} - \tau_{2j-1})}{\prod_{i,j}(\tau_{2i} - \tau_{2j-1})}\right]^{2[(\pi+\delta)/\pi]^2}. \quad (3.8)$$

Since $J/2$ appears with each $S_\pm s(0)_\pm$ operator, we have to multiply this last expression by $(J/2)^{2n}$.

IV. The Method of Schotte and Schotte

Schotte and Schotte [11] transformed the Fermi gas into a Bose gas, using the method due to Tomonaga [12].

Since the interaction terms (H_1 and H_2) act on the Fermi gas at the origin, we are only interested in s electrons; those with higher orbital angular momentum do not interact with S. The s electrons form a one-dimensional gas; in the neighborhood of the Fermi surface, the energy E is a linear function of the momentum k. This is where we again make the asymptotic assumption that J is small, and that we need treat only electrons near the Fermi surface accurately.

7. ASYMPTOTICALLY EXACT METHODS IN THE KONDO PROBLEM

If we consider density wave operators in this one-dimensional gas

$$\rho_q = \sum_{E_k < E_F} a^\dagger_{k+q} a_k$$

$$E_{k+q} > E_F,$$

we find

$$[H, \rho_q] = vq\rho_q, \tag{4.1}$$

where v is the Fermi velocity. The ρ_q obey Bose statistics so long as $q \ll k_F$ and the Fermi gas is only weakly excited.

Having transformed the Fermi gas into a Bose gas, we have to transform the electron creation and annihilation operators (due to flips of S_z) into Boson operators. Since a change of π in the scattering phase shift causes an extra bound state to appear (or disappear), without otherwise affecting the scatterer, we expect such a change in the scattering phase shift at the origin to imitate the creation and annihilation operators fairly well. This does indeed happen [11], and the effect of the electron creation and annihilation operators is to replace the weak scatterer $\pm J$, giving a phase shift $\pm \delta$, by a much stronger scatterer with a phase shift of $\pm(\pi \pm \delta)$.

Now that we have a time-dependent potential exciting an array of harmonic oscillators, we can use Feynman's original expression [5] for the resulting amplitude due to each oscillator, and find the amplitude for any path $S_z(t)$ by multiplying them together. Since these amplitudes are each the exponential of a bilinear expression in the perturbing potential, we will find $(\pi \pm \delta)^2$ appearing as a part of an exponent in the expression for the amplitude. Summing up over the oscillators, Schotte and Schotte find an amplitude

$$(J_\pm/2)^{2n} \left[\frac{\prod_{i<j} (\tau_{2i} - \tau_{2j})(\tau_{2i-k} - \tau_{2j-1})}{\prod_{i,j} (\tau_{2i} - \tau_{2j-1})} \right]^{2[(\pi+\delta)/\pi]^2} \tag{4.2}$$

as in (2.3) $[(J_\pm/2)^{2n}$ being again due to the $(J_\pm/2)S_\pm \cdot s_\mp(0)$ term in $H_2]$.

V. Summing Up over the Paths

We now have an amplitude for each path of S_z (over imaginary time). These amplitudes are all real and positive.† Moreover, the amplitudes are products of terms

$$(\tau_i - \tau_j)^{\pm 2[(\pi+\delta)/\pi]},$$

---

† This only holds if we work on the imaginary-time line, which is why we work there.

involving two times each. Therefore we associate an "energy" E with each path, so as to make its amplitude equal to[†] $e^{-\beta E}$, and we find that E is a sum of two-body logarithmic interactions between the flips. That is, we may write

$$E = \sum_{i,j} (-)^{i-j} \cdot (2 - \varepsilon) \ln(\tau_i - \tau_j). \tag{5.1}$$

If we now regard the amplitudes as if they were probabilities, we find the Kondo problem equivalent to the statistical mechanics of a gas of *classical* particles (the flips at times t_i) on a straight line, interacting via a logarithmic potential, and with a chemical potential $\ln(J_{\pm/2})$ (since a path with $2n$ flips has a factor $J_\pm^{2n}$ in its amplitude).

Integrating by parts twice, the interaction "energy"

$$\iint_{|\tau-\tau'|>\tau_0} d\tau\, d\tau'\, \ln(\tau-\tau')\, \frac{dS_z}{d\tau}\, \frac{dS_z}{d\tau'} \tag{5.2}$$

between flips is equivalent, so long as no two flips occur within a time τ_0 of each other, to an interaction

$$\iint_{|\tau-\tau'|>\tau_0} d\tau\, d\tau' \cdot \frac{1}{(\tau-\tau')^2} \cdot S_z(\tau) \cdot S_z(\tau') \tag{5.3}$$

between the spins S_z at various (imaginary) times. Thus instead of a logarithmic interaction between the spin flips, we have an inverse-square interaction between the spins $S_z(t)$ at any two times. If we had let τ_i, τ_j approach each other arbitrarily close (corresponding to a really infinite band), we might have had trouble about convergence, since the integral $\iint d\tau\, d\tau'[-(2+\varepsilon)\ln(\tau-\tau)']$ has a divergence for $\tau-\tau' \ll 1$ if $\varepsilon > 0$. If we impose a cutoff, say that $|\tau_i - \tau_j|$ must always be greater than a minimum t_0 (which is approximately the inverse bandwidth, because of the uncertainty relationship $\Delta t\, \Delta E > h$), these divergences disappear, and we have a finite integral; instead of diverging, it gives us a contribution that behaves (in the region of interest) like $\ln t_0$. We recall that the Kondo problem arose because of difficulties in perturbation theory for arbitrarily small J, that is, in the region where flips are rare, and where $\varepsilon = \delta/\pi + 2(\delta^2/\pi^2)$ is very small, so that the logarithmic divergences at long range dominate the behavior of the classical gas.

Logarithmically divergent integrals depend only very weakly on the detailed shape of the cutoff, and we therefore expect the thermodynamics

---

[†] β^{-1} is a "temperature" which we shall choose to be unity.

of the system to remain essentially unaltered if, instead of the τ_i being continuous variables subject to the condition $|\tau_i - \tau_j| > t_0$, we make all the τ_i different integer multiples of t_0. Let us consider then a one-dimensional Ising chain specified by

$$E = \sum_{i,j} \frac{J_{LR}}{(i-j)^2} S_i S_j + \sum_i J_{NN} S_i S_{i+1}. \tag{5.4}$$

It is shown [13] that this is mathematically equivalent, as far as the long-range interactions between spin reversals along the chain are concerned, to the hard-rod model above, and thus also the spin $\frac{1}{2}$ Kondo problem. The formulas giving the correspondence are

$$\left|\frac{J_{\pm} t_0}{2}\right| = \exp(-\beta[J_{NN} + (1+C) J_{LR}])$$

$$2\beta J_{LR} = 2 - \varepsilon \cong 2 - 2 J_z t_0, \tag{5.5}$$

where C is Euler's constant.

Unfortunately, this Ising problem remains the subject of controversy in the literature, although a solution based on our work on the Kondo problem is now beginning to be accepted.

VI. Finite Temperatures

The methods used in the last two sections to study the ground-state behavior of the Kondo problem can also be used for the finite-temperature case. The approach is the same with both methods: We impose the requirement that the (imaginary-time) behavior be periodic modulo $i\beta = i/kT$. Another way of formulating it is to limit the spin flips, or the Ising spins, to a line of length $i\beta$, and have the interaction behave as

$$\log\left(\frac{\beta}{\pi} \sin \pi \frac{\tau_i - \tau_j}{\beta}\right)$$

instead of $\log(\tau_i - \tau_j)$ between the flips, or as

$$\frac{\beta^2/\pi^2}{\sin^2(\pi/\beta)(\tau_i - \tau_j)}$$

instead of $1/(\tau_i - \tau_j)^2$ (between the Ising spins). We can summarize the equivalence between the Kondo system and the two classical systems as shown in Table I. Since no finite system can exhibit a rigorous phase

TABLE I

COMPARISON OF THE KONDO SYSTEM WITH THE TWO CLASSICAL SYSTEMS

Kondo system	Charged-rod model of the flips	Ising model
β	System is on circle of circumference β	System is on the circle of circumference β
J_z	Logarithmic interaction proportional to $1 + J_z t_0$	Inverse-square interaction proportional to $1 + J_z t_0$
$J_\pm$	Chemical potential proportional to $\ln \mid J_\pm \mid$	Nearest-neighbor interaction proportional to $\ln \mid J_\pm \mid$
Scaling factor in the interactions	Temperature	Temperature

transition, we only expect nonanalytic dependence on J_z, $J_\pm$ in the properties of the ground state (corresponding to an infinite system), and not at any finite temperature (corresponding to a finite system).

The finite-temperature behavior of the Kondo problem is thus equivalent to the thermodynamics of a *finite* classical system, as against the infinite system, found for the ground-state behavior. Since it takes a system of a certain size to tell whether we are above or below the phase transition "temperature," we expect to find a sharp difference between the antiferromagnetic Kondo system (for which we shall show the classical system to be uncondensed) and the ferromagnetic system (for which the classical system is condensed) only at $T = 0$. One way of defining the Kondo temperature is as the temperature at which large differences between the two systems show up.

VII. Numerical Results

Using Monte Carlo or other computational methods, it is possible to go directly from the classical models to properties of the Kondo system. One such calculation has been made by K. D. Schotte [14], who has been kind enough to let us present his results. The magnetic susceptibility at any temperature may be directly related to the $\langle S_z(0) S_z(\tau) \rangle$ correlation function, which is calculable in terms of charged rods on a ring. One approximation is made: a canonical rather than grand canonical distribution of the number of spin flips; but this is increasingly unimportant as the temperature is lowered. The basic limitation on computation time is that as $\beta \to \infty$ ($T \to 0$) the ring becomes infinitely large; in practice no more than 50 spin flips were treated.

Three separate values of $J\tau_0$ were treated ranging from 0.175 [$T_K = \exp(-1/J\tau_0) = 3.3 \times 10^{-3}$] to 0.225 ($T_K = 12.4 \times 10^{-3}$). Since

7. ASYMPTOTICALLY EXACT METHODS IN THE KONDO PROBLEM

these do not differ greatly, we have placed them on the same graph (Fig. 2) along with results of Ting [15] from the Suhl equations, for contrast between the exact and the approximate methods. The two agree very well at T_K. Above T_K the deviation comes about because Ting's formulas neglect $J\rho \cong J\tau_0$ in the sense that terms of relative order $J\tau_0 \ln(T/T_K)$ are explicitly neglected, whereas no essential approximation occurs in the Schotte results.[†]

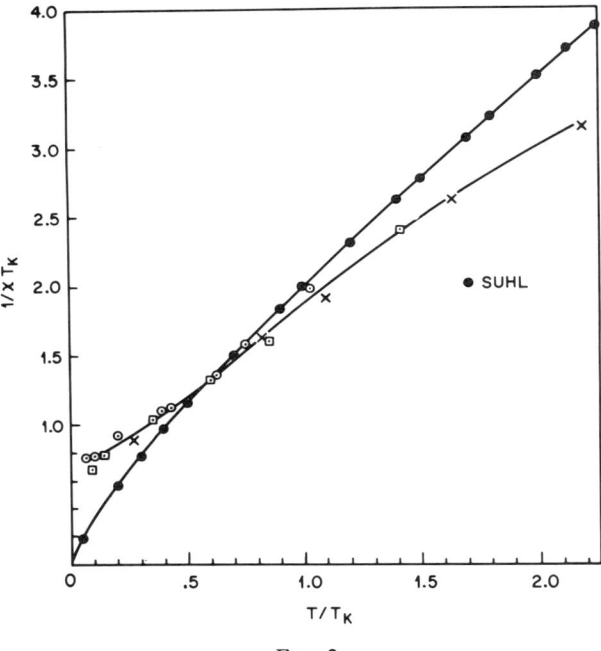

FIG. 2.

Below T_K the Suhl results, and those of almost all other methods as well which are valid at high T, lead to a divergent susceptibility, and thus $1/\chi \to 0$, usually with vertical slope. Schotte's results curve upward instead, as they must if a bound state is to be reached and Nernst's theorem is valid (as it need not necessarily be!). It is noticeable that the region of interest near T_K is almost linear, in agreement with the majority of experimental results, which all agree that

$$\chi \simeq \frac{C}{T + \text{``}T_K\text{''}}. \tag{7.1}$$

[†] Mr. John Armytage has found that actually, Schotte's results are not accurate at higher T because of the microcanonical assumption, so that only the lower T differences are germane.

This is a consequence of the inflection point which must separate negative curvature at high T/T_K from positive at low T/T_K.

VIII. The Scaling Method

From previous work on the Kondo problem (in particular Mattis [16]) it seems clear that the behavior of the system is entirely different in the ferromagnetic case ($J > 0$) and in the antiferromagnetic case ($J < 0$). For such a dramatic change to occur in the thermodynamics of the classical gases discussed in Section V, we have to have a phase transition separating the $J > 0$ and $J < 0$ regions. The Kondo problem, which is really a problem of very low J values, is thus closely related to the critical-point behavior of the equivalent classical system.

In order to look for such a phase transition, we want to normalize away the fluctuations due to pairs of flips near each other (Fig. 3). This renormalization process turns out to be the crucial point of the solution[†]: We use the model in which the τ_i are continuous variables with $|\tau_i - \tau_j| < t_0$, and consider the pairs of flips (we shall call them *close pairs*) with $t_0 < |\tau_i - \tau_j| < t_0 + dt$. These close pairs are very rare (since dt is infinitesimal), and thus they do not interact with each other. Since they are small, and the other spins are usually far away from them[‡], it is a good approximation to assume that they appear uniformly along the t line. Therefore, we can ignore (i.e., renormalize away) all these pairs, so long as we adjust the interaction of the flips that remain; this adjustment is necessary, because a region that looks (neglecting the close

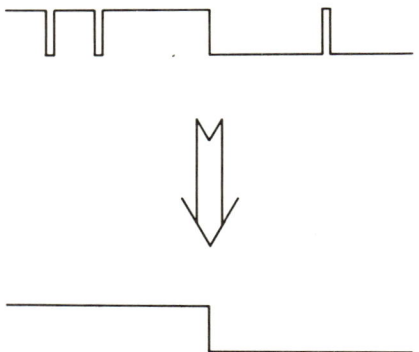

Fig. 3. Renormalizing the fluctuations away.

[†] See Anderson et al. [17] for details of the mathematics.
[‡] The interaction between a close pair and any other flips is weak, because the two flips in the pair give terms that nearly cancel each other out.

pairs) as if $S_z = \frac{1}{2}$ all along it has a finite probability of having in it an $S_z = -\frac{1}{2}$ region of length t_0; thus all interactions between the flips are multiplied by a factor.

In this way, we can renormalize t_0 upward in infinitesimal steps,[†] and in the process change the values of the various interactions in the classical system. To follow the behavior of the classical gas under the renormalization any further, it is easier to consider isotropic and anisotropic Kondo models at the same time; J appears twice in the Hamiltonian, in $H_1 = J_z S_z \cdot \sum \langle \rangle$ and in $H_2 = J_{\pm}/2 \cdot (S_+ \sum \langle \rangle + \text{h.c.})$. If $J_\pm \neq J_z$, we have two parameters to replace J: In the classical gas, J_z gives the strength of the interaction between the spin flips, while $J_\pm$ gives their chemical potential.

When we change t_0 continuously in the renormalization process, $|J_\pm|$[‡] and J_z move over a two-dimensional diagram as in Fig. 4

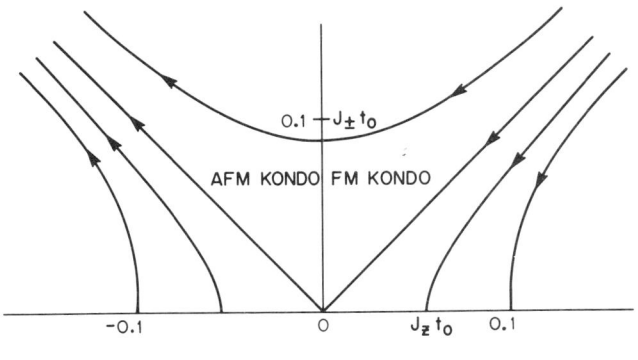

FIG. 4.

(this figure is reversed with regard to the figures of Anderson et al. [17]).

Thus, if we start in the region $J_z > |J_\pm|$, the renormalization process carries the system into one on the X axis, on which $|J_\pm| = 0$. Such a system cannot have any spin flips, and corresponds to a completely condensed gas. For $J_z < |J_\pm|$, renormalization will ultimately bring us to a region where $|J_\pm|$ grows larger and larger; therefore, in that region the flips become more and more frequent (on the scale of the renormal-

---

[†] Since t_0 is approximately the inverse bandwidth, this renormalization is equivalent to a downward renormalization of the bandwidth. The latter can be performed without reference to the path-integral method; see Section XI.
[‡] The sign of $J_\pm$ has no effect on the mathematics.

ized t_0), and the system is clearly uncondensed.[†] The phase boundary is on the isotropic ferromagnetic line $J_z = |J_\pm|$, but states on this line behave as if they were below the transition.

The phase transition in this model (and in the Ising model of Section V) has a discontinuity in the order parameter (i.e., the magnetization). Thouless [18] has already shown that such a discontinuity must occur here if there is a phase transition at all; we find that the order parameter just below the transition has precisely the minimum value it can have according to Thouless. Dyson [19] has rigorously shown such a phase transition for a special case of his hierarchical model. Several other properties of the phase transition are also anomalous. In our previous work [13, 17] we have discussed the thermal properties of the infinite Ising model, which correspond to the ground-state energy of the Kondo problem, and demonstrated that its analytic behavior agrees with that calculated from perturbation theory by Kondo.

IX. The Kondo Temperature

We have previously mentioned that the Kondo effect has a characteristic temperature scale—the Kondo temperature. It should be emphasized that the Kondo temperature is *not* the temperature of a phase transition; it is only a characteristic temperature for a gradual change in the properties of the system. Using the scaling method, this characteristic temperature can be given a precise meaning.

When we renormalize τ_0 upward, β/τ_0 scales downward. As we have seen in [17], there is a precise equivalence between two Kondo systems with different values of β/τ_0, if J is scaled accordingly.[‡] Using this equivalence in the opposite direction, two Kondo systems with different values of J (but the same sign) are equivalent to each other if we consider them at two *different* temperatures, the ratio between which depends on the J values. There is thus a natural unit of temperature, depending on the system (a Kondo temperature), such that in terms of this unit, all ferromagnetic Kondo systems have the same behavior, and so do all antiferromagnetic systems. This Kondo temperature is found to be

$$T_K = \exp(-1/J\tau_0).$$

[†] Indeed, if we continue the scaling sufficiently far into the left-hand side of the diagram, we reach a point where the interaction between flips gives a $(\tau_i - \tau_j)^{\pm 1}$ term in the amplitude [rather than $(\tau_i - \tau_j)^{\pm 2}$ in the $J \to 0$ limit]. As Toulouse pointed out, this point corresponds to the path-integral problem for the spin-up electron tunneling in and out of a resonant state, without any spin-down electrons, which is exactly soluble.

[‡] A change of both β and τ by the same factor has no effect on the mathematics, if $J\tau$ is kept constant.

X. Physical Implications

We have found out that, in the path-integral formalism, the ferromagnetic and antiferromagnetic Kondo systems correspond to the statistical mechanics of a classical system just above and just below its phase transition, respectively. To see the implication of this for the original quantum mechanical problem, we consider the long-range order.

In the ferromagnetic case, there is infinite-range order, and if $S_z > 0$ at one point, the expectation value of S_z will remain positive infinitely far from this point. Since the space dimension in the classical system is the time line, this means that S_z has a memory for arbitrarily long time intervals. This agrees with a triplet ground state: If the total angular momentum points in the positive z direction, S_z is always more likely than not to be positive.

In the antiferromagnetic case, S_z has no infinite-time memory; this agrees with a singlet ground state, which is nondegenerate, and has no preferred direction in space. Since, for small J, the system is just above its transition, we have correlations (i.e., spin memory) extending over extremely long regions (i.e., time intervals), and above the temperature corresponding to the correlation range (which is again the Kondo temperature), we can ignore the singlet nature of the ground state.

Far enough below the temperature, the system to which the physical one is equivalent is one with a *large* $J\tau_0$ and a singlet ground state. Near zero temperature it behaves, then, simply like a localized spin fluctuation (LSF), but so far the types of theories which have been applied to that problem are incapable of handling the essential stage of connecting together the LSF and magnetic regimes, as we can.

XI. "Renormalization Group" Methods in the Kondo Problem

Another group of methods have been applied to the Kondo problem which also have the (as yet incompletely exploited) possibility of retaining intact the analytical nature of the problem, and which lead at first to a very similar scaling procedure to that already derived. Since the results do not seriously differ (insofar as they remain exact) from those already discussed, this section is rather brief and merely serves to call attention to these other approaches.

The papers we discuss here are those of Anderson [20] and Fowler and Zawadowski [21]. These papers have different emphasis and points of view, although there is a close initial similarity. Anderson's is very much simpler mathematically and easier to follow; Fowler and Zawadowski use the full panoply of the renormalization group method and try to achieve a complete solution, but make certain approximations which Anderson avoids in principle if not in practice.

Let us first sketch the simple Anderson method. The idea here was to carry out a *projection* of the problem with an upper cutoff energy D onto the set of states appropriate to a problem with lower cutoff energy $D - \varepsilon$. When the problem is reexpressed in terms of the exact resolvent operator $G = (E - H)^{-1}$, this projection can be done by summing over perturbation theory diagrams involving the high-energy states between D and $D - \varepsilon$, which will be of order $(\varepsilon^n/D^n)J^n$ for a diagram of nth order. Clearly only the lowest order is relevant since we choose ε at will.

In a general problem such a procedure will lead to a small change in the effective interaction $V \to V + VP_\varepsilon V/D$, which is not useful because it is more complicated than the original V. Here, however, we find that to a very good approximation the new interaction is almost exactly like the old one. This is what makes it possible again to write down scaling differential equations like those obtained by the fluctuation renormalization method:

$$d(J_\pm \tau) = (J_\pm \tau)(J_z \tau) \frac{d\tau}{\tau},$$

$$d(Et\tau) = \left(\frac{J_\pm \tau}{2}\right)^2 \frac{d\tau}{\tau^2},$$

(11.1)

and so on (where $d\tau = \varepsilon$, $\tau = D^{-1}$).

Unfortunately, this procedure is not quite exact because the change $d(J\tau)$ depends to some extent on the energy of the initial and final states of the scattering process. This energy dependence becomes severe when the effective exchange integral itself becomes large. In fact, if we follow the equations (11.1) to their logical conclusion, eventually $J\tau$ increases without limit: but only for electrons right at the Fermi level.

Anderson's solution [20] to this problem was to suggest that this continuous scaling be stopped at a definite point where $(J\tau)$ has become reasonably large, and that this resulting strong coupling problem be treated by entirely separate methods. This is the problem discussed by Mattis [16], for instance, of a spin strongly coupled to the nearest Wannier function, with relatively small transfer matrix elements (given by the cutoff band-width $1/\tau = D$). Perturbation theory in D/J should give correct answers for this problem. But Anderson did not show that $J\tau$ could be scaled to values large enough to make this perturbation theory converge rapidly, nor did he carry out that calculation. Nonetheless, this method is a very useful simple visualization of the essentially more exact results of the space-time methods.

Fowler and Zawadowski [21], on the other hand, specifically emphasized the frequency dependence of the effective interactions. They worked from the start with Abrikosov's [22] pseudo-Fermion technique and conventional Green's functions. Studying the dependence of the renormalized Green's functions on the physical input parameters J, D, and the frequency variable ω in the scattering process, they observed that different values of input parameters could give the same renormalized scattering amplitudes. One then looks for the group of transformations on the inputs which gives the same physical results: This is the "renormalization group."

There are two regions of the parameters where the renormalizations are very small. First, if we set the input J nearly equal to the scattering matrix T itself (which in this theory is called the "invariant coupling") and the frequency and cutoff energies are very small, we have very little renormalization. Second, in the case where D is large but ω is at some sufficiently large value also, then all perturbation denominators are very large (there is a regularization procedure here introduced by Bogoliubov), and for the physical J also renormalization is small. These two regimes are equivalent under the renormalization group, and can be scaled into each other by use of a differential equation involving only the latter regime. Here, however, it becomes clear that ω dependence plays a crucial role, so that we improve the accuracy by taking into account higher diagrams. In lowest order, one gets the conventional Abrikosov results; in the next order, the very accurate approximation of Noziéres [23].

By the nature of the method it is still only a very sophisticated way of generating approximations. But again it has the great advantage of bringing out clearly the fact that the antiferromagnetic Kondo problem scales into a strong coupling problem for arbitrarily small values of J, and that the factor giving the scaling is the conventional "Kondo temperature" energy.

Acknowledgment

We would like to acknowledge the help of Mr. John Armytage, especially in preparing Fig. 3 and in the preparation of Section XI.

References

1. J. Kondo, *Progr. Theor. Phys.* **32**, 37 (1964).
2. P. W. Anderson, *Phys. Rev.* **124**, 41 (1961).
3. G. D. Mahan, *Phys. Rev.* **163**, 612 (1967).

4. P. Noziéres and C. de Dominicis, *Phys. Rev.* **178**, 1097 (1969).
5. R. P. Feynman, Ph. D. Thesis, Princeton, Univ. Princeton, New Jersey, 1942.
6. P. W. Anderson and G. Yuval, *Phys. Rev. Lett.* **23**, 89 (1969).
7. G. Yuval and P. W. Anderson, *Phys. Rev. B* **1**, 1522 (1970).
8. G. Polya and G. Szegö, "Aufgabe Und Lehrsätze aus der Analyse." Dover, New York, 1945.
9. N. I. Muskhelishvili and D. A. Kveselava, *Tr. Tbilis. Mat. Inst.* **11**, 141 (1942).
10. P. W. Anderson, *Phys. Rev.* **164**, 352 (1967).
11. K. D. Schotte and U. Schotte, *Phys. Rev.* **182**, 479 (1969).
12. S. Tomonaga, *Prog. Theor. Phys.* **5**, 544 (1950).
13. P. W. Anderson and G. Yuval, *J. Phys. C* **4**, 607 (1971).
14. K. D. Schotte and U. Schotte, *Phys. Rev. B* **4**, 2228 (1971).
15. C. S. Ting, *J. Phys. Chem. Solids* **31**, 777 (1970).
16. D. C. Mattis, *Phys. Rev. Lett.* **19**, 1478 (1967).
17. P. W. Anderson, G. Yuval, and P. R. Hamann, *Phys. Rev. B* **1**, 4464 (1970); *Solid State Commun.* **8**, 1033 (1970).
18. D. Thouless, *Phys. Rev.* **187**, 732 (1969).
19. F. J. Dyson, *Commun. Math. Phys.* **21**, 269 (1971).
20. P. W. Anderson, *J. Phys. C.* **3**, 2436 (1970).
21. M. Fowler and A. Zawadowski, *Solid State Commun.* **9**, 471 (1971).
22. A. A. Abrikosov, *Physics (Long Island City, N.Y.)* **2**, 5 (1965).
23. P. Nozières, unpublished, 1970; also A. A. Abrikosov and A. A. Migdal, *J. Low Temp. Phys.* **3**, 519 (1970).

8. Functional Integral Methods in the Magnetic Impurity Problem

D. R. Hamann

*Bell Telephone Laboratories, Inc.
Murray Hill, New Jersey*

J. R. Schrieffer

*Department of Physics
University of Pennsylvania
Philadelphia, Pennsylvania*

The principal problem presented by the theory of dilute alloys is understanding the mechanism through which the interacting electron gas can exhibit a susceptibility ranging continuously from enhanced Pauli paramagnetism to Curie law behavior. This variation takes place either when the exchange interactions on the impurities are increased in strength for fixed temperature or when the temperature is sufficiently increased for fixed interaction. The divergences associated with the Kondo effect [1] play an important role in determining some aspects of this variation, and a complete theory must take this into account. It is clearly desirable to formulate the problem in a way which will most fully exploit what intuitive understanding of local moment formation we do have. One easily established characteristic of local moment systems is that large, long-lived fluctuations of the electron spin density must occur on the impurities [2]. If the fluctuations were not locally large, neither saturation in an applied field nor a Curie law could be observed. This follows from considering an exact expression for the total susceptibility, $\chi = \mu_B{}^2 \langle |\mathbf{S}|^2 \rangle / 3 k_B T$ where $\mathbf{S}$ is the total spin. For free electrons, the numerator increases linearly with T until T is of the order of the Fermi energy, yielding the Pauli susceptibility. In the alloy, the thermally driven fluctuations on the impurities must saturate at a much lower temperature to yield a Curie law contribution to χ. This indicates that most of the oscillator strength of the susceptibility must be at energies below this temperature. On the basis of the uncertainty principle, the impurity spin relaxation time will then be long compared to that of a host atom, whose characteristic relaxation time is ε_F^{-1}.

This discussion suggests that it will be advantageous to single out the spin density on the impurity site as the dynamical object of primary interest. Furthermore it will be necessary to consider the nonlinear response of this object.

Divergences similar to those associated with the Kondo effect have recently been found in studying the theory of x-ray absorption in metals [3, 4]. The problem is solved by casting it in the form of a local time-dependent potential acting on the electron gas. This solution suggests that the Kondo divergence occurs in considering the response of the spin-up electron gas to the low-frequency components of the local potential exerted on it by the spin-down electron gas, and vice versa. Of course these potentials are self-consistently determined by the responses, but it would be advantageous to separate this self-consistency problem from the response problem [5–7].

A convenient formulation which gives explicit expression to this point of view is provided by a transformation of the many-body problem introduced by Stratonovich and Hubbard [8].[†] To be explicit, we will use Anderson's model of a single magnetic impurity, whose Hamiltonian is, in the standard notation [10],

$$H_0 = \sum_{k\sigma} \varepsilon_{k\sigma} n_{k\sigma} + \sum_{\sigma} \varepsilon_{d\sigma} n_{d\sigma} + \sum_{k\sigma} (V_{kd} C^\dagger_{k\sigma} C_{d\sigma} + V^*_{kd} C^\dagger_{d\sigma} C_{k\sigma}), \quad (1)$$

$$H_1 = U n_{d+} n_{d-}. \quad (2)$$

The partition function can be expressed as

$$Z = Z_0 \left\langle T_\tau \exp\left[-\int_0^\beta d\tau H_1(\tau)\right] \right\rangle, \quad (3)$$

where Z_0 is the partition function for H_0, H_1 is in the interaction representation defined by H_0, T_τ is the ordering operator with respect to τ, and $\langle \ \rangle$ is the thermal average with respect to H_0 [11]. The transformation requires that H_1 be a diagonal quadratic form of the operators. We shall choose the form

$$H_1 = (U/4)(n_{d+} + n_{d-})^2 - (U/4)(n_{d+} - n_{d-})^2. \quad (4)$$

In Eq. (3), the operators in H_1 may be treated as if they commute for

---

[†] Application of this technique to the Anderson model was investigated by Mühlschlegel [9].

different τ as long as they are in the domain of T_τ. Therefore, for each interval $d\tau$ we may use the identity

$$\exp(a^2) = \int_{-\infty}^{\infty} dx \exp(-\pi x^2 - 2\pi^{1/2}ax) \tag{5}$$

to replace the exponential of the squared operators, represented by a, by an integral in which the operators only appear linearly in the exponent. This yields

$$Z = Z_0 \int \delta x\, \delta y \left\langle T_\tau \exp\left\{-\int_0^\beta d\tau \left(\frac{\pi x^2}{\beta} + \frac{\pi y^2}{\beta} + \tilde{H}_1\right)\right\}\right\rangle, \tag{6}$$

where

$$\tilde{H}_1 = (\pi U/\beta)^{1/2} \{x(\tau)[n_{d+}(\tau) - n_{d-}(\tau)] + iy(\tau)[n_{d+}(\tau) + n_{d-}(\tau)]\}. \tag{7}$$

In Eq. (7), x and y are explicitly τ-dependent c-number functions, and $n_{d\sigma}$ are time dependent through the interaction representation. The integrals in Eq. (6) run over all functions $x(\tau)$ and $y(\tau)$.

The form of Eq. (7) shows that the function $x(\tau)$ acts as an effective exchange field coupled to the z component of the localized spin density, while $iy(\tau)$ acts as a potential coupled to the localized particle density. For any particular pair of functions x and y, Eq. (6) factors into a product of two terms of the form

$$Z_\sigma = \left\langle T_\tau \exp\left\{-\int_0^\beta d\tau\, v_\sigma(\tau)\, n_{d\sigma}(\tau)\right\}\right\rangle, \tag{8}$$

one for each spin system, where

$$v_\sigma(\tau) = (\pi U/\beta)^{1/2} [\sigma x(\tau) + iy(\tau)], \qquad \sigma = \pm 1. \tag{9}$$

Thus the interacting electron system reduces to a gas of free electrons moving in an external potential. The problem now separates into two steps. The first is calculating the response of each spin system to the fluctuating potential. The second is selecting the physically important aspects of $v_\sigma(\tau)$, and carrying out the functional average. It is the functional average which reinstates the interaction between the two spin systems.

To evaluate the quantum average in Eq. (8) we multiply v_σ by a coupling constant g. Differentiating with respect to g we find

$$\frac{\partial \log Z_\sigma}{\partial g} = -\int_0^\beta v_\sigma(\tau)\langle n_{d\sigma}(\tau)\rangle_g\, d\tau, \tag{10}$$

where for any operator A

$$\langle A \rangle_g = \frac{\langle T_{\tau'} A \exp[-g \int_0^\beta v(\tau') n_{d\sigma}(\tau') \, d\tau'] \rangle}{\langle T_{\tau'} \exp[-g \int_0^\beta v(\tau') n_{d\sigma}(\tau') \, d\tau'] \rangle}. \tag{11}$$

The average $\langle n_{d\sigma}(\tau) \rangle_g$ is the limit $\tau' = \tau^+$ of the localized state Green's function

$$G_{d\sigma}(\tau, \tau') = -\langle T_\tau C_{d\sigma}(\tau) C_{d\sigma}^\dagger(\tau') \rangle_g, \tag{12}$$

which satisfies

$$G_{d\sigma}(\tau, \tau') = G_{d\sigma}^0(\tau - \tau') + g \int_0^\beta G_{d\sigma}^0(\tau - \tau'') v_\sigma(\tau'') G_{d\sigma}(\tau'', \tau') \, d\tau''. \tag{13}$$

G^0 is the localized state Green's function for H_0. Its Fourier transform is [10]

$$G_{d\sigma}^0(\omega_n) = \frac{1}{i\omega_n - \varepsilon_{d\sigma} + i\varDelta(\omega_n/|\omega_n|)},$$

$$\omega_n = \frac{(2n+1)\pi}{\beta}, \quad n = 0, \pm 1, \pm 2,..., \tag{14}$$

for a Lorentzian virtual state of width $\varDelta$. From Eq. (10) we see that Z_σ is given by

$$Z_\sigma = \exp\left[-\int_0^1 dg \int_0^\beta v_\sigma(\tau) G_{d\sigma}(\tau, \tau^+) \, d\tau\right]. \tag{15}$$

The complete partition function may be expressed as

$$Z = Z_0 \int Z_+(x, y) Z_-(x, y) \exp\left[-\int_0^\beta (\pi x^2 + \pi y^2) \, d\tau\right] \delta x \, \delta y. \tag{16}$$

Thus the probability distribution of the fields x and y acting on the +spin is weighted by Z_-, and vice versa.

The relation of this formalism and conventional perturbation theoretic diagrams is established by writing $Z_+ Z_-$ as a power series in v. The Gaussian functional average of each term in the series is zero unless each v is paired with another v at the same τ, since v's at different τ's are independent and have zero average value. A factor $U^{1/2}$ enters at each vertex so that paired vertices contribute a factor of U corresponding to the conventional two-body interaction [5, 12].

It is convenient at this point to introduce a simplifying approximation in carrying out the functional integral. It is plausible that local charge

8. MAGNETIC IMPURITY PROBLEM FUNCTIONAL INTEGRAL METHODS

density fluctuations have high energy and do not have large amplitude, as do the low-energy spin fluctuations. We shall neglect the contribution of these fluctuations to the partition function, since they should only give a temperature-independent term in the free energy. This is equivalent to making an extremal approximation to the y integral. For the symmetric case $\varepsilon_d = -U/2$, the extremal y function is

$$y(\tau) = y_0 = -\frac{i}{2}\left(\frac{\beta U}{\pi}\right)^{1/2}. \tag{17}$$

When this value of y is substituted into (7) it may be combined with the ε_d term in H_0 and the shifted virtual level is centered at the Fermi surface. We could avoid this approximation by using another form of H_1,

$$H_1 = -\frac{U}{2}(n_{d+} - n_{d-})^2 + \frac{U}{2}(n_{d+} + n_{d-}), \quad n_{d\sigma}^2 = n_{d\sigma}, \tag{18}$$

and introducing a single variable z to carry out the transformation [6, 8]. While this scheme is in principle simpler to handle than the two-variable scheme, extremal paths in z do not correspond to the Hartree–Fock approximation as they do in the x, y scheme, a fact which makes the latter scheme preferable in the Kondo regime where the HF states play a special role [5, 12].

The ratio $U/\pi\varDelta$ measures the strength of the two-body interaction, U. An approximation which is exact in the weak coupling limit ($U \to 0$), and when the strong coupling limit is approached keeping the temperature well above the Kondo temperature, is given by restricting $x(\tau)$ to τ-independent functions (the static approximation). This approximation provides a smooth interpolation between these limits. Then x enters to shift the up- and down-spin virtual levels in opposite directions and one finds from Eq. (13) for the symmetric case,

$$G_{d\sigma}(\tau, \tau^+) = \tfrac{1}{2} - (1/\pi)\tan^{-1} g\sigma\xi, \tag{19}$$

where

$$\xi = (\pi U/\beta\varDelta^2)^{1/2} x. \tag{20}$$

Substituting this result into (15) one finds that Z is expressed as an average of the statistical weight for each ξ,

$$Z = Z_0 \int_{-\infty}^{\infty} e^{-\beta V(\xi)} (\beta\varDelta^2/\pi U)^{1/2} d\xi, \tag{21}$$

where

$$V = \frac{\varDelta^2}{U}\xi^2 - \frac{2\varDelta}{\pi}[\xi \tan^{-1}\xi - \tfrac{1}{2}\log(1+\xi^2)]. \tag{22}$$

The effective free energy V is plotted in Fig. 1 for several values of $U/\pi\varDelta$. For weak coupling, the free energy has a minimum at $\xi = 0$ corresponding to the most probable value of the local moment being zero. Values of ξ near zero are most heavily weighted so the rms ampli-

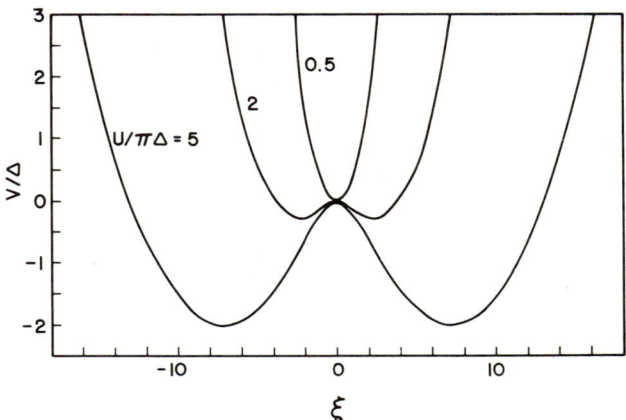

FIG. 1. The effective free energy $V(\xi)$ for a time-independent path ξ with $U/\pi\varDelta = $ 0.5, 2, and 5, corresponding to weak, intermediate, and strong coupling, respectively.

tude of the local spin density is small. This is the region of enhanced Pauli paramagnetism. As $U/\pi\varDelta$ approaches unity, the curvature at the origin changes sign and two minima develop, corresponding to the spin-up and spin-down states of a weak localized moment. For $U/\pi\varDelta \gg 1$, the minima deepen and a full localized spin $\frac{1}{2}$ is achieved. Thus a smooth transition from nonmagnetic to magnetic behavior is predicted within this simple approximation. We note that the minima occur at the values predicted by the HF approximation [10, 13].

The dynamics of the spin fluctuations, which are neglected in the static approximation, can be treated by expanding the exponential in Eq. (15) to lowest nonvanishing order in v, that is, second order. This is accomplished by substituting the first iteration of Eq. (13) into Eq. (15). The argument of the exponent of the total functional integral becomes a quadratic form in $x(\tau)$ within this approximation. This quadratic form can be transformed to its diagonal representation by Fourier transformation,

$$x(\tau) = \sum_\nu x_\nu \exp(-i\Omega_\nu \tau), \quad \Omega_\nu = 2\pi\nu/\beta, \quad \nu = 0, \pm 1, \pm 2,.... \quad (23)$$

8. MAGNETIC IMPURITY PROBLEM FUNCTIONAL INTEGRAL METHODS

The partition function becomes

$$Z = Z_0 \int \delta x \exp\left[-\pi \sum_\nu [1 - U\chi_0(\Omega_\nu)] \mid x_\nu \mid^2\right], \quad (24)$$

where χ_0 is the bubble diagram shown in Fig. 2a,

$$\chi_0(\Omega) = \frac{\Delta}{\pi \mid \Omega \mid (\mid \Omega \mid + 2\Delta)} \log\left(1 + \frac{\mid \Omega \mid (\mid \Omega \mid + 2\Delta)}{\Delta^2}\right). \quad (25)$$

We can evaluate Eq. (24) directly, or observe that the expansion of the exponential gives a sum of an arbitrary number of unlinked loops which

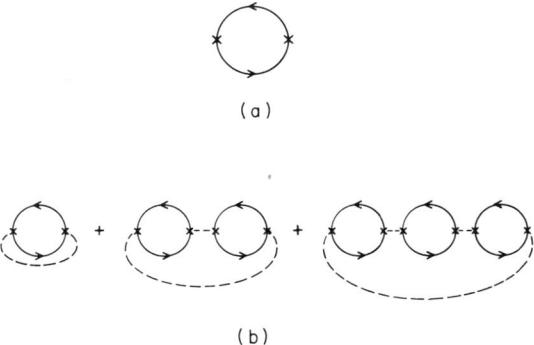

FIG. 2. (a) Diagrammatic representation of χ_0. (b) Diagrammatic sum for the free energy in the random phase approximation.

are connected together by interaction lines on performing the functional integral. Either way one obtains the conventional bubble chain of Fig. 2b and

$$\log\left(\frac{Z}{Z_0}\right) = -\frac{1}{2} \sum_\nu \log[1 - U\chi_0(\Omega_\nu)]. \quad (26)$$

This approximation diverges for $U/\pi\Delta > 1$. Since quartic terms in $V(\xi)$ are necessary for convergence in the static approximation for $U/\pi\Delta > 1$, it is reasonable that similar anharmonic effects will stabilize the unstable modes in Eq. (24).

The leading anharmonic effects arise from the v^4 terms in vG, as shown in Fig. 3a for the linked cluster diagram and in Fig. 3b for a few of the diagrams after the functional integral has been carried out [7]. The general term involves a product of four x_ν's. This corresponds to an interaction between spin fluctuation modes defined in the quadratic

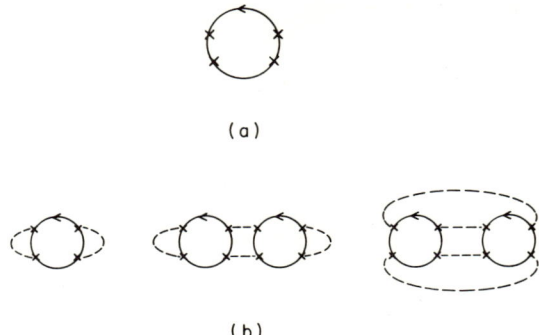

FIG. 3. (a) Diagrammatic representation of the coefficient of the quartic terms in Eq. (27). (b) Low-order diagrams contributing to the free energy in the expansion of the quartic approximation, Eq. (27). The dashed lines represent the propagator of the spin fluctuations given by the quadratic approximation, $\chi_0/(1 - U\chi_0)$.

approximation. There are a large number of important modes ($\sim \beta\Delta$) and the correlation between these modes should not be strong in the intermediate coupling regime. In this case we expect that interactions between four distinct modes will be less important than when modes occur in pairs. This approximation leads to the functional,

$$\frac{Z}{Z_0} = \int \delta x \exp\left[-\pi \sum_\nu [1 - U\chi_0(\Omega_\nu)] \mid x_\nu \mid^2 - \sum_{\nu_1 > \nu_2 \geq 0} a_{\nu_1,\nu_2} \mid x_{\nu_1} \mid^2 \mid x_{\nu_2} \mid^2 \right.$$
$$\left. - \sum_{\nu \geq 0} b_\nu \mid x_\nu \mid^4 \right]. \qquad (27)$$

For ν_1 and $\nu_2 \ll \beta\Delta$, $a_{\nu_1,\nu_2} = \pi U^2/\beta\Delta^3$ and $b_\nu = a_{\nu\nu}/4$. These coefficients, like $\chi_0(\Omega_\nu)$, drop off with increasing Ω_ν on the scale of Δ. Even though $U/\pi\Delta > 1$, the quartic terms lead to a convergent result.

To compare these approximations we consider the dynamic local susceptibility, which can be shown to be given by

$$\chi_{loc}(\Omega_\nu) = \frac{2\mu_B^2}{U}[2\pi\langle \mid x_\nu \mid^2 \rangle - 1], \qquad (28)$$

where $\langle \mid x_\nu \mid^2 \rangle$ is given by averaging $\mid x_\nu \mid^2$ over x_ν with a weight given by the integrand of Eq. (27). For the random phase approximation (RPA), one has

$$\chi_{loc}(\Omega_\nu) = \frac{2\mu_B^2 \chi_0(\Omega_\nu)}{1 - U\chi_0(\Omega_\nu)}. \qquad (29)$$

The exchange enhancement is largest for zero frequency, and χ diverges as $U/\pi\Delta \to 1$ for $\Omega_\nu = 0$, due to the breakdown of the RPA. This divergence is not present in either the static or quartic approximations. A plot of $\chi_{\text{loc}}(0)$ as a function of $\beta\Delta$ is given in Fig. 4. For $U/\pi\Delta \ll 1$,

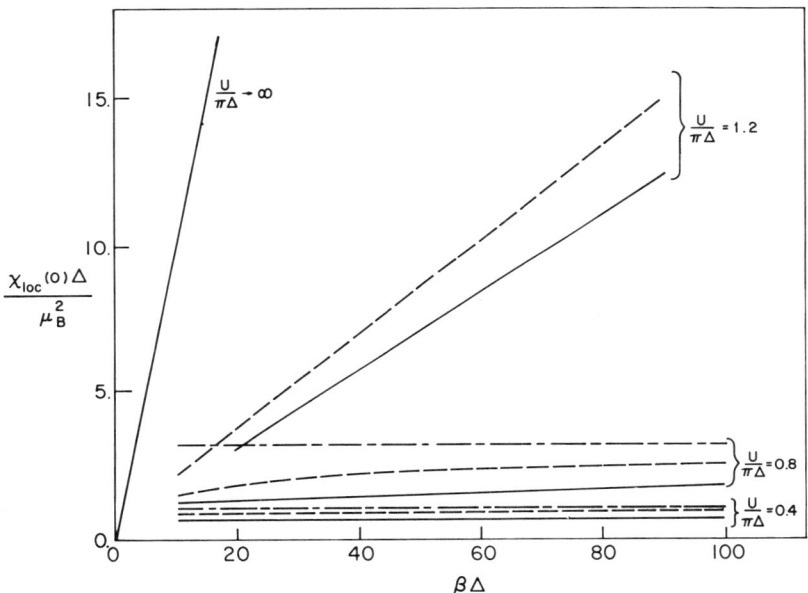

FIG. 4. Plot of zero-frequency local susceptibility $\chi_{\text{loc}}(0)$ as a function of $\beta\Delta$. Solid, dashed, and dash-dotted lines give the results of the static, quartic, and quadratic approximations, respectively. The quadratic approximation diverges for $U/\pi\Delta \geq 1$.

one has temperature-insensitive Pauli paramagnetism. For intermediate coupling, the RPA result diverges while the quartic result is somewhat reduced from that static result. For $U/\pi\Delta \to \infty$ with T greater than the Kondo temperature (which goes to zero in this limit), the static result is exact.

In Fig. 5, $\chi_{\text{loc}}(\Omega_\nu)$ is plotted as a function ν for $U/\pi\Delta = 0.8$ and 1.2 with $\beta\Delta = 40$. The plot shows that the susceptibility drops off rapidly with frequency on the scale of Δ.

When the Coulomb interaction U becomes sufficiently large compared to Δ, there is no longer any justification for keeping a limited number of powers of x in the exponent in Eq. (15). However, in this case we expect that the Anderson model may behave like the s–d exchange model [14], and have its most important spin fluctuations occur with

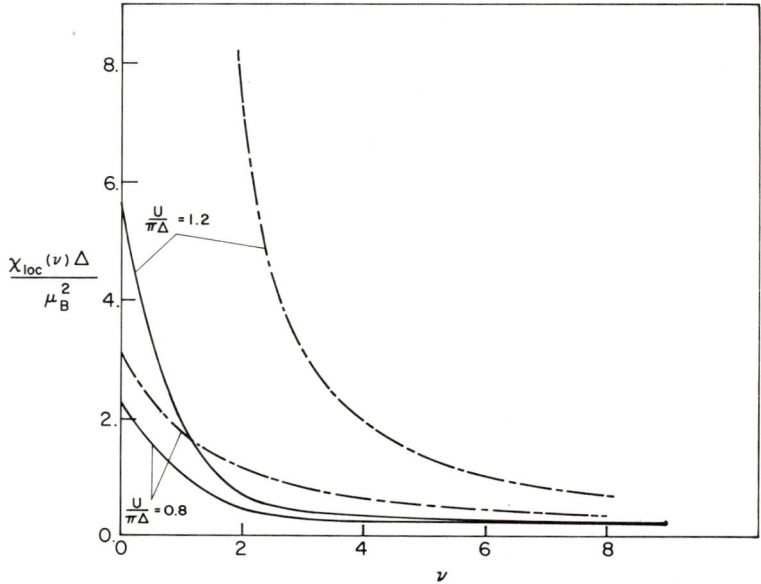

FIG. 5. The local susceptibility $\chi_{\text{loc}}(\Omega_l)$ plotted as a continuous function of frequency index ν. The solid and dash-dotted curves correspond to the quartic and quadratic approximations, respectively. Note the divergence in the latter near $\nu = 2$ for $U/\pi\Delta = 1.2$.

a frequency of the order of the Kondo temperature. In this regime, the asymptotic approximation introduced by Nozières and de Dominicis [4] should be highly accurate.

To derive the corresponding asymptotic approximation for this problem [5], we first consider the exact expression for $G_d^0(\tau)$,

$$G_d^0(\tau) = \int_{-\infty}^{\infty} \frac{d\varepsilon}{\pi} \frac{\Delta}{(\varepsilon - \varepsilon_d)^2 + \Delta^2} e^{-\varepsilon\tau}[f(\varepsilon) - \theta(\tau)], \tag{30}$$

where θ is the unit step function and f is the Fermi function. When $|\tau|$ is large compared to ε_d or Δ, the first factor in the integrand is slowly varying and can be replaced by its value at $\varepsilon = 0$. The integral of the remaining factors is proportional to $[\sin(\pi\tau/\beta)]^{-1}$. A well-defined function which agrees with this asymptotic result and which gives the correct value for its integral around $\tau = 0$ is

$$G_d^0 \approx \frac{\Delta}{\varepsilon_d^2 + \Delta^2} \left[\frac{P}{\beta \sin(\pi\tau/\beta)} + \frac{\varepsilon_d}{\Delta} \delta(\tau) \right], \tag{31}$$

8. MAGNETIC IMPURITY PROBLEM FUNCTIONAL INTEGRAL METHODS

where P denotes principal value. Because we incorporate the extremal value of the density fluctuation field in H_0 and specialize to the case $\varepsilon_d = U/2$, the effective ε_d in $G_d{}^0$ is actually zero.

If Eq. (31) is substituted in Eq. (13), the resulting equation can be solved in closed form using techniques similar to those of Muskhelishvili [18] for any arbitrary fluctuating potential $\xi(\tau)$. The resulting expression for the Green's function contains two groups of terms,

$$G_d(\tau, \tau') = G_d{}^A(\tau, \tau') + G_d{}^T(\tau, \tau'), \tag{32}$$

which are distinguished by their behavior as $\tau' \to \tau$. In this limit, which must be taken in using Eq. (15) to compute the contribution to the partition function, $G_d{}^A$ is singular. However, it can be shown to have precisely the form of the approximate $G_d{}^0$, Eq. (31), but with ε_d/Δ replaced by the instantaneous potential $\xi(\tau)$. It is clear that this is a contribution to the Green's function which represents the system adiabatically following the fluctuating potential, and that we should evaluate the equal-time limit from the exact expression for $G_d{}^0$, Eq. (19). The remaining term is well behaved in the equal-time limit, and is zero if $\xi(\tau)$ is a constant. This term can be associated with the transient response of the system. While $G_d{}^T$ is an unmanageably complicated functional of ξ, the τ integral of the equal-time limit appearing in Eq. (15) can be simplified considerably, and the coupling constant integration can be carried out explicitly.

The result of these calculations is a relatively simple closed-form functional,

$$Z(\xi) = \exp[-\beta(V + T)], \tag{33}$$

where

$$V = \beta^{-1} \int_0^\beta d\tau\, V[\xi(\tau)], \tag{34}$$

and

$$T = \frac{P}{\pi\beta^2} \int_0^\beta d\tau\, d\tau' \cot\left[\frac{\pi(\tau - \tau')}{\beta}\right] \xi(\tau) \frac{d\xi(\tau')}{d\tau'}$$
$$\times \frac{1}{\xi^2(\tau) - \xi^2(\tau')} \ln \frac{1 + \xi^2(\tau)}{1 + \xi^2(\tau')}. \tag{35}$$

$V(\xi)$ in Eq. (34) is just the function Eq. (22), which was previously identified as an effective free energy of a τ-independent path and is shown in Fig. 1. In the present calculation, V arises from the "adiabatic" part of the Green's function and the Gaussian weighting factor. The term T arises from the "transient" part. It is zero for a τ-independent path, and appears to be positive semidefinite, although this has only been verified for some specific classes of functions. Extending the previous interpretation slightly, V and T could be identified as potential and kinetic contributions to an effective path energy.

To check the functional for small U/Δ, where $V(\xi)$ will confine the fluctuations in ξ to small values, we can expand the functional to second order in ξ. This reproduces the RPA result, but with $\chi_0(\Omega)$ [Eq. (25)] approximated by the first two terms in its expansion about zero frequency.

The real motivation for obtaining the closed-form functional on the basis of the asymptotic approximation was to be able to identify important classes of paths. We first note that the minima of $V(\xi)$ become arbitrarily deep and widely separated as U/Δ increases. However, T is independent of U/Δ and the value of its integrand saturates when $\xi(\tau)$ and $\xi(\tau')$ both become large. Therefore we expect that V will be the dominant part of the functional at large U/Δ. This requires that the important paths spend most of their time at or near a minimum of $V(\xi)$. However, a path need not remain near a single minimum. It may hop back and forth between the minima at only a small cost in increased "potential energy" if it spends only a short time τ_0 crossing the maximum. Such a path is sketched in Fig. 6. The "entropy" gained in summing $Z(\xi)$ over all such functions is easily seen to outweigh the energy lost in departing from the extremal solution.

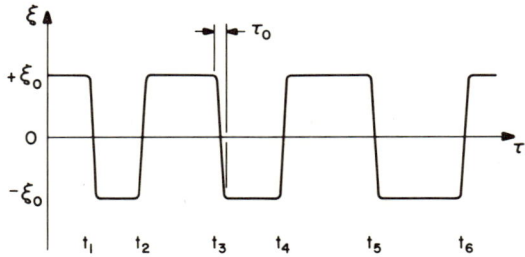

FIG. 6. Typical "hopping" path contributing to the functional integral for large $U/\pi\Delta$. The $\pm\xi_0$ are the positions of the minima of $V(\xi)$, and the positions of the hops t_i are arbitrary, except for the requirement $t_{i+1} - t_i > \tau_0$.

To evaluate T for an arbitrary hopping path, we first perform an integration by parts on τ in Eq. (35),

$$T = -\frac{P}{\pi\beta^2}\int_0^\beta d\tau\, d\tau' \ln\left(\sin\left[\frac{\pi(\tau-\tau')}{\beta}\right]\right)\frac{d\xi(\tau')}{d\tau'}$$

$$\times \frac{d}{d\tau}\left[\frac{\xi(\tau)}{\xi^2(\tau)-\xi^2(\tau')}\ln\frac{1+\xi^2(\tau)}{1+\xi^2(\tau')}\right]. \quad (36)$$

8. MAGNETIC IMPURITY PROBLEM FUNCTIONAL INTEGRAL METHODS

The derivative of ξ is zero outside the hops, so the integrand is zero unless both τ and τ' are within hops. The integral can thus be written

$$T = \sum_{i \neq j} \int_{t_i}^{t_i + \tau_0} d\tau \int_{t_j}^{t_j + \tau_0} d\tau' \, F(\tau, \tau')$$

$$+ \sum_i \int_{t_i}^{t_i + \tau_0} d\tau \, d\tau' \, F(\tau, \tau'), \tag{37}$$

where F is the integrand in Eq. (36). Since we expect the spacing $t_i - t_j$ to be large compared to τ_0 in most cases, we can evaluate the "interhop" term without assumption about the shape of ξ within the hop. To evaluate the "intrahop" term, we assume a linear shape within the hop, but argue that reasonable changes in this only produce a numerical contribution which is negligible for large ξ_0. After evaluating the various terms, we vary τ_0 to minimize $T + V$, and find that the minimum is at $\tau_0 = 6/U$,[†] independent, except for the numerical factor, of any details of the path. This time is near the limit of the minimum for which we can hope to use the asymptotic approximation. Therefore numerical factors depending on details within each hop are outside the accuracy of the closed-form functional.

The sum of $Z(\xi)$ over the hopping paths is given by

$$Z = \sum_{n=0}^{\infty} \xi_0^{-2n} \int_0^\beta \frac{dt_{2n}}{\tau_0} \int_0^{t_{2n} - \tau_0} \frac{dt_{2n-1}}{\tau_0} \cdots \int_0^{t_2 - \tau_0} \frac{dt_1}{\tau_0}$$

$$\times \exp\left[\left(\frac{2 \tan^{-1} \xi_0}{\pi}\right)^2 \sum_{i \neq j} (-1)^{i+j} \ln \left|\left(\frac{\beta}{\pi \tau_0}\right) \sin \left[\frac{\pi(t_i - t_j)}{\beta}\right]\right|\right]. \tag{38}$$

This expression has the form of the grand partition function of a gas of classical particles moving in one dimension and interacting via both a hard-core and a long-range logarithmic potential which is alternately attractive and repulsive. The hops may be thought of as particles, and the t_i as their coordinates. The length of the system is β.

In studying the s–d exchange model, Anderson and Yuval were able to show its correspondence to the same problem in classical mechanics [16]. Furthermore the correspondence between the coefficients in

---

[†] The value $\tau_0 = \Delta^{-1}$ which was assumed prior to the complete evaluation of the "intrahop" term [5, first paper] was shown to lead to erroneous conclusions by Ramakrishnan [15].

their result and Eq. (38) is just that predicted by the lowest order canonical transformation relating the models [14]. This correspondence breaks down as $U/\pi\Delta$ approaches the intermediate coupling regime, as can be seen from the following argument. To complete the functional integral, we could add fluctuations to each straight section of the hopping paths. It is clear that these would be confined to the vicinity of each minimum, and would lower the free energy by an amount that is essentially independent of the positions of the hops. As U/Δ diminished, however, the wells of V would become shallow and the fluctuations would grow to the point that they destroyed the identity of the hops. At this point the hopping approximation fails, and the correspondence between the Anderson model and the s–d exchange model breaks down. The approximations described earlier in this chapter, which are appropriate to the intermediate coupling regime, have no parallel in the s–d exchange model.

The hopping path expression for the partition function, while it represents a great simplification of the original problem of interacting electrons, is itself a difficult problem in classical statistical mechanics. An approximate analysis of this model has been carried out [17] and is described elsewhere in this volume [16].

The approximate analysis shows that the partition function, Eq. (38), displays a phase transition. This is to be understood in the following sense: The coefficient of the logarithmic interaction and the "fugacity" prefactor ξ_0^{-1} are treated as independent variables. The size of the system β is let go to infinity (i.e., the physical temperature is set equal to zero). Then the correlation function

$$\langle \xi(\tau)\, \xi(0) \rangle \equiv Z^{-1} \int \delta\xi\, \xi(\tau)\, \xi(0)\, Z(\xi) \tag{39}$$

is studied in the limit $\tau \to \infty$. For small values of interaction coefficient, there is no long-range correlation. For large values, the correlation function has a finite limit, indicating long-range order. In the actual problem, this coefficient is bounded. Its maximum, obtained by letting $\xi_0 \to \infty$, is precisely on the phase transition line. The regime of physical interest, ξ_0 large but finite, is thus in the "critical region." The range of the correlation function Eq. (39) can be shown to be $\tau_K = \tau_0 \exp(\pi U/8\Delta)$ in this critical region, which is just the inverse of the Kondo temperature. The typical paths have sections of length τ_K in which $\xi(\tau)$ equals, say, $+\xi_0$ for most τ. The occasional hops to $-\xi_0$ are closely paired with hops back to $+\xi_0$. A fuzzy area in which neither valley is clearly preferred separates this section from the next, where $\xi(\tau) = -\xi_0$ predominates.

8. MAGNETIC IMPURITY PROBLEM FUNCTIONAL INTEGRAL METHODS 251

These results permit us to understand the qualitative differences in magnetic alloy behavior above and below the Kondo temperature. For $T > T_K$, the corresponding classical system is small compared to the range of its correlations. The dominant paths are either mostly $+\xi_0$ or mostly $-\xi_0$, and the partition function becomes, effectively, that of a two-state system like a spin $\frac{1}{2}$. The susceptibility is a Curie law to the first approximation, with a moment slightly reduced from the full free spin value. It would be legitimate to describe the system as a statistical superposition of the two HF states. As the temperature is lowered, the classical system becomes larger than the range of its correlations, so that paths which spend significantly more time in one valley than the other become increasingly less probable. Thus the effective magnetic moment falls smoothly with decreasing temperature. In the low-temperature limit, the susceptibility goes to a constant, which is easily shown to be of order μ_B^2/T_K by making use of the range of the correlation function. More detailed considerations in the solution of the classical model indicate that the approach to zero temperature is smooth, and that the physical properties are analytic functions of T at $T = 0$. Thus the magnetic impurity below its Kondo temperature behaves like a nonmagnetic impurity, but with a characteristic energy renormalized by the Kondo exponential.

The impurity problem in a metal is distinguished from most many-body problems in that a single degree of freedom, the local spin density on the impurity, contains all the dynamics of the system. The functional integral technique is tractable because a single set of functions of one variable is adequate to describe the interactions. It has permitted us to deal with the nonlinear dynamics of the local spin density in considerable detail. While this approach has not yielded simple formulas for the physical properties of the system valid for all ranges of the parameters, it has produced a satisfactory explanation of the full range of non-magnetic to magnetic behavior. Furthermore, it has enabled us to discuss rather thoroughly the motivation for the approximation schemes employed in various parameter regions.

It is not easy to assess the possible future role of the functional integral technique in a broader context of quantum mechanical many-body problems. It has proved useful here in treating the nonlinear dynamics of a single variable, in contrast to the well-studied examples of quasi-linear dynamics of many variables. The challenging unsolved problem of quantum mechanical systems in the critical region appears to involve both nonlinear dynamics and many variables. The present work only broadens the base for an attack on such problems.

References

1. J. Kondo, *Progr. Theor. Phys.* **32**, 37 (1964).
2. P. Lederer and D, L. Mills, *Solid State Commun.* **5**, 131 (1967); *Phys. Rev. Lett.* **20**, 1036 (1967); N. Rivier and M. J. Zuckermann, *Ibid.* **21**, 904 (1968); H. Suhl, *Ibid.* **19**, 442 (1967); M. Levine and H. Suhl, *Phys. Rev.* **171**, 567 (1968); M. Levine, T. V. Ramakrishnan, and R. A. Weiner, *Phys. Rev. Lett.* **20**, 1370 (1968); D. R. Hamann, *Phys. Rev.* **186**, 549 (1969); Y. Kuroda, *Progr. Theor. Phys.* **43**, 870 (1970); M. T. Béal-Monod and D. L. Mills, *Phys. Rev. Lett.* **24**, 225 (1970).
3. G. D. Mahan, *Phys. Rev.* **163**, 612 (1967); B. Roulet, J. Gavoret, and P. Nozières, *Ibid.* **178**, 1072 (1969); P. Nozières, J. Gavoret, and B. Roulet, *Ibid.* **178**, 1084 (1969).
4. P. Nozières and C. T. de Dominicis, *Phys. Rev.* **178**, 1097 (1969).
5. D. R. Hamann, *Phys Rev. Lett.* **23**, 95 (1969); *Phys. Rev. B* **2**, 1373 (1970); *J. Phys. (Paris)* **32**, C1-207 (1971).
6. S. Q. Wang, W. E. Evenson, and J. R. Schrieffer, *Phys. Rev. Lett.* **23**, 92 (1969); *J. Appl. Phys.* **41**, 1199 (1970).
7. J. R. Schrieffer, W. E. Evenson, and S. Q. Wang, *J. Phys. (Paris)* **32**, C1-19 (1971); S. Q. Wang, Thesis, Univ. of Pennsylvania, Philadelphia, Pennsylvania, 1970, unpublished.
8. R. L. Stratonovich, *Dokl. Akad. Nauk. SSSR* **115**, 1097 (1957); J. Hubbard, *Phys. Rev. Lett.* **3**, 77 (1959).
9. B. Mühlschlegel, unpublished lecture notes, Univ. of Pennsylvania, Philadelphia, Pennsylvania, 1965.
10. P. W. Anderson, *Phys. Rev.* **124**, 41 (1961).
11. A. A. Abrikosov, L. P. Gorkov, and I. E. Dzyaloshinski, "Methods of Quantum Field Theory in Statistical Physics" translated by R. A. Silverman p. 130. Prentice-Hall, Englewood Cliffs, New Jersey 1963.
12. H. Keiter, *Phys. Rev. B* **2**, 3777 (1970).
13. B. Kjøllerstrøm, *Phys. Stat. Solidi.* **43**, 203 (1971).
14. J. R. Schrieffer and P. A. Wolff, *Phys. Rev.* **149**, 491 (1966).
15. T. V. Ramakrishnan, *Phys. Rev. B* **1**, 3881 (1970).
16. P. W. Anderson and G. Yuval, *Phys. Rev. Lett.* **23**, 89 (1969); *Phys. Rev. B* **1**, 1552 (1970); *in* "Magnetism" (H. Suhl ed.), Vol. V, p. 217–236. Academic Press, New York, 1971.
17. P. W. Anderson, G. Yuval, and D. R. Hamann, *Solid State Commun.* **8**, 1033 (1970); *Phys. Rev. B* **1**, 4464 (1970).
18. N. I. Muskhelishvili, "Singular Integral Equations," translated by J. R. M. Radok. P. Noordhoff, Groningen, Holland, 1953.

9. The Ground State of the s–d Model

Kei Yosida and Akio Yoshimori

Institute for Solid State Physics
University of Tokyo, Roppongi
Tokyo, Japan

I. Introduction 253
II. Perturbation Theoretic Approach for the Singlet Ground State 258
III. Bound State for the Anisotropic Exchange Interaction 264
IV. Charge, Spin Polarization, and Spin Correlation Densities 270
V. Bound State in the Presence of Magnetic Field 274
VI. Local Electron Distributions and Magnetoresistance 278
VII. Concluding Remarks 284
References 285

I. Introduction

This chapter deals with the nature of the ground state of the system consisting of the conduction electrons and a localized spin embedded in an otherwise pure metal which are coupled by the s–d exchange interaction.

The s–d exchange interaction, no matter how weak it may be, has so strong an effect on this system that it may change the free state of the localized spin drastically. This nature of the s–d exchange interaction bears a resemblance to that of the attraction between conduction electrons in superconductors. The logarithmic divergence appearing in the perturbation expansion of the scattering t matrix presumably originates in this particular character of this interaction. The logarithmic divergence was found by Kondo [1], and was a milestone in the researches of localized moments in metals.

The s–d exchange Hamiltonian for a single localized spin situated at the origin is written as

$$H_{sd} = -\frac{J}{2N} \sum_{kk'\alpha\alpha'} a^\dagger_{k\alpha} \sigma_{\alpha\alpha'} a_{k'\alpha'} \cdot S, \quad (1.1)$$

where $a^\dagger_{k\alpha}$ and $a_{k\alpha}$ represent, respectively, the creation and annihilation operators for the conduction electron with wave vector k and spin α; S is the localized spin whose magnitude is assumed to be $\frac{1}{2}$, σ are the Pauli matrices, and N is the number of lattice points. The exchange coupling constant, J, is assumed to be independent of k and k'.

For iron group impurities with which we are concerned this s–d exchange interaction is regarded as an effective Hamiltonian derived from the more fundamental Anderson Hamiltonian [2] for a single d orbital,

$$H_A = \sum_{k\alpha} \varepsilon_k a^\dagger_{k\alpha} a_{k\alpha} + \sum_\alpha E_d a^\dagger_{d\alpha} a_{d\alpha}$$

$$+ \sum_{k,\alpha} [V_{kd} a^\dagger_{k\alpha} a_{d\alpha} + V_{dk} a^\dagger_{d\alpha} a_{k\alpha}] + U n_{d\uparrow} n_{d\downarrow}, \quad (1.2)$$

in the limiting case where the intra-Coulomb integral U between two d electrons and the energy depth of the d level $|E_d|$ are large compared with the s–d mixing rate $\rho |V|^2$, ρ being the density of states at the Fermi surface.

As it is well known [3–5], in the limit of large U and large $|E_d|$, Eq. (1.2) becomes equivalent to the sum of the impurity potential and the s–d exchange, Eq. (1.1). The impurity potential vanishes in the neutral case of $U + 2E_d = 0$, and in this case the exchange coupling constant J is given by

$$\frac{J}{N} \simeq -2|V|^2 \left[\frac{1}{U + E_d - \varepsilon_{k'}} + \frac{1}{|E_d| + \varepsilon_k} \right] \simeq -8 \frac{|V|^2}{U}. \quad (1.3)$$

Therefore the ground state of the s–d system coupled with the s–d exchange Hamiltonian Eq. (1.1) is also the ground state for the Anderson Hamiltonian in its limit of large U.

In the usual perturbation theoretic approach applied to the present system, the wave function of the two separated free systems of the conduction electrons and the localized spin is taken as an unperturbed wave function

$$\psi_{0\beta} = \chi_\beta \psi_\nu, \quad (1.4)$$

where $\chi_\beta (\beta = \uparrow, \downarrow)$ represents the state of the localized spin whose z component is $S_z = \frac{1}{2}$ or $-\frac{1}{2}$, and ψ_v is the Fermi vacuum for the conduction electrons. In the following, β denotes the localized spin state. Thus this unperturbed state is doubly degenerate, corresponding to two possible components of the localized spin. This degeneracy will not be removed, in general, by the finite-order perturbation of the s–d exchange because this interaction conserves the total spin of the system.

The usual finite-temperature perturbation calculation [6] gives rise to the following result for the expectation value, thermally averaged with respect to the conduction-electron state, of S_z in the perturbed state, which can be interpreted as the magnitude of the localized spin,

$$\langle S_z \rangle = S_z \left[1 + \frac{\frac{1}{2}(J\rho/N)^2 \log(kT/D)}{1 - (J\rho/N)\log(kT/D)} \right] \quad (1.5)$$

in which the density of states of the conduction electrons is assumed to be constant ρ for $-D \leqslant \varepsilon \leqslant D$ and otherwise zero, the Fermi energy being taken as the origin of energy, and the most divergent terms are only retained in each order of perturbation.

The total spin polarization of the conduction electrons $\langle \sigma \rangle$ is associated with this expectation value of the localized spin by the relation

$$\langle \sigma \rangle = \frac{J\rho}{2N} \langle S \rangle, \quad (1.6)$$

which holds generally for the local spin $\langle S \rangle$ of any origin. These two results immediately lead to the susceptibility of the localized spin,

$$\chi = \frac{C}{T} \left\{ 1 + \frac{J\rho/N}{1 - (J\rho)/N \log(kT/D)} \right\}, \quad (1.7)$$

which can also be derived by direct calculation [7]. Here C is the Curie constant for a free spin.

As Eqs. (1.5) and (1.6) show, the magnitude of the localized spin nearly keeps its free value accompanied by the spin polarization of $(J\rho/2N)S$ at high temperatures. For ferromagnetic coupling ($J > 0$), it decreases monotonically as the temperature is lowered, and at $T = 0$ it tends to $S(1 - J\rho/2N)$. This diminution of the localized spin is exactly compensated by the spin polarization of the conduction electrons insofar as a linear term in J is concerned. This means that the total spin associated with the localized spin returns to the original free value and that the total spin is conserved at $T = 0$.

On the other hand, for antiferromagnetic coupling ($J < 0$), the diminution is large and $\langle S \rangle$ goes through zero and diverges at $T = T_K$,

$$T_K = D\, e^{N/J\rho}. \tag{1.8}$$

This behavior of the localized spin obtained by the perturbation calculation suggests that at the absolute zero of temperature both the localized spin and the induced spin polarization associated with it vanish and the system will be in a singlet ground state. As mentioned earlier, the perturbation calculation at zero temperature conserves the total spin. Therefore, in order to obtain a singlet ground state by perturbation method, we should start with an unperturbed singlet state.

A singlet state of the present system may be described by

$$\psi_{\text{singlet}} = (1/\sqrt{2})[\psi_\uparrow \chi_\uparrow - \psi_\downarrow \chi_\downarrow], \tag{1.9}$$

where $\psi_\uparrow (\psi_\downarrow)$ denotes the wave function of the conduction electrons which possesses one excess electron with $\downarrow (\uparrow)$ spin compared with the electrons with $\uparrow (\downarrow)$ spin which may be localized around the impurity and coupled with the localized spin of the impurity.

Two approaches have so far been attempted in order to construct such wave functions, ψ_β ($\beta = \uparrow, \downarrow$). One approach is variational and the other is perturbative. In the variational approach adopted by Anderson [8], the ψ_β are, in the version of the s–d model, approximated by the Slater determinants constructed by one-particle unrestricted wave functions $\phi^\beta_{n\uparrow}$ and $\phi^\beta_{n\downarrow}$ which are treated as variational functions. Anderson pointed out that a difficulty arises here that unless the differences between the phase shifts of $\phi^\uparrow_{n\uparrow}$ and $\phi^\downarrow_{n\uparrow}$ and that of $\phi^\uparrow_{n\downarrow}$ and $\phi^\downarrow_{n\downarrow}$ are, respectively, equal to $-\pi$ and $+\pi$ at the Fermi surface, two components $\psi_\uparrow$ and $\psi_\downarrow$ cannot be connected by the s–d exchange interaction. This conservation of the local number of electrons is of essential importance in the Kondo problem and is fully discussed by Anderson [9] in his lecture at Les Houches. Thus, in order to avoid this difficulty, called "infrared catastrophe" [10], the phase shifts of $\phi^\beta_{n\uparrow}$ and $\phi^\beta_{n\downarrow}$ are chosen as shown in Fig. 1. In Anderson's practical calculation the energy dependence of this phase shift is varied so as to minimize the expectation value of the total energy.

A simpler restricted variational function for a singlet wave function has been used by Kondo [11]. His singlet wave function is taken as

$$\psi_{\text{Kondo}} = (1/\sqrt{2})(a^\dagger_{0\downarrow}\chi_\uparrow - a^\dagger_{0\uparrow}\chi_\downarrow) \prod a^\dagger_{n\uparrow} a^\dagger_{n\downarrow} \psi_0, \tag{1.10}$$

where the zero orbital represents a one-particle bound-state wave function which is treated as a variational function, one-electron scattered

9. THE GROUND STATE OF THE s–d MODEL

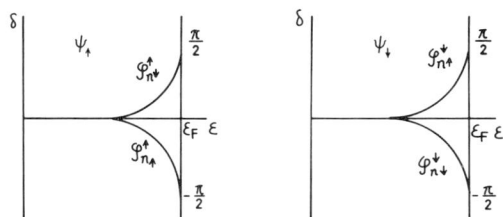

FIG. 1. The phase shifts of one-particle wave functions in $\psi_\uparrow$ and $\psi_\downarrow$.

state occupied by a pair of up and down spin electrons is determined by the orthogonality conditions, and ψ_0 denotes the true vacuum. This variational function has some weak points because of its simple form, one of which is that it leads to a spurious bound-state solution for ferromagnetic exchange $J > 0$.

In the perturbational approach [12–17], an unperturbed initial singlet wave function

$$\psi_\uparrow{}^0 = \sum_k \Gamma_k a_{k\downarrow}^\dagger \psi_\mathrm{v}, \qquad \psi_\downarrow{}^0 = \sum_k \Gamma_k a_{k\uparrow}^\dagger \psi_\mathrm{v} \qquad (1.11)$$

is taken, and $\psi_\uparrow$ and $\psi_\downarrow$ are calculated straightforwardly by the perturbation method by starting from this initial singlet. Therefore these are expressed by a series of the wave functions with excited electron–hole pairs besides a one-particle excitation; Γ_k and higher order amplitudes for these wave functions with pair excitations are determined by the Schrödinger equation. In this initial singlet, the electron orbitals below the Fermi level are simply the plane waves. One electron above the Fermi level is coupled with the localized spin, forming a bound state, and therefore, the phase shift at the Fermi energy in the Anderson sense is π for $\phi_\downarrow{}^\uparrow$ and $\phi_\uparrow{}^\downarrow$ and zero for $\phi_\uparrow{}^\uparrow$ and $\phi_\downarrow{}^\downarrow$, respectively. Thus this initial singlet wave function satisfies the Anderson condition, but violates charge neutrality (or electron–hole symmetry). However, since charge neutrality is restored in the perturbed final state, as is shown later, it would be unnecessary to force charge neutrality on the initial state in the perturbation theoretic approach if it is correctly carried through. Actually, for the purpose of removing shortcomings inherent to Kondo's variational function, which satisfies charge neutrality, Appelbaum and Kondo [18, 19] employed Eq. (1.10) in place of Eq. (1.11) as a starting singlet in perturbation calculation. However, this modification turned out to result in making actual calculation formidable and moreover had the disadvantage of losing the expansion parameter [20]. Heeger and Jensen [21] constructed a ground-state wave function possessing charge neutrality, but it violates the condition of particle number conservation and stays at the initial singlet.

In this chapter, the ground state of the *s–d* system is described and discussed mainly on the basis of the perturbation theoretic approach, because this approach has some advantages over the variational one. The perturbation theoretic approach is closely related to the scattering *t* matrix, the electronic structure of the ground state can more easily be demonstrated, calculation can be done symmetrically with respect to the transverse and longitudinal components of the *s–d* exchange interaction, and so on.

II. Perturbation Theoretic Approach for the Singlet Ground State

In the perturbation approach, the singlet ground-state wave functions $\psi_\uparrow$ and $\psi_\downarrow$ are calculated in an expansion form with respect to the exchange coupling J by starting from the initial singlet given by Eq. (1.11). The wave functions $\psi_\uparrow$ and $\psi_\downarrow$ are expanded as follows:

$$\psi_\uparrow = \Big[\sum_1 \Gamma_1 a^\dagger_{1\downarrow} + \sum_{123}(\Gamma^{\uparrow,\downarrow}_{12,3} a^\dagger_{1\downarrow} a^\dagger_{2\downarrow} a_{3\downarrow} + \Gamma^{\uparrow,\uparrow}_{12,3} a^\dagger_{1\uparrow} a^\dagger_{2\downarrow} a_{3\uparrow})$$

$$+ \sum_{12345}(\Gamma^{\uparrow,\downarrow\downarrow}_{123,45} a^\dagger_{1\downarrow} a^\dagger_{2\downarrow} a^\dagger_{3\downarrow} a_{4\downarrow} a_{5\downarrow}$$

$$+ \Gamma^{\uparrow,\downarrow\uparrow}_{123,45} a^\dagger_{1\downarrow} a^\dagger_{2\downarrow} a^\dagger_{3\uparrow} a_{4\downarrow} a_{5\uparrow} + \Gamma^{\uparrow,\uparrow\uparrow}_{123,45} a^\dagger_{1\downarrow} a^\dagger_{2\uparrow} a^\dagger_{3\uparrow} a_{4\uparrow} a_{5\uparrow})$$

$$+ \cdots \Big] \psi_v . \tag{2.1}$$

$\psi_\downarrow$ = expression derived from $\psi_\uparrow$ by exchanging up and down spins in the suffixes of annihilation and creation operators,

where 1, 2, and 3 are used for k_1, k_2, and k_3 for simplicity. On account of the singlet condition, some relations hold among the amplitudes $\Gamma_{12,3}$, $\Gamma_{123,45}$, and so on. For example,

$$\Gamma^{\uparrow,\downarrow}_{[12],3} = \Gamma^{\uparrow,\uparrow}_{12,3} - \Gamma^{\uparrow,\uparrow}_{21,3}, \tag{2.2}$$

where $\Gamma_{[12],3}$ means the antisymmetrized sum of $\Gamma_{12,3}$, though these relations are automatically satisfied by the solutions of the Schrödinger equation.

If we insert this expansion form in the Schrödinger equation

$$(H - E)\psi_{\text{singlet}} = 0, \tag{2.3}$$

9. THE GROUND STATE OF THE s–d MODEL 259

where H consists of the kinetic energy and the s–d exchange interaction given by Eq. (1.1), we obtain a hierarchy of equations which connect the amplitudes for the states with n particle–hole excitations to the amplitudes for those with $n \pm 1$ particle–hole excitations. The first two equations of the hierarchy are shown below:

$$\Gamma_1(\varepsilon_1 - E) + \frac{3J}{4N}\sum_2 \Gamma_2 + \frac{3J}{4N}\sum_{23}\Gamma^{\uparrow,\uparrow}_{12,3} = 0, \qquad (2.4)$$

$$\Gamma^{\uparrow,\uparrow}_{12,3}(\varepsilon_1 + \varepsilon_2 - \varepsilon_3 - E) + \frac{J}{4N}(2\Gamma_1 + \Gamma_2) + \frac{J}{4N}\sum_4 (\Gamma^{\uparrow,\uparrow}_{14,3} - \Gamma^{\uparrow,\uparrow}_{42,3} + \Gamma^{\uparrow,\uparrow}_{12,4}$$

$$- 2\Gamma^{\uparrow,\downarrow}_{[41],3} + 2\Gamma^{\uparrow,\uparrow}_{21,4}) + \frac{J}{4N}\sum_{45}(\Gamma^{\uparrow,\downarrow\uparrow}_{[24]1,53} + \Gamma^{\uparrow,\uparrow\uparrow}_{2[14],[53]}$$

$$- 2\Gamma^{\uparrow,\downarrow\uparrow}_{[41]2,35}) = 0, \qquad (2.5)$$

where wave number subscripts situated before a comma represent the states above the Fermi energy and those after a comma the states below the Fermi energy. Here we express higher order amplitudes in terms of the lower order amplitudes by iterative procedure, express $\Gamma^{\uparrow,\uparrow}_{12,3}$ in a power series in J which includes only Γ_1, and eliminate it from the first equation.

Thus we obtain the integral equation for Γ_1. This equation is somewhat complicated but its structure is as simple as

$$\Gamma_1(\varepsilon_1 - \tilde{E}) + \frac{3J}{4N}\sum_2 \Gamma_2 = \sum_2 \Gamma_2 K(\varepsilon_1, \varepsilon_2; \tilde{E}), \qquad (2.6)$$

where $\tilde{E}$ is defined by

$$E = \tilde{E} + \Delta E. \qquad (2.7)$$

ΔE denotes the energy shift of the state with one excited electron above the Fermi sea and is given by the following perturbation series:

$$\Delta E = -6\left(\frac{J}{4N}\right)^2 \sum_{23} \frac{1}{\varepsilon_2 - \varepsilon_3 + (\varepsilon_1 - \tilde{E})} + 12\left(\frac{J}{4N}\right)^3 \sum_{234} \frac{1}{\varepsilon_2 - \varepsilon_3 + (\varepsilon_1 - \tilde{E})}$$

$$\times \left[\frac{1}{\varepsilon_4 - \varepsilon_3 + (\varepsilon_1 - \tilde{E})} + \frac{1}{\varepsilon_2 - \varepsilon_4 + (\varepsilon_1 - \tilde{E})}\right] + \cdots. \qquad (2.8)$$

This ΔE is approximated by its limiting value of $\varepsilon_1 - \tilde{E} \to 0$ which is equal to the normal energy shift of the doublet state. This approximation corresponds to neglecting a quantity

$$\frac{3}{8}(\varepsilon - \tilde{E})\left(\frac{J\rho}{N}\right)^2 \log\frac{\varepsilon - \tilde{E}}{D}\left[1 + \frac{J\rho}{N}\log\frac{\varepsilon - \tilde{E}}{D} + \left(\frac{J\rho}{N}\log\frac{\varepsilon - \tilde{E}}{D}\right)^2 + \cdots\right], \quad (2.9)$$

which converges insofar as $\varepsilon - \tilde{E} > T_K$ ($-\tilde{E}$ will turn out to be given by T_K). If one adds this correction in the first terms of Eq. (2.6), $\Gamma_1(\varepsilon_1 - \tilde{E})$ is multiplied by an extra factor

$$a = 1 - \frac{3}{8}\left(\frac{J\rho}{N}\right)^2 \log\frac{\varepsilon - \tilde{E}}{D}\left[1 + \frac{J\rho}{N}\log\frac{\varepsilon - \tilde{E}}{D} + \cdots\right]. \quad (2.10)$$

Here, the second term is a next divergent quantity of order $J\rho/N$, because we are considering $(J\rho/N)\log(-\tilde{E}/D)$ to be a quantity of order unity. Therefore, this can be neglected against unity, if calculation is intended to be exact in the weak coupling limit $J\rho/N \to 0$. Incidentally, in the Kondo Brillouin–Wigner perturbation approach [22] to the ground state which starts from one of the doublet states, the anomalous energy-lowering $\tilde{E}$ is determined by the condition of $a = 0$ in which ε is put as zero. However, it will never be allowed to put less divergent quantities as equal to unity from the standpoint of the present theory.

By neglecting next divergent terms associated with $D_{12,3}$, and so on, arising from the same origin as (2.9), the integration kernel $K(\varepsilon_1, \varepsilon_2; \tilde{E})$ in Eq. (2.6) can be expressed in the following series in J:

$$K(\varepsilon_1, \varepsilon_2; \tilde{E}) = \frac{3}{16}\left(\frac{J}{N}\right)^2 \left[\sum_3 \frac{1}{D_{12,3}} + \frac{J}{4N}\sum_{3,4}\left(\frac{2}{D_{24,3}D_{14,3}}\right.\right.$$
$$\left.\left. + \frac{1}{D_{12,3}D_{42,3}} + \frac{1}{D_{14,3}D_{12,3}} - \frac{5}{D_{12,3}D_{12,4}}\right) + \cdots\right], \quad (2.11)$$

$$D_{12,3} = \varepsilon_1 + \varepsilon_2 - \varepsilon_3 - \tilde{E}. \quad (2.12)$$

The J^2 term in Eq. (2.11) corresponds to Fig. 2a and the four J^3 terms to (b1)–(b4) in Fig. 2, where solid lines represent an electron line to the right and a hole line to the left and the dashed lines represent a localized spin line.

Among these five terms, the first J^3 term (b1) does not give the most divergent term after integration over ε_3 and ε_4 and can be neglected consistently to the approximation in which (2.9) is neglected. The

9. THE GROUND STATE OF THE s–d MODEL

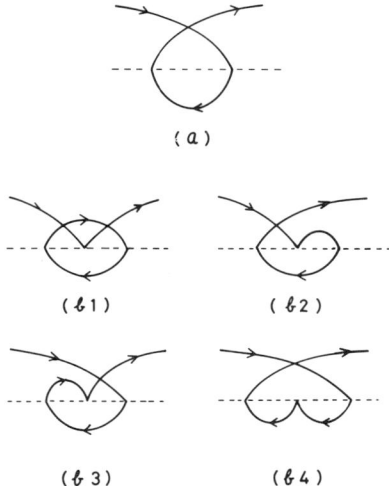

FIG. 2. Diagrams for the terms of J^2 and J^3 in the integration kernel. The solid line represents a conduction electron and the dashed line the localized spin.

other terms give the most divergent contributions after integration, and Eq. (2.11) is calculated in this approximation as

$$K(\varepsilon_1, \varepsilon_2; \tilde{E}) = -\frac{3}{16}\frac{J}{N}\left(\frac{J\rho}{N}\right)\log\frac{\varepsilon_1 + \varepsilon_2 - \tilde{E}}{D}$$
$$\times \left[1 + \frac{J\rho}{N}\log\frac{\varepsilon_1 + \varepsilon_2 - \tilde{E}}{D} + \cdots\right]. \quad (2.13)$$

Similarly, J^4 and J^5 terms can be calculated to give

$$[(J\rho/N)\log\{(\varepsilon_1 + \varepsilon_2 - \tilde{E})/D\}]^2 \quad \text{and} \quad [(J\rho/N)\log\{(\varepsilon_1 + \varepsilon_2 - \tilde{E})/D\}]^3,$$

respectively. Thus it is confirmed that the most divergent terms in $K(\varepsilon_1, \varepsilon_2; \tilde{E})$ form a geometric series. The graphical representation for the integration kernel described in Fig. 2 tells us that the kernel has the same structure as that of the vertex function $\Lambda_2(\omega)$ introduced by Abrikosov [23], as noted by Nakajima [24].

With the use of this form of the integration kernel, the integral equation for the amplitude $\Gamma(\varepsilon)$ is brought into a closed form as

$$(\varepsilon - \tilde{E})\Gamma(\varepsilon) + \frac{3J\rho}{4N}\int_0^D \Gamma(\varepsilon')\,d\varepsilon' + \frac{3}{16}\left(\frac{J\rho}{N}\right)^2\int_0^D d\varepsilon'\,\Gamma(\varepsilon')\log\frac{\varepsilon + \varepsilon' - \tilde{E}}{D}$$
$$\times \left[1 - \frac{J\rho}{N}\log\frac{\varepsilon + \varepsilon' - \tilde{E}}{D}\right]^{-1} = 0. \quad (2.14)$$

To solve this equation we introduce a function defined by

$$G(\varepsilon) = \int_0^\varepsilon \Gamma(\varepsilon') \, d\varepsilon' \tag{2.15}$$

and insert it into Eq. (2.14). Then the third term can be manipulated by partial integration as

$$-\frac{3}{16}\left(\frac{J\rho}{N}\right)^2 \int_0^D d\varepsilon' \frac{G(\varepsilon')}{\varepsilon + \varepsilon' - \tilde{E}} \left(1 - \frac{J\rho}{N} \log \frac{\varepsilon + \varepsilon' - \tilde{E}}{D}\right)^{-2}. \tag{2.16}$$

This can be replaced in the weak coupling limit by

$$-\frac{3}{16}\left(\frac{J\rho}{N}\right)^2 \int_\varepsilon^D d\varepsilon' \frac{G(\varepsilon')}{\varepsilon' - \tilde{E}} \left(1 - \frac{J\rho}{N} \log \frac{\varepsilon' - \tilde{E}}{D}\right)^{-2}. \tag{2.17}$$

This replacement enables us to transform Eq. (2.14) into a differential equation for $G(\varepsilon)$ as

$$(\varepsilon - \tilde{E})^2 \frac{d^2 G}{d\varepsilon^2} + (\varepsilon - \tilde{E}) \frac{dG}{d\varepsilon} + \frac{3}{16}\left(\frac{J\rho}{N}\right)^2 G(\varepsilon) \left[1 - \frac{J\rho}{N} \log \frac{\varepsilon - \tilde{E}}{D}\right]^{-2} = 0. \tag{2.18}$$

This can readily be solved and we obtain a solution satisfying the boundary conditions $G(0) = 0$ and

$$\left[(\varepsilon - \tilde{E}) \frac{dG}{d\varepsilon}\right]_{\varepsilon = D} = -\frac{3J\rho}{4N} G(D), \tag{2.19}$$

as

$$G(\varepsilon) = G(D)[1 - \tfrac{1}{3}(1-x)^{1/2}]^{-1}$$
$$\times \left[\left(1 - \frac{J\rho}{N} \log \frac{\varepsilon - \tilde{E}}{D}\right)^{3/4} - (1-x)^{1/2} \left(1 - \frac{J\rho}{N} \log \frac{\varepsilon - \tilde{E}}{D}\right)^{1/4}\right]. \tag{2.20}$$

The amplitude $\Gamma(\varepsilon)$ is derived from this as

$$\Gamma(\varepsilon) = \frac{1}{\varepsilon - \tilde{E}} \left[\left(1 - \frac{J\rho}{N} \log \frac{\varepsilon - \tilde{E}}{D}\right)^{-1/4} - \frac{1}{3}(1-x)^{1/2} \left(1 - \frac{J\rho}{N} \log \frac{\varepsilon - \tilde{E}}{D}\right)^{-3/4}\right], \tag{2.21}$$

where x is defined by

$$x = (J\rho/N) \log(-\tilde{E}/D) \tag{2.22}$$

9. THE GROUND STATE OF THE s–d MODEL

and determined by the eigenvalue equation

$$1 - \frac{1-(1-x)^{1/2}}{1-(1-x)^{1/2}/3} = 0. \tag{2.23}$$

The anomalous part of the energy eigenvalue is then given by $x = 1$, the end of the branch cut,

$$\tilde{E} = -De^{N/J\rho} = -T_K, \tag{2.24}$$

which provides us with a physical meaning of the Kondo temperature T_K at which the scattering amplitude diverges.

These results are exact in the most divergent approximation. It should be remarked that the second term of Eq. (2.21) for $\Gamma(\varepsilon)$ cannot be dropped, because it gives a quantity of the same order as the first term at $\varepsilon = 0$, and that the normalization integral for the present singlet wave function with the use of this $\Gamma(\varepsilon)$ diverges in this weak coupling limit as

$$\langle \psi | \psi \rangle = \frac{\rho}{-\tilde{E}} \frac{8}{9} (1-x)^{-1/2}, \tag{2.25}$$

where contributions from the higher order amplitudes are omitted because they are less divergent. We believe that this divergence could in principle be removed by taking into account the less divergent terms and does not imply any difficulty in the present theory. As a matter of fact, we can calculate physical quantities without any ambiguity with the use of this diverging normalization integral, as is shown later.

The same calculation as above can also be carried out by starting from a triplet wave function

$$\psi_0^t = \sum_k \Gamma_k (a_{k\downarrow}^\dagger \chi_\uparrow + a_{k\uparrow}^\dagger \chi_\downarrow) \psi_v, \tag{2.26}$$

which is degenerate in energy with the other two wave functions

$$\sum_k \Gamma_k a_{k\uparrow}^\dagger \chi_\uparrow \psi_v, \quad \sum_k \Gamma_k a_{k\downarrow}^\dagger \chi_\downarrow \psi_v. \tag{2.27}$$

The eigenvalue equation for this case is obtained in place of Eq. (2.23) as

$$1 + \frac{1-(1-x)^{3/2}}{5+(1-x)^{3/2}} = 0. \tag{2.28}$$

This equation has no solution for either sign of J, although for positive J the wave functions (2.26) and (2.27) have a bound-state solution in the zero approximation in which the states with excited electron–hole pairs are neglected. An energy gain made by forming a collective bound state survives in the final state only when the lowest bound state in the zero approximation is not degenerate.

The eigenvalue equation for a general value of S can also be derived in a similar way, although this case seems to be unphysical in the present specialized Hamiltonian (1.1), as

$$1 - \frac{1 - (1-x)^S}{1 - [(1-S)/(1+S)](1-x)^S} = 0 \quad \text{for} \quad S - \frac{1}{2}, \quad (2.29)$$

$$1 + \frac{1 - (1-x)^{S+1}}{(S+2)/S + (1-x)^{S+1}} = 0 \quad \text{for} \quad S + \frac{1}{2}. \quad (2.30)$$

Equation (2.29) has one solution for $x = 1$. For $S \to \infty$, $J \to 0$, keeping $JS = $ constant, this becomes

$$1 - \tanh(Sx/2) = 0, \quad (2.31)$$

to show that in this classical limit a bound-state solution disappears

Thus it is concluded that the ground-state energy of the s–d system is given by

$$E_g = \Delta E - D e^{N/J\rho} \quad (2.32)$$

for $J < 0$. The anomalous part denoted by $\tilde{E}$ may be interpreted as a binding energy of the singlet bound state. In this state, which is not a one-particle bound state (though it was in the starting approximation), but is collective in its nature, the spin polarization amounting to one electron spin is bound by the impurity. ΔE is defined by the power series in J given by Eq. (2.8) and is common in its form to both the ferromagnetic and antiferromagnetic exchange. This normal part does not include any logarithmic singular terms in the limit of $\varepsilon - \tilde{E} = 0$ [25, 26].

III. Bound State for the Anisotropic Exchange Interaction

The theory of singlet ground state by the perturbation theoretic approach is closely related to the scattering amplitude by Abrikosov as mentioned in the preceding section. This relation will also be maintained in more general cases for the anisotropic s–d exchange interaction.

For the anisotropic exchange Eq. (1.1) is extended by referring to the principal axes x, y, and z as

$$H_{sd} = -\frac{1}{2N} \sum_i \sum_{kk'\alpha\alpha'} J_i a_{k\alpha}^\dagger \sigma_{\alpha\alpha'}^i a_{k'\alpha'} \cdot S_i, \quad (3.1)$$

9. THE GROUND STATE OF THE s–d MODEL

where i represents x, y, and z components. For this Hamiltonian the zero-approximation wave function is described by a linear combination of the following four wave functions:

$$\psi_0^{\alpha\beta} = \sum_k \Gamma_k^{\alpha\beta} a_{k\alpha}^\dagger \chi_\beta \psi_v , \tag{3.2}$$

where as before α denotes the spin state of an excited electron and β the localized spin state.

When J_x, J_y, and J_z are all different, a bound state with the largest binding energy is nondegenerate in the zero approximation; but for a uniaxial case in which two of J, for example, J_x and J_y, are equal, there is a region in which two lowest bound states are degenerate, the hatched region and the negative J_z axis in Fig. 3. The existence of this boundary which separates two regions in the $J_\perp$–J_z plane leading to essentially different ground states has also been pointed out and discussed by Anderson et al. [27] on the basis of their scaling law.

Starting with the zero-approximation wave function given by Eq. (3.2) or their linear combination, we can proceed similarly in principle to the case of the isotropic case. This extension has been made by Shiba [28]. We describe his theory in this section.

The integral equation for the lowest amplitudes $\Gamma_{\alpha\beta}(\varepsilon)$ can readily be written down as

$$(\varepsilon - \tilde{E}) \Gamma_{\alpha\beta}(\varepsilon) - \frac{1}{2N} J_i \sigma_{\alpha\alpha'}^i S_{\beta\beta'}^i \rho \int_0^D d\varepsilon' \, \Gamma_{\alpha'\beta'}(\varepsilon')$$
$$- \rho \int_0^D d\varepsilon' \, K_{\alpha\beta\alpha'\beta'}(\varepsilon + \varepsilon' - \tilde{E}) \Gamma_{\alpha'\beta'}(\varepsilon') = 0, \tag{3.3}$$

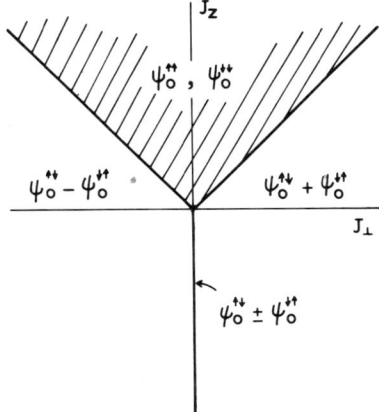

FIG. 3. The bound-state wave functions in the zero approximation and their degeneracy for the s–d exchange of uniaxial symmetry.

where here and in the following, repeated superscripts and subscripts are summed. However, it is difficult to bring the most divergent terms in the integration kernel $K_{\alpha\beta\alpha'\beta'}$ in a closed form by inspecting first three or four perturbation terms in this series. This difficulty can be overcome as follows. We observe that the integration kernel in the isotropic case has the same structure as the vertex function Λ_2 introduced by Abrikosov and therefore that $K(\omega)$ can be identified with this Λ_2. The integration kernel for the singlet amplitude, Eq. (2.11), is given by its singlet component

$$K_{\text{singlet}}(\omega) = \Lambda_{2,\downarrow\uparrow\downarrow\uparrow} - \Lambda_{2,\downarrow\uparrow\uparrow\downarrow}. \quad (3.4)$$

Abrikosov's Λ_2 is connected to his vertex function or the spin-flip amplitude, $\Gamma^A(\omega)$, in the approximation so that less divergent terms are neglected, by the relation

$$\Lambda_{2\alpha\beta\alpha'\beta'}(\omega) = -\rho \int_{|\omega|}^{D} \frac{d\omega_1}{\omega_1} \Gamma^A_{\alpha\beta''\alpha''\beta'}(\omega_1) \Gamma^A_{\alpha''\beta\alpha'\beta''}(\omega_1). \quad (3.5)$$

$\Gamma^A(\omega)$ satisfies the following integral equation in the same approximation:

$$\Gamma^A_{\alpha\beta\alpha'\beta'}(\omega) = \frac{J}{2N} (\sigma \cdot S)_{\alpha\beta\alpha'\beta'} + \rho \int_{|\omega|}^{D} \frac{d\omega_1}{\omega_1} [\Gamma^A_{\alpha\beta\alpha''\beta''}(\omega_1) \Gamma^A_{\alpha''\beta''\alpha'\beta'}(\omega_1)$$

$$- \Gamma^A_{\alpha\beta''\alpha''\beta'}(\omega_1) \Gamma^A_{\alpha''\beta\alpha'\beta''}(\omega_1)]. \quad (3.6)$$

Putting

$$\Gamma^A_{\alpha\beta\alpha'\beta'}(\omega) = \Gamma^A(\omega)(\sigma \cdot S)_{\alpha\beta\alpha'\beta'}, \quad (3.7)$$

we obtain the integral equation for $\Gamma^A(\omega)$ as

$$\Gamma^A(\omega) = \frac{J}{2N} - 2\rho \int_{|\omega|}^{D} \frac{d\omega_1}{\omega_1} \Gamma^A(\omega_1)^2. \quad (3.8)$$

Inserting the solution of this equation

$$\Gamma^A(\omega) = \frac{J}{2N} \left[1 - \frac{J\rho}{N} \log \left| \frac{\omega}{D} \right| \right]^{-1} \quad (3.9)$$

into Eq. (3.5), we obtain $K_{\text{singlet}}(\omega)$ from Eq. (3.4) as

$$K_{\text{singlet}}(\omega) = -\frac{3}{16} \frac{J}{N} \left(\frac{J\rho}{N} \right) \log \left| \frac{\omega}{D} \right| \left[1 - \frac{J\rho}{N} \log \left| \frac{\omega}{D} \right| \right]^{-1}. \quad (3.10)$$

This is identified with the integration kernel of Eq. (2.14) derived directly.

9. THE GROUND STATE OF THE s–d MODEL

For the anisotropic s–d exchange, the inhomogeneous term of Eq. (3.6) is simply replaced by

$$\frac{1}{2N} J_i \sigma^i_{\alpha\alpha'} S^i_{\beta\beta'} . \tag{3.11}$$

Corresponding to this modification, Eq. (3.7) is now generalized as

$$\Gamma^A_{\alpha\beta\alpha'\beta'}(\omega) = \Gamma^A_i(\omega) \sigma^i_{\alpha\alpha'} \cdot S^i_{\beta\beta'} . \tag{3.12}$$

Then, the integral equation for $\Gamma^A_i(\omega)$ becomes

$$\Gamma^A_i(\omega) = \frac{J_i}{2N} - 2N \int_0^t ds\, \Gamma^A_j(s)\, \Gamma^A_k(s), \tag{3.13}$$

where t is defined by $t = (\rho/N) \log(D/|\omega|)$ and i, j, and k represent x, y, and z in their cyclic order.

We assume $J_x^2 \geq J_y^2 \geq J_z^2$ without loss of generality. Then the solution of these integral equations is expressed by

$$\Gamma^A_x(t)^2 = \frac{1}{4N^2}(J_x^2 - J_z^2) \frac{1}{\operatorname{sn}^2[(t - t_c)(J_x^2 - J_z^2)^{1/2}, k]},$$

$$\Gamma^A_y(t)^2 = \frac{1}{4N^2}(J_x^2 - J_z^2) \frac{\operatorname{dn}^2[(t - t_c)(J_x^2 - J_z^2)^{1/2}, k]}{\operatorname{sn}^2[(t - t_c)(J_x^2 - J_z^2)^{1/2}, k]}, \tag{3.14}$$

$$\Gamma^A_z(t)^2 = \frac{1}{4N^2}(J_x^2 - J_z^2) \frac{\operatorname{cn}^2[(t - t_c)(J_x^2 - J_z^2)^{1/2}, k]}{\operatorname{sn}^2[(t - t_c)(J_x^2 - J_z^2)^{1/2}, k]},$$

where $\operatorname{sn}(z)$, $\operatorname{cn}(z)$, and $\operatorname{dn}(z)$ are the Jacobian elliptic functions [29] and related to each other by

$$\operatorname{cn}^2(z) = 1 - \operatorname{sn}^2(z), \qquad \operatorname{dn}^2(z) = 1 - k^2 \operatorname{sn}^2(z). \tag{3.15}$$

The modulus k and t_c are given, respectively, by

$$k = \left(\frac{J_x^2 - J_y^2}{J_x^2 - J_z^2}\right)^{1/2}, \qquad 0 \leq k \leq 1, \tag{3.16}$$

$$t_c = \pm \frac{1}{(J_x^2 - J_z^2)^{1/2}} \int_0^{(J_x^2 - J_z^2)^{1/2}/|J_x|} \frac{dt}{(1 - t^2)^{1/2}(1 - k^2 t^2)^{1/2}}, \tag{3.17}$$

where the $\pm$ sign is chosen according as $J_x J_y J_z \lessgtr 0$.

The vertex functions given by Eq. (3.14) have poles at zeros of sn function. Therefore, in general they have diverging points on the positive real axis of t ($\infty > t \geq 0$). An exceptional case is that in which the real period $4K$ of sn function becomes infinite and in addition t_c is negative. This condition is given by

$$J_x J_y J_z \geq 0,$$
$$J_x^2 > J_y^2 = J_z^2, \quad \text{or} \quad k = 1, \tag{3.18}$$

and satisfied in the hatched region and on the negative J_z axis in Fig. 3, coinciding with the condition that the lowest bound state in the zero approximation is spin degenerate. The first pole occurs at $t = t_c$ for $t_c > 0$ and at $t = 2K/(J_x^2 - J_z^2)^{1/2} + t_c$ for $t_c < 0$, where K is given by

$$K = \int_0^{\pi/2} (1 - k^2 \sin^2 \phi)^{-1/2} d\phi. \tag{3.19}$$

Corresponding to this, the largest value of ω at which the scattering amplitude diverges is given by

$$|\omega_c| = D \exp\left(\frac{-t_c}{\rho/N}\right) \quad \text{for} \quad t_c > 0,$$
$$|\omega_c| = D \exp\left[-\frac{2K/(J_x^2 - J_z^2)^{1/2} + t_c}{\rho/N}\right] \quad \text{for} \quad t_c < 0. \tag{3.20}$$

For the uniaxial case, $J_x^2 = J_y^2 = J_\perp^2 > J_z^2$, where k becomes zero, we have

$$\operatorname{sn}(z, 0) = \sin z, \quad \operatorname{cn}(z, 0) = \cos z, \quad \operatorname{dn}(z, 0) = 1, \tag{3.21}$$

and t_c is calculated as

$$t_c = -\frac{\operatorname{sgn} J_z}{(J_\perp^2 - J_z^2)^{1/2}} \operatorname{Sin}^{-1} \frac{(J_\perp^2 - J_z^2)^{1/2}}{|J_\perp|}, \tag{3.22}$$

and therefore, the scattering amplitudes or vertex functions are obtained as

$$\Gamma_\perp^A(t) = -\frac{1}{2N}(J_\perp^2 - J_z^2)^{1/2} \frac{\operatorname{sgn}(J_\perp J_z)}{\sin\{\operatorname{Tan}^{-1}[(J_\perp^2 - J_z^2)^{1/2}/-J_z] - (J_\perp^2 - J_z^2)^{1/2}t\}}, \tag{3.23}$$

$$\Gamma_z^A(t) = -\frac{1}{2N}(J_\perp^2 - J_z^2)^{1/2} \cot\left[\operatorname{Tan}^{-1}\frac{(J_\perp^2 - J_z^2)^{1/2}}{-J_z} - (J_\perp^2 - J_z^2)^{1/2}t\right].$$

For $J_x^2 > J_y^2 = J_z^2 = J_\perp^2$ where $k = 1$, we have

$$t_c = \frac{1}{(J_x^2 - J_\perp^2)^{1/2}} \tanh^{-1} \frac{(J_x^2 - J_\perp^2)^{1/2}}{-J_x} \tag{3.24}$$

9. THE GROUND STATE OF THE s–d MODEL

and

$$\Gamma_\perp{}^A(t) = -\frac{1}{2N}(J_x{}^2 - J_\perp{}^2)^{1/2} \frac{\text{sgn}(J_\perp J_x)}{\sinh\{\tanh^{-1}[(J_x{}^2 - J_\perp{}^2)^{1/2}/-J_x] - (J_x{}^2 - J_\perp{}^2)^{1/2}t\}},$$

(3.25)

$$\Gamma_x{}^A(t) = \frac{-1}{2N}(J_x{}^2 - J_\perp{}^2)^{1/2} \coth\left[\tanh^{-1}\frac{(J_x{}^2 - J_\perp{}^2)^{1/2}}{-J_x} - (J_x{}^2 - J_\perp{}^2)^{1/2}t\right].$$

Since $\Gamma_i{}^A(t)$ have been obtained as Eqs. (3.23) and (3.25), we can solve the integral equation (3.3) for $\Gamma_{\alpha\beta}(\varepsilon)$ for the anisotropic s–d exchange interaction of uniaxial symmetry with the use of $\Lambda_2(\omega)$ given by the relation (3.5) for $K(\omega)$. The results are as follows. In the region where the scattering amplitude has no pole on the real positive axis of t, that is, the hatched region and the negative J_z axis, a bound-state solution does not appear; but in other regions a nondegenerate bound-state solution appears whose binding energy, denoted by $\tilde{E}$, is given in Table I.

TABLE I

BINDING ENERGY $\tilde{E}$ FOR THE ANISOTROPIC s–d EXCHANGE OF UNIAXIAL SYMMETRY[a]

	$J_z < 0$	$J_z > 0$
$\|J_\perp\| > \|J_z\|$	$-D \exp\left[\frac{-N}{(J_\perp{}^2 - J_z{}^2)^{1/2}\rho} \times \text{Tan}^{-1}\frac{(J_\perp{}^2 - J_z{}^2)^{1/2}}{-J_z}\right]$	$-D \exp\left\{\frac{-N}{(J_\perp{}^2 - J_z{}^2)^{1/2}\rho} \times \left[\pi + \text{Tan}^{-1}\frac{(J_\perp{}^2 - J_z{}^2)^{1/2}}{-J_z}\right]\right\}$
$\|J_\perp\| < \|J_z\|$	$-D \exp\left[\frac{-N}{(J_z{}^2 - J_\perp{}^2)^{1/2}\rho} \times \tanh^{-1}\frac{(J_z{}^2 - J_\perp{}^2)^{1/2}}{-J_z}\right]$	0

[a] Shiba [28].

Thus it can be concluded that the essential function of the s–d exchange interaction is to remove the degeneracy of the localized spin. A nondegenerate bound state is generally formed by this interaction, and the cases in which the spin degeneracy survives may be regarded as rather exceptional.

IV. Charge, Spin Polarization, and Spin Correlation Densities

In this section, we return to the isotropic case and consider the electronic structure of the singlet ground state.

The wave function for the singlet ground state can be written down in terms of the lowest amplitude $\Gamma(\varepsilon)$, and therefore various physical quantities can be calculated. The anomalous part of the kinetic and exchange energy can be obtained with the use of the relation

$$J \frac{d\tilde{E}}{dJ} = \tilde{E}_{\text{ex}} \tag{4.1}$$

as

$$\tilde{E}_{\text{ex}} = \frac{-\tilde{E}}{J\rho/N},$$

$$\tilde{E}_{\text{kin}} = \frac{\tilde{E}}{(J\rho/N)} + \tilde{E}. \tag{4.2}$$

The spin density of the conduction electrons at the impurity site in the $\psi_\uparrow$ and $\psi_\downarrow$ components of the ground-state wave function,

$$\sigma_\uparrow(0) = -\sigma_\downarrow(0) = \frac{1}{2V} \sum_{k'k} \frac{\langle \psi_\uparrow | a^\dagger_{k'\uparrow} a_{k\uparrow} - a^\dagger_{k'\downarrow} a_{k\downarrow} | \psi_\uparrow \rangle}{\langle \psi_\uparrow | \psi_\uparrow \rangle} \tag{4.3}$$

can be derived from $\tilde{E}_{\text{ex}}$ as

$$\sigma_\uparrow(0) = -\frac{2}{3} \frac{\rho}{V} \frac{-\tilde{E}}{(J\rho/N)^2}. \tag{4.4}$$

On the other hand, as for the charge density of the conduction electrons at the impurity site in each component,

$$\rho_\uparrow(0) = \rho_\downarrow(0) = \frac{1}{V} \sum_{kk'} \frac{\langle \psi_\uparrow | a^\dagger_{k'\uparrow} a_{k\uparrow} + a^\dagger_{k'\downarrow} a_{k\downarrow} | \psi_\uparrow \rangle}{\langle \psi_\uparrow | \psi_\uparrow \rangle}, \tag{4.5}$$

direct calculation is needed with the use of the ground-state wave function. For this purpose, we express $\langle \psi_\uparrow | a^\dagger_{k'\uparrow} a_{k\uparrow} | \psi_\uparrow \rangle$ and $\langle \psi_\uparrow | a^\dagger_{k'\downarrow} a_{k\downarrow} | \psi_\uparrow \rangle$ in terms of the amplitudes Γ_1, Γ_{123}, and so on, using Eq. (2.1), as

$$\langle \psi_\uparrow | a^\dagger_{k'\downarrow} a_{k\downarrow} | \psi_\uparrow \rangle_{\text{ee}} = \Gamma_{k'} \Gamma_k + \sum_{12} (\Gamma^{\uparrow,\downarrow}_{[k'1],2} \Gamma^{\uparrow,\downarrow}_{[k1],2}$$

$$+ \Gamma^{\uparrow,\uparrow}_{1k',2} \Gamma^{\uparrow,\uparrow}_{1k,2}) + \cdots, \tag{4.6}$$

9. THE GROUND STATE OF THE s–d MODEL

$$\langle \psi_\uparrow | a_{k'\downarrow}^\dagger a_{k\downarrow} | \psi_\uparrow \rangle_{\text{eh}} = -\sum_1 \Gamma_1 \Gamma_{[k'1],k}^{\uparrow,\downarrow} + \cdots, \tag{4.7}$$

$$\langle \psi_\uparrow | a_{k'\downarrow}^\dagger a_{k\downarrow} | \psi_\uparrow \rangle_{\text{hh}} = -\frac{1}{2}\sum_{12} \Gamma_{[12],k'}^{\uparrow,\downarrow} \Gamma_{[12]k}^{\uparrow,\downarrow} + \cdots, \tag{4.8}$$

$$\langle \psi_\uparrow | a_{k'\uparrow}^\dagger a_{k\uparrow} | \psi_\uparrow \rangle_{\text{ee}} = \sum_{12} \Gamma_{k'1,2}^{\uparrow,\uparrow} \Gamma_{k1,2}^{\uparrow,\uparrow} + \cdots, \tag{4.9}$$

$$\langle \psi_\uparrow | a_{k'\uparrow}^\dagger a_{k\uparrow} | \psi_\uparrow \rangle_{\text{eh}} = -\sum_1 \Gamma_1 \Gamma_{k'1k}^{\uparrow,\uparrow} + \cdots, \tag{4.10}$$

$$\langle \psi_\uparrow | a_{k'\uparrow}^\dagger a_{k\uparrow} | \psi_\uparrow \rangle_{\text{hh}} = -\sum_{12} \Gamma_{12,k'}^{\uparrow,\uparrow} \Gamma_{12,k}^{\uparrow,\uparrow} + \cdots, \tag{4.11}$$

where the subscripts e and h denote the electron and hole states for k' and k, and in Eqs. (4.8) and (4.11) the contributions from the Fermi vacuum are omitted.

With the use of these expressions, $\rho_\uparrow(0)$ can be obtained as

$$\rho_\uparrow(0) = \frac{1}{V}\frac{1}{\langle \psi_\uparrow | \psi_\uparrow \rangle} \Big[\sum_{12}\Gamma_1\Gamma_2 + 2\sum_{123}\Gamma_1(\Gamma_{12,3}^{\uparrow,\uparrow} - 2\Gamma_{21,3}^{\uparrow,\uparrow})$$

$$+ \sum_{1234}\Gamma_{12,3}^{\uparrow,\uparrow}(\Gamma_{21,4}^{\uparrow,\uparrow} - 2\Gamma_{12,4}^{\uparrow,\uparrow} + 2\Gamma_{14,3}^{\uparrow,\uparrow} - \Gamma_{41,3}^{\uparrow,\uparrow} + 2\Gamma_{42,3}^{\uparrow,\uparrow} - \Gamma_{24,3}^{\uparrow,\uparrow}) + \cdots\Big]. \tag{4.12}$$

By expressing the higher order amplitudes $\Gamma_{12,3}$ in terms of the lowest order amplitude $\Gamma(\varepsilon)$ and retaining the most divergent terms among those appearing after integration over k_3, k_4, and so on, we obtain

$$\rho_\uparrow(0) = \frac{1}{V}\frac{1}{\langle \psi_\uparrow | \psi_\uparrow \rangle}\Big[\Big(\sum_1 \Gamma_1\Big)^2 + \frac{3}{2}\sum_{12}\Gamma_1\Gamma_2 \log\Big(1 - \frac{J\rho}{N}\log\frac{\varepsilon_1 + \varepsilon_2 - \tilde{E}}{D}\Big)\Big]. \tag{4.13}$$

If we use $\Gamma(\varepsilon)$ as given by Eq. (2.21) in Eq. (4.13), we can see that Eq. (4.13) vanishes. Furthermore, combining this result and Eq. (4.4) we have the relation

$$\sum_{k'k}\frac{\langle \psi_\uparrow | a_{k'\uparrow}^\dagger a_{k\uparrow} | \psi_\uparrow \rangle}{\langle \psi_\uparrow | \psi_\uparrow \rangle} = -\sum_{k'k}\frac{\langle \psi_\uparrow | a_{k'\downarrow}^\dagger a_{k\downarrow} | \psi_\uparrow \rangle}{\langle \psi_\uparrow | \psi_\uparrow \rangle}$$

$$= -\frac{2}{3}\rho\frac{-\tilde{E}}{(J\rho/N)^2}. \tag{4.14}$$

This result that the charge density at the impurity site vanishes in each component suggests that the total charge bound by the impurity also vanishes in each component, although one electron charge is localized in the initial state from which we started.

The situation in which the total bound charge vanishes in the final stage can most easily be understood by the following consideration. First we observe the relation

$$\langle \psi | [H, a^\dagger_{k'\uparrow} a_{k\uparrow} + a^\dagger_{k'\downarrow} a_{k\downarrow}] | \psi \rangle = 0, \tag{4.15}$$

which generally holds for an eigenfunction of the Hamiltonian. Applying this relation to the present singlet wave function which represents the ground state for the s–d system, we obtain the following relation:

$$\frac{\langle \psi_\uparrow | a^\dagger_{k'\uparrow} a_{k\uparrow} + a^\dagger_{k'\downarrow} a_{k\downarrow} | \psi_\uparrow \rangle}{\langle \psi_\uparrow | \psi_\uparrow \rangle} = -\frac{3J}{4N} \frac{M_{k'} - M_k}{\varepsilon_{k'} - \varepsilon_k}, \tag{4.16}$$

where M_k is defined by

$$M_k = \sum_l \frac{\langle \psi_\uparrow | a^\dagger_{l\uparrow} a_{k\uparrow} - a^\dagger_{l\downarrow} a_{k\downarrow} | \psi_\uparrow \rangle}{\langle \psi_\uparrow | \psi_\uparrow \rangle}. \tag{4.17}$$

The localized charge in $\psi_\uparrow$ (or $\psi_\downarrow$) is given by

$$\sum_k \lim_{k' \to k} \frac{\langle \psi_\uparrow | a^\dagger_{k'\uparrow} a_{k\uparrow} + a^\dagger_{k'\downarrow} a_{k\downarrow} | \psi_\uparrow \rangle}{\langle \psi_\uparrow | \psi_\uparrow \rangle}. \tag{4.18}$$

When $\varepsilon_{k'} > 0$, $\varepsilon_k > 0$ and $\varepsilon_{k'} < 0$, $\varepsilon_k < 0$, this is calculated as

$$(n_{\uparrow,\uparrow} + n_{\uparrow,\downarrow})_{ee} + (n_{\uparrow,\uparrow} + n_{\uparrow,\downarrow})_{hh}$$

$$= -\frac{3J\rho}{4N} \left(\int_0^D \frac{dM_k}{d\varepsilon_k} d\varepsilon_k + \int_{-D}^0 \frac{dM_k}{d\varepsilon_k} d\varepsilon_k \right)$$

$$= -\frac{3J\rho}{4N} \{[M(D) - M(-D)] - [M(0_+) - M(0_-)]\}, \tag{4.19}$$

and when $\varepsilon_{k'} > 0$, $\varepsilon_k < 0$ and $\varepsilon_{k'} < 0$, $\varepsilon_k > 0$, as

$$(n_{\uparrow,\uparrow} + n_{\uparrow,\downarrow})_{eh}$$

$$= -\frac{3J}{4N} (M(0_+) - M(0_-)) \sum_k \lim_{k' \to k} \left[\left(\frac{1}{\varepsilon_{k'} - \varepsilon_k}\right)_{\substack{\varepsilon_{k'}>0 \\ \varepsilon_k<0}} - \left(\frac{1}{\varepsilon_{k'} - \varepsilon_k}\right)_{\substack{\varepsilon_{k'}<0 \\ \varepsilon_k>0}} \right]$$

$$= -\frac{3J\rho}{4N} (M(0_+) - M(0_-)). \tag{4.20}$$

9. THE GROUND STATE OF THE s-d MODEL

For a symmetrical band, $M(D) = M(-D)$ and the two terms arising from the discontinuity of $M(\varepsilon_k)$ at $\varepsilon_k = 0$ in Eqs. (4.19) and (4.20) cancel each other. Thus we can see the situation in which the total bound charge vanishes in each component. This result can also be directly derived with the use of the present ground-state wave function. For the total localized spin in each component, direct calculation is needed.

The electron–electron and hole–hole contributions to the total charge and the total spin are given by

$$(n_{\uparrow,\uparrow} \pm n_{\uparrow,\downarrow})_{ee} + (n_{\uparrow,\uparrow} \pm n_{\uparrow,\downarrow})_{hh} = \pm 1. \tag{4.21}$$

The electron–hole contribution to them will come from the discontinuity at the Fermi surface,

$$n_{\uparrow\pm} = \lim_{k' \to k} \left[\frac{\langle \psi_\uparrow | a^\dagger_{k'\uparrow} a_{k\uparrow} \pm a^\dagger_{k'\downarrow} a_{k\downarrow} | \psi_\uparrow \rangle_{eh}}{\langle \psi_\uparrow | \psi_\uparrow \rangle} \right] (\varepsilon_{k'} - \varepsilon_k). \tag{4.22}$$

Direct calculation actually shows, in the most divergent approximation,

$$n_{\uparrow+} = -1/\rho, \qquad n_{\uparrow-} = 0. \tag{4.23}$$

These results confirm that the total bound charge vanishes and that the spin density of the total amount equal to one electron spin is bound. Therefore we can see that in the $\psi_\uparrow$ component of the singlet ground state half an electron with down spin and half a hole with up spin are bound by the impurity. This situation is schematically shown in Fig. 4.

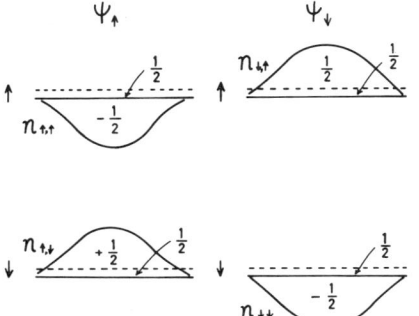

FIG. 4. The electron distributions around the impurity.

In this way, the present perturbation theoretic approach automatically leads us to the state which satisfies Anderson's condition

$$\delta_{\uparrow,\uparrow(\downarrow)} = (\mp)\pi/2, \qquad \delta_{\downarrow,\uparrow(\downarrow)} = (\pm)\pi/2, \tag{4.24}$$

where $\delta_{\uparrow,\uparrow(\downarrow)}$ and $\delta_{\downarrow,\uparrow(\downarrow)}$ represent the phase shift of the electron with up and down spin at the Fermi surface in $\psi_\uparrow$ and $\psi_\downarrow$ components. This is a physical ground for the unitarity-limit value of the scattering t matrix derived by the scattering theory [30–32].

V. Bound State in the Presence of Magnetic Field

The effects of magnetic field on the singlet bound state have been studied by a few authors [33–35]. In the following the results by Ishii [35] are given in some detail. An external field is assumed to be applied only to the localized spin. It can be proved that when conduction electrons are also subjected to the magnetic field, the Pauli paramagnetism is merely added. We have the Zeeman term

$$H_Z = 2\Delta S_z, \qquad \Delta = \tfrac{1}{2}g\mu_B H, \tag{5.1}$$

in addition to the s–d exchange Hamiltonian. The starting wave function is given by

$$\psi_0 = \sum_1 (\Gamma_1^\uparrow a_{1\downarrow}^\dagger \chi_\uparrow + \Gamma_1^\downarrow a_{1\uparrow}^\dagger \chi_\downarrow)\psi_v. \tag{5.2}$$

The part of the excited electron–hole pairs in $\psi_\uparrow$ and $\psi_\downarrow$ is the same as Eq. (2.1), but one must discriminate between $\Gamma_{12,3}^{\uparrow\uparrow,\uparrow}$ and $\Gamma_{12,3}^{\downarrow\downarrow,\downarrow}$, and so on. The equation for $\Gamma_1^\uparrow$ turns out to be

$$\Gamma_1^\uparrow(\varepsilon_1 - \tilde{E} + \Delta) + \frac{J}{4N}\sum_2 (\Gamma_2^\uparrow - 2\Gamma_2^\downarrow)$$
$$- \frac{J}{4N}\sum_{23}(\Gamma_{[21],3}^{\uparrow,\downarrow} + \Gamma_{12,3}^{\uparrow,\uparrow} - 2\Gamma_{21,3}^{\downarrow,\downarrow}) = 0, \tag{5.3}$$

and the equation for $\Gamma_1^\downarrow$ can be obtained by interchanging $\uparrow$ and $\downarrow$ and by changing the sign of Δ. The equations for $\Gamma_{12,3}^{\uparrow\uparrow,\uparrow}$ and other amplitudes can be written similarly, but are omitted here. The iterative procedure

9. THE GROUND STATE OF THE s-d MODEL

as before leads to the following two integral equations for $\Gamma_\uparrow(\varepsilon_1)$ and $\Gamma_\downarrow(\varepsilon_1)$:

$$(\varepsilon - \tilde{E} + \Delta) \Gamma_\uparrow(\varepsilon) = \frac{J\rho}{4N} \int_0^D d\varepsilon_1 \, [2\Gamma_\downarrow(\varepsilon_1) - \Gamma_\uparrow(\varepsilon_1)]$$
$$+ \frac{1}{3}\rho \int_0^D d\varepsilon_1 \, [\Gamma_\uparrow(\varepsilon_1) + 4\Gamma_\downarrow(\varepsilon_1)] K(\varepsilon + \varepsilon_1 - \tilde{E}), \quad (5.4)$$

$$(\varepsilon - \tilde{E} - \Delta) \Gamma_\downarrow(\varepsilon) = \frac{J\rho}{4N} \int_0^D d\varepsilon_1 \, [2\Gamma_\uparrow(\varepsilon_1) - \Gamma_\downarrow(\varepsilon_1)]$$
$$+ \frac{1}{3}\rho \int_0^D d\varepsilon_1 \, [\Gamma_\downarrow(\varepsilon_1) + 4\Gamma_\uparrow(\varepsilon_1)] K(\varepsilon + \varepsilon_1 - \tilde{E}),$$

where the integration kernel is given by Eq. (2.13).

These rather simple equations are derived on the basis of the following two approximations. The first one is related to the energy shifts ΔE_β [see Eqs. (2.7) and (2.8)]; $\Delta E_\uparrow$ is given by

$$\Delta E_\uparrow = -2 \left(\frac{J}{4N}\right)^2 \sum_{12} \left(\frac{1}{\varepsilon_1 + \varepsilon - \varepsilon_2 + \Delta - \tilde{E}} + \frac{2}{\varepsilon_1 + \varepsilon - \varepsilon_2 - \Delta - \tilde{E}} \right) + \cdots$$

$$= -2\left(\frac{J\rho}{4N}\right)^2 \left[6D \log 2 + (\varepsilon - \tilde{E} + \Delta) \log \frac{\varepsilon + \Delta - \tilde{E}}{D} \right.$$
$$\left. + 2(\varepsilon - \Delta - \tilde{E}) \log \frac{\varepsilon - \Delta - \tilde{E}}{D} \right] + \cdots, \quad (5.5)$$

and $\Delta E_\downarrow(\Delta) = \Delta E_\uparrow(-\Delta)$. Then, the logarithmic terms are neglected, and only the normal terms in each order of $J\rho/N$ are retained, so that we have

$$\Delta E_\uparrow \cong \Delta E \cong \Delta E_\downarrow, \quad (5.6)$$

where ΔE is given by Eq. (2.8) with $\varepsilon_1 - \tilde{E} = 0$. This is allowed under the condition of

$$-\tilde{E} - \Delta \gg 2\left(\frac{J\rho}{4N}\right)^2 (-\tilde{E} + \Delta) \left| \log \frac{-\tilde{E} + \Delta}{D} \right| \quad (5.7)$$

By using the result of $-\tilde{E} = (\tilde{E}_0^2 + \Delta^2)^{1/2}$ which will be obtained after solving Eq. (5.4), this becomes

$$\Delta \ll |\tilde{E}_0| (-J\rho/N)^{-1/2}. \quad (5.8)$$

In the weak coupling limit, this is valid up to even much larger values of Δ than the binding energy $|\tilde{E}_0|$ in the absence of external field.

The second approximation gives a less stringent condition than Eq. (5.8), but it is a pivotal approximation to simplify the integration kernel greatly. This enables us to neglect the field dependence in the argument of the logarithmic functions in the kernel. The approximation is allowed under the condition of

$$\left|\log\left(\frac{-\tilde{E}_0}{D}\right)\right| \gg \left|\log\left\{\frac{(-\tilde{E} \pm \Delta)}{-\tilde{E}_0}\right\}\right|$$

and the condition for Δ is written, with the use of $-\tilde{E} = (\tilde{E}_0^2 + \Delta^2)^{1/2}$ again, in the form

$$\log(\Delta/-\tilde{E}_0) \ll -N/J\rho. \qquad (5.9)$$

It is easy to see that this condition is weak and holds up to very large values of Δ compared with $|\tilde{E}_0|$ for $-J\rho/N \ll 1$. This is due to the nature of logarithmic function, in which H dependence is so weak that it is destroyed easily by the existence of $-\tilde{E}_0$.

Equation (5.4) can be solved in a paralled way to the case in the absence of the external field. After converting the integral equations into simultaneous differential equations, one finds a solution,

$$G_\uparrow(\varepsilon) = -A_3 G_3(\varepsilon, -\Delta) - A_1 G_1(\varepsilon, -\Delta) + A_5 G_5(\varepsilon, -\Delta) + A_{-1} G_{-1}(\varepsilon, -\Delta),$$

$$G_\downarrow(\varepsilon) = A_3 G_3(\varepsilon, \Delta) + A_1 G_1(\varepsilon, \Delta) + A_5 G_5(\varepsilon, \Delta) + A_{-1} G_{-1}(\varepsilon, \Delta), \qquad (5.10)$$

where $G_\beta(\varepsilon)$ are defined by

$$G_\beta(\varepsilon) = \int_0^\varepsilon d\varepsilon' \Gamma_\beta(\varepsilon') \qquad (5.11)$$

and $G_{4\gamma}$ are given as

$$G_{4\gamma}(\varepsilon, \Delta) = \left(1 - \frac{J\rho}{N} \log \frac{\varepsilon - \tilde{E}_0}{D}\right)^\gamma - \gamma \left(\frac{J\rho}{N}\right) \left(1 - \frac{J\rho}{N} \log \frac{\varepsilon - \tilde{E}_0}{D}\right)^{\gamma-1}$$

$$\times \log \frac{\varepsilon - \tilde{E} - \Delta}{\varepsilon - \tilde{E}_0}. \qquad (5.12)$$

Integration constants $A_{4\gamma}$ and energy eigenvalue $\tilde{E}$ are determined by $G_\beta(0) = 0$ from Eq. (5.11) and by Eq. (5.4). Note that the solution is correct in the highest order of logarithm in each order of J for its deviation from the singlet solution of $\Delta = 0$.

The energy eigenvalue $\tilde{E}$ is found to be

$$\tilde{E} = -(\tilde{E}_0^2 + \Delta^2)^{1/2}, \qquad (5.13)$$

and, by differentiating $G_\beta(\varepsilon)$ with respect to ε, $\Gamma_\beta(\varepsilon)$ are derived:

$$\Gamma_\uparrow(\varepsilon) = \frac{-1}{\varepsilon - \tilde{E} + \varDelta}\left[\left(1 - \frac{J\rho}{N}\log\frac{\varepsilon - \tilde{E}_0}{D}\right)^{-1/4}\right.$$
$$\left. - \frac{1}{3}(1-x)^{1/2}\left(1 - \frac{J\rho}{N}\log\frac{\varepsilon - \tilde{E}_0}{D}\right)^{-3/4}\right],$$
$$\Gamma_\downarrow(\varepsilon) = \frac{1}{\varepsilon - \tilde{E} - \varDelta}\left[\left(1 - \frac{J\rho}{N}\log\frac{\varepsilon - \tilde{E}_0}{D}\right)^{-1/4}\right.$$
$$\left. - \frac{1}{3}(1-x)^{1/2}\left(1 - \frac{J\rho}{N}\log\frac{\varepsilon - \tilde{E}_0}{D}\right)^{-3/4}\right],$$
(5.14)

besides the normalization constant, where x denotes $(J\rho/N)\log(-\tilde{E}_0/D)$ as before. From Eq. (5.13) the expectation value of the localized spin in the presence of magnetic field is shown to be

$$\langle S_z \rangle = (\tfrac{1}{2})\frac{d\tilde{E}}{d\varDelta} = -(\tfrac{1}{2})\varDelta(\tilde{E}_0^2 + \varDelta^2)^{-1/2}. \quad (5.15)$$

As is seen in Fig. 5, the magnitude of the induced spin moment increases with field and tends to the saturated value of $\tfrac{1}{2}$ at high fields. The

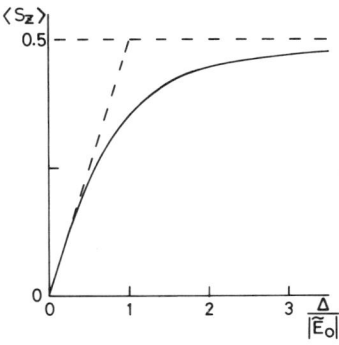

FIG. 5. The magnitude of the localized spin induced by magnetic field.

susceptibility at weak fields is given by the initial slope of Eq. (5.15),

$$\chi = \frac{(g\mu_B/2)^2}{-\tilde{E}_0}. \quad (5.16)$$

The same expression as Eqs. (5.13), (5.15), and (5.16) are obtained in the case of the anisotropic s–d exchange interaction by Okiji et al. [36], where $\tilde{E}_0$ is replaced by the values in Table I. Putting aside the g factor, one has the isotropic susceptibility in spite of the anisotropic interaction in the approximation of the most divergent terms.

The constant susceptibility or the Δ^2 dependence in energy at small Δ is in agreement with the predicted result due to the scaling-law argument by Anderson et al. [27] and is in sharp contrast with the results by the Green's function method on the Nagaoka decoupling. By the Green's function approach, the $(T \log T)^{-1}$ dependence for the susceptibility, divergent at $T \to 0$, results besides the negatively divergent T^{-1} part [32, 37]. The expression (5.16) has been derived by Nam and Woo [34] in the approximation of the zeroth order in the bound-state approach. The anomalous Green's function approach [38] gives the constant susceptibility at 0°K similar to Eq. (5.16).

VI. Local Electron Distributions and Magnetoresistance

In Section IV we discussed the local electron distributions in the singlet ground state. The variation of these distributions due to external field, which has been analyzed by Ishii [39] and will give further insights to the problem, is the subject of this section.

In order to find electron distributions in $\psi_\uparrow$ and $\psi_\downarrow$ separately as before, we provide first the normalization integrals for ψ_β. By the use of Eq. (5.14), $\langle \psi_\uparrow | \psi_\uparrow \rangle$ is calculated as

$$\langle \psi_\uparrow | \psi_\uparrow \rangle = (\tfrac{4}{9}) \rho(-\tilde{E} + \Delta)^{-1} (1-x)^{-1/2}. \tag{6.1}$$

Similarly, $\langle \psi_\downarrow | \psi_\downarrow \rangle$ can be calculated as

$$\langle \psi_\downarrow | \psi_\downarrow \rangle = (\tfrac{4}{9}) \rho(-\tilde{E} - \Delta)^{-1} (1-x)^{-1/2}. \tag{6.2}$$

The value of $\langle S_z \rangle$ is given generally by

$$\langle S_z \rangle = \frac{1}{2} \frac{\langle \psi_\uparrow | \psi_\uparrow \rangle - \langle \psi_\downarrow | \psi_\downarrow \rangle}{\langle \psi | \psi \rangle}, \quad \langle \psi | \psi \rangle = \langle \psi_\uparrow | \psi_\uparrow \rangle + \langle \psi_\downarrow | \psi_\downarrow \rangle, \tag{6.3}$$

so that one has general relations,

$$\begin{aligned} \frac{\langle \psi_\uparrow | \psi_\uparrow \rangle}{\langle \psi | \psi \rangle} &= \frac{1}{2} + \langle S_z \rangle \\ \frac{\langle \psi_\downarrow | \psi_\downarrow \rangle}{\langle \psi | \psi \rangle} &= \frac{1}{2} - \langle S_z \rangle. \end{aligned} \tag{6.4}$$

Equations (6.1) to (6.3) reproduce Eq. (5.15) automatically.

9. THE GROUND STATE OF THE s–d MODEL

The electron distributions are calculated from $\langle \psi_\uparrow | a_{k\alpha}^\dagger a_{k'\alpha} | \psi_\uparrow \rangle$ and $\langle \psi_\downarrow | a_{k\alpha}^\dagger a_{k'\alpha} | \psi_\downarrow \rangle$. A simple symmetry relation can be easily proved from the equations for Γ_k^β and higher order amplitudes so that the expression of $\langle \psi_\downarrow | a_{k\alpha}^\dagger a_{k'\alpha} | \psi_\downarrow \rangle$ is obtained from $\langle \psi_\uparrow | a_{k-\alpha}^\dagger a_{k'-\alpha} | \psi_\uparrow \rangle$ by changing the sign of Δ, or vice versa.

The charge and spin densities at the impurity spin site in the components $\psi_\uparrow$ are given, respectively, by

$$n_{\uparrow,\uparrow}(0) + n_{\uparrow,\downarrow}(0) = \frac{1}{V} \frac{1}{\langle \psi_\uparrow | \psi_\uparrow \rangle} \sum_{kk'} \langle \psi_\uparrow | a_{k\uparrow}^\dagger a_{k'\uparrow} + a_{k\downarrow}^\dagger a_{k'\downarrow} | \psi_\uparrow \rangle, \quad (6.5)$$

$$\frac{1}{2}[n_{\uparrow,\uparrow}(0) - n_{\uparrow,\downarrow}(0)] = \frac{1}{2V} \frac{1}{\langle \psi_\uparrow | \psi_\uparrow \rangle} \sum_{kk'} \langle \psi_\uparrow | a_{k\uparrow}^\dagger a_{k'\uparrow} - a_{k\downarrow}^\dagger a_{k'\downarrow} | \psi_\uparrow \rangle, \quad (6.6)$$

where V denotes the total volume of the crystal. The right-hand side of Eq. (6.5) can be expressed in terms of Γ_k^β by using the iterative procedure. After a straightforward manipulation, one has

$$n_{\uparrow,\uparrow}(0) + n_{\uparrow,\downarrow}(0) = (V \langle \psi_\uparrow | \psi_\uparrow \rangle)^{-1} \left[\left(\sum \Gamma_1^\uparrow \right)^2 - \sum_{12} \Gamma_1^\uparrow \left(\frac{1}{2} \Gamma_2^\uparrow - \Gamma_2^\downarrow \right) \log \left(1 - \frac{J\rho}{N} \log \frac{\varepsilon_1 + \varepsilon_2 - \tilde{E}_0}{D} \right) \right]. \quad (6.7)$$

Using $\Gamma_\beta(\varepsilon)$ given by Eq. (5.14), one finds finally that Eq. (6.7) vanishes. Obviously, one has also

$$n_{\downarrow,\uparrow}(0) + n_{\downarrow,\downarrow}(0) = 0. \quad (6.8)$$

That is, the charge density at the impurity site disappears. In a similar way, the spin polarization and the z-component part of the spin correlation are calculated with the use of Eq. (6.6) and the symmetry relation between $\langle \psi_\uparrow | a_{k\alpha}^\dagger a_{k'\alpha} | \psi_\uparrow \rangle$ and $\langle \psi_\downarrow | a_{k-\alpha}^\dagger a_{k'-\alpha} | \psi_\downarrow \rangle$ as follows:

$$\frac{1}{2V \langle \psi | \psi \rangle} \sum_{kk'} \langle \psi | a_{k\uparrow}^\dagger a_{k'\uparrow} - a_{k\downarrow}^\dagger a_{k'\downarrow} | \psi \rangle$$

$$= \frac{1}{2}\left[(n_{\uparrow,\uparrow}(0) - n_{\uparrow,\downarrow}(0)) \left(\frac{1}{2} + \langle S_z \rangle \right) + (n_{\downarrow,\uparrow}(0) - n_{\downarrow,\downarrow}(0)) \left(\frac{1}{2} - \langle S_z \rangle \right) \right]$$

$$= \langle S_z \rangle \frac{J}{VN} \sum_{kk'} \frac{1}{\varepsilon_k - \varepsilon_{k'}}, \quad (6.9)$$

$$\frac{1}{2V\langle\psi|\psi\rangle} \sum_{kk'} \langle\psi|(a^\dagger_{k\uparrow}a_{k'\uparrow} - a^\dagger_{k\downarrow}a_{k'\downarrow}) S_z|\psi\rangle$$

$$= \frac{1}{4}\left[(n_{\uparrow,\uparrow}(0) - n_{\uparrow,\downarrow}(0))\left(\frac{1}{2} + \langle S_z\rangle\right) - (n_{\downarrow,\uparrow}(0) - n_{\downarrow,\downarrow}(0))\left(\frac{1}{2} - \langle S_z\rangle\right)\right]$$

$$= -\frac{\rho}{3V}\left(\frac{N}{J\rho}\right)\frac{\tilde{E}_0^2}{-\tilde{E}} + \frac{J}{4VN}\sum_{kk'}\frac{1}{\varepsilon_k - \varepsilon_{k'}}, \quad (6.10)$$

where the factors $(\frac{1}{2} \pm \langle S_z \rangle)$ account for the correct weights and the normalization. The result of Eq. (6.9) is the value given by the Ruderman Kittel terms [40, 41]. It is noted here that the z-component part of the spin correlation density at the impurity spin site, Eq. (6.10) remains as one-third of $(1/V)\langle\sigma(0) \cdot S\rangle$ obtained from the derivative of the ground-state energy with respect to J, though the magnitude of the anomalous part decreases with increasing Δ as a whole.

Now the electron densities $n_{\beta,\alpha}(0)$ are given by

$$n_{\uparrow,\uparrow}(0) = -n_{\uparrow,\downarrow}(0) = -\frac{2\rho}{3V}\left(\frac{N}{J\rho}\right)^2\frac{\tilde{E}_0^2}{-\tilde{E}-\Delta} + \frac{J}{2VN}\sum_{kk'}\frac{1}{\varepsilon_k - \varepsilon_{k'}} \quad (6.11)$$

$$n_{\downarrow,\uparrow}(0) = -n_{\downarrow,\downarrow}(0) = +\frac{2\rho}{3V}\left(\frac{N}{J\rho}\right)^2\frac{\tilde{E}_0^2}{-\tilde{E}+\Delta} - \frac{J}{2VN}\sum_{kk'}\frac{1}{\varepsilon_k - \varepsilon_{k'}}.$$

With increasing Δ, $-n_{\uparrow,\uparrow}(0)$ and $n_{\uparrow,\downarrow}(0)$ increase, while $n_{\downarrow,\uparrow}(0)$ and $-n_{\downarrow,\downarrow}(0)$ decrease.

The total localized electron densities are defined as before:

$$n_{\uparrow,\alpha} = \sum_k \lim_{k'\to k} \frac{\langle\psi_\uparrow|a^\dagger_{k\alpha}a_{k'\alpha}|\psi_\uparrow\rangle}{\langle\psi_\uparrow|\psi_\uparrow\rangle},$$

$$n_{\downarrow,\alpha} = \sum_k \lim_{k'\to k} \frac{\langle\psi_\downarrow|a^\dagger_{k\alpha}a_{k'\alpha}|\psi_\downarrow\rangle}{\langle\psi_\downarrow|\psi_\downarrow\rangle}. \quad (6.12)$$

We can calculate these quantities by expressing them in terms of Γ_k^β, as in the case of vanishing field. Instead, here we discuss the problem from a more general standpoint. Using the relation $\langle\psi|[H, a^\dagger_{k\alpha}a_{k'\alpha}]|\psi\rangle = 0$ which holds for any eigenstate of H, and assuming ψ nondegenerate, we can derive the following equation:

$$\frac{1}{\langle\psi|\psi\rangle}\langle\psi|a^\dagger_{k\uparrow}a_{k'\uparrow} - a^\dagger_{k\downarrow}a_{k'\downarrow}|\psi\rangle = -\frac{J}{2N}\frac{L_k - L_{k'}}{\varepsilon_k - \varepsilon_{k'}}, \quad (6.13)$$

9. THE GROUND STATE OF THE s–d MODEL

where L_k is defined by

$$L_k = \sum_l \frac{\langle \psi |(a^\dagger_{k\uparrow}a_{l\uparrow} + a^\dagger_{k\downarrow}a_{l\downarrow}) S_z - a^\dagger_{k\downarrow}a_{l\uparrow}S_+ + a^\dagger_{k\uparrow}a_{l\downarrow}S_- | \psi\rangle}{\langle\psi|\psi\rangle}. \quad (6.14)$$

It is easily shown from Eqs. (6.4) and (6.13) that the total localized spin polarization of the conduction electrons, $\langle \sigma_z \rangle$, is given by

$$\langle \sigma_z \rangle = \frac{1}{2}\left[(n_{\uparrow,\uparrow} - n_{\uparrow,\downarrow})\left(\frac{1}{2} + \langle S_z \rangle\right) + (n_{\downarrow,\uparrow} - n_{\downarrow,\downarrow})\left(\frac{1}{2} - \langle S_z \rangle\right)\right]$$

$$= -\frac{J\rho}{4N}[L(D) - L(-D)], \quad (6.15)$$

where $L(\pm D)$ means L_k at $\varepsilon_k = \pm D$; L_k is expressed in terms of Γ^B and $L(\pm D)$ can be shown to be

$$L(D) = 0, \quad L(-D) = 2\langle S_z\rangle. \quad (6.16)$$

Then we have

$$\langle \sigma_z \rangle = (J\rho/2N)\langle S_z\rangle. \quad (6.17)$$

That is, the relation of the Ruderman–Kittel term given by Eq. (1.6) is also proved. The total localized charge is proved to disappear in a similar way, so that we have

$$(n_{\uparrow,\uparrow} + n_{\uparrow,\downarrow})(\tfrac{1}{2} + \langle S_z\rangle) + (n_{\downarrow,\uparrow} + n_{\downarrow,\downarrow})(\tfrac{1}{2} - \langle S_z\rangle) = 0. \quad (6.18)$$

Equations (6.17) and (6.18) can be regarded as quite general ones from their derivation.

As it is argued in the case of no magnetic field, the transverse component of the s–d exchange interaction changes local numbers of electrons by -1 for up spin and by $+1$ for down spin when it operates on $\psi_\downarrow \chi_\downarrow$, and since we have a finite matrix element between $\psi_\uparrow\chi_\uparrow$ and $\psi_\downarrow\chi_\downarrow$ even in the presence of magnetic field, the theorem by Anderson [10] suggests the relations*

$$n_{\uparrow,\uparrow} - n_{\downarrow,\uparrow} = -1, \quad n_{\uparrow,\downarrow} - n_{\downarrow,\downarrow} = 1. \quad (6.19)$$

---

* These relations are actually satisfied in the starting approximation (5.2). The calculated results [39] of $n_{\beta,\alpha}$, by including the higher order electron–hole pair effects in a similar way to those in Section IV, do not satisfy the relation (6.19), though they satisfy the relations (6.17) and (6.18). Further investigation is necessary on this point.

Now there are four equations, (6.17)–(6.19), sufficient to determine $n_{\beta,\alpha}$. We have

$$-n_{\uparrow,\uparrow} = n_{\uparrow,\downarrow} = \frac{1}{2} - \langle S_z \rangle \left(1 + \frac{J\rho}{2N}\right),$$

$$n_{\downarrow,\uparrow} = -n_{\downarrow,\downarrow} = \frac{1}{2} + \langle S_z \rangle \left(1 + \frac{J\rho}{2N}\right).$$

(6.20)

The phase shifts, $\delta_{\beta,\alpha}$, for α-spin electrons at the Fermi level in ψ_β are obtained, respectively, as

$$-\frac{\delta_{\uparrow,\uparrow}}{\pi} = \frac{\delta_{\uparrow,\downarrow}}{\pi} = \frac{1}{2} - \langle S_z \rangle \left(1 + \frac{J\rho}{2N}\right),$$

$$\frac{\delta_{\downarrow,\uparrow}}{\pi} = -\frac{\delta_{\downarrow,\downarrow}}{\pi} = \frac{1}{2} + \langle S_z \rangle \left(1 + \frac{J\rho}{2N}\right).$$

(6.21)

From each of these, one can calculate a contribution to the resistivity. The result from each $\delta_{\beta,\alpha}$ turns out to be identical, so that one has a general formula for the magnetoresistance at $0°K$,

$$R = R_0 \cos^2 \left[\pi \langle S_z \rangle \left(1 + \frac{J\rho}{2N}\right)\right],$$

(6.22)

where R_0 denotes the s-wave unitarity limit value for resistivity. This result is given by Ishii [39]. This is also given by Hamann [42] in the functional integral approach and by Kurata [43] on the Takano–Ogawa model.

The expression for $\langle S_z \rangle$, Eq. (5.15), gives curves for R vs. Δ shown in Fig. 6 in linear and logarithmic scales of Δ. The decrease of R is proportional to Δ^2 for small Δ; that is, R shows quite normal behavior, too. The logarithmic dependence of Δ disappears, but it is interesting for the R–log Δ curve to show a considerable wide part of straight line around H_K, which is defined as $|\tilde{E}_0|/g\mu_B$. With increasing Δ, R approaches the value in the Born approximation, $R_0(\pi^2/16)(J\rho/N)^2$. The results for the spin polarization of conduction electrons at the impurity and for the total one suggest that it is entirely normal, that is, given by the Ruderman–Kittel term (RK term). In fact, the calculation by Ishii [44] shows that it is true in the whole range of distance of primary importance. The

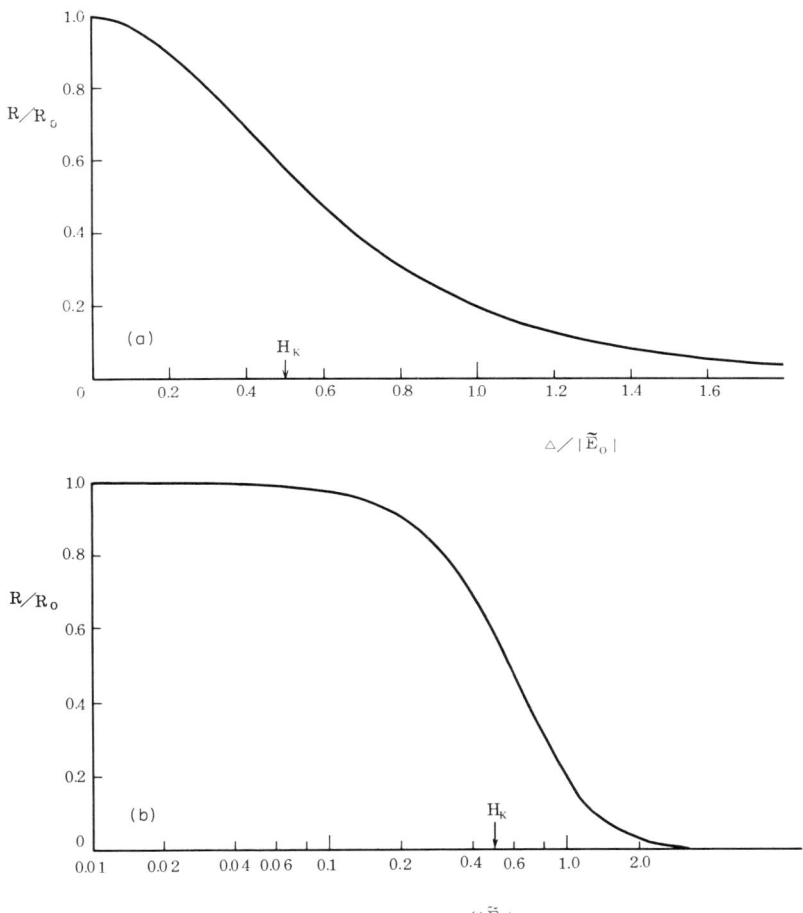

FIG. 6. Curves for magnetoresistance versus magnetic field (a) in the linear scale and (b) in the logarithmic scale.

spin polarization density, $\langle \sigma_z(r) \rangle$, is calculated with the use of Eq. (6.13) as

$$\langle \sigma_z(r) \rangle = -\frac{J}{2NV} \sum_q \sum_k \frac{L_{k+q} - L_k}{\varepsilon_{k+q} - \varepsilon_k} e^{iq \cdot r}, \tag{6.23}$$

$$= \frac{-1}{4\pi} \left(\frac{J\rho}{N} \right) \langle S_z \rangle \frac{\cos 2k_F r}{r^3}, \quad r \ll \tilde{r}(-J\rho/N) \tag{6.24}$$

$$= -\frac{1}{2\pi} \langle S_z \rangle \frac{\cos 2k_F r}{r^3}, \quad r \gg \tilde{r} \tag{6.25}$$

where $\varDelta$ is assumed to be small, L_k is estimated in the approximation of the most divergent terms, k_F is the Fermi wave vector, and $\tilde{r}$ is defined by

$$k_F \tilde{r} = e^{-N/J\rho}. \tag{6.26}$$

The expression (6.24) is the RK term, of which the asymptotic form for $r \gg k_F^{-1}$ is given for simplicity. In the range of $\tilde{r}(-J\rho/N) \lesssim r \lesssim \tilde{r}$, the RK term and the anomalous term are comparable to each other. Since in the present calculation the weak coupling limit is assumed, the critical distance $(-J\rho/N)\tilde{r}$ is very large, so that the spin polarization ceases to be appreciable beyond $(-J\rho/N)\tilde{r}$, whether it is normal or anomalous, and the RK term covers the whole range of the appreciable polarization. However, it seems interesting that there exists the anomalous term of a simple form [Eq. (6.25)] when $r \gg \tilde{r}$ although its magnitude is very small.

VII. Concluding Remarks

The important physical aspects of the singlet bound state, some of which are common to the Anderson theory, can be summarized as follows: The spin density of conduction electrons is bound around the localized spin ($S = \frac{1}{2}$), amounts to one-half in total, and forms a local singlet with the localized spin. This gives the binding energy, which is equal to $k_B T_K$, in addition to the energy shift due to the perturbation to the free spin state. Phase shifts of conduction electrons at the Fermi level are concluded to be $\pm \pi/2$ and lead to the unitarity limit value of the resistivity. Under an applied magnetic field, the bound state remains to exist beyond H_K, and the susceptibility, $\langle S_z \rangle$, the spin polarization of conduction electrons, and the magnetoresistance are found to be normal in their behavior. In other words, no logarithmic dependence of H is involved but the quite normal H-power laws appear.

Strictly speaking, the simple formulas for $\langle S_z \rangle$, the binding energy, and others are valid under the condition $H \ll (2 \mid \tilde{E}_0 \mid /g\mu_B)(-J\rho/N)^{-1/2}$. However, the fact that the bound state survives beyond H_K and does not make a sudden transition at H_K is emphasized here. This is in accordance with a general requirement that no phase transition should occur in an essentially small system of a localized spin in a metal. It is interesting that the expressions like $\langle S_z \rangle$ go smoothly to the expected limiting values. The whole calculation except some based on general arguments has been carried out in the approximation of the most divergent terms. This approximation, valid in the weak coupling limit, is further supported by the reasonable behavior of the calculated physical quantities and the satisfactory physical picture in spite of the divergent normalization. The important role played by binding energy $\tilde{E}$ in arguments of logarithmic functions to wipe out logarithmic field dependence in the results should also be emphasized. In this sense, it is natural that the bound state exhibits the nonsingular behavior in the magnetic field dependence and also probably in the temperature dependence.

The low-lying excited states above the ground state must be those with an electron–hole pair excited far from the impurity spin and form a continuous energy spectrum, which is inevitably affected by the formation of the bound state. It will be misleading simply to say that the bound-state theory gives a gap in an excitation energy, although it will cost $-\tilde{E}$ in energy to release the localized spin from the bound state. No attempt to calculate the excited states or the free energy of the bound state has been successful yet. The bound-state theory at finite temperatures is awaited.

References

1. J. Kondo, *Progr. Theor. Phys.* **32**, 37 (1964).
2. P. W. Anderson, *Phys. Rev.* **124**, 41 (1961).
3. P. W. Anderson and A. M. Clogston, *Bull. Amer. Phys. Soc.* **6**, 124 (1961).
4. J. Kondo, *Progr. Theor. Phys.* **28**, 846 (1962).
5. J. R. Schrieffer and P. A. Wolff, *Phys. Rev.* **149**, 491 (1966).
6. K. Yosida and A. Okiji, *Progr. Theor. Phys.* **34**, 505 (1965).
7. H. Miwa, *Progr. Theor. Phys.* **34**, 1040 (1965).
8. P. W. Anderson, *Phys. Rev.* **164**, 352 (1967).
9. P. W. Anderson, *in* "Many Body Physics" (C. Dewitt and R. Balian, eds.), pp. 229–295. Gordon & Breach, New York, 1967.
10. P. W. Anderson, *Phys. Rev. Lett.* **18**, 1049 (1967).
11. J. Kondo, *Progr. Theor. Phys.* **36**, 429 (1966).
12. K. Yosida, *Phys. Rev.* **147**, 223 (1966).
13. K. Yosida, *Progr. Theor. Phys.* **36**, 875 (1966).
14. A. Okiji, *Progr. Theor. Phys.* **36**, 712 (1966).
15. A. Yoshimori, *Phys. Rev.* **168**, 493 (1968).
16. A. Yoshimori and K. Yosida, *Progr. Theor. Phys.* **39**, 1413 (1968).
17. K. Yosida and A. Yoshimori, *Progr. Theor. Phys.* **42**, 753 (1969).
18. J. A. Appelbaum and J. Kondo, *Phys. Rev. Lett.* **19**, 906 (1967).
19. J. A. Appelbaum and J. Kondo, *Phys. Rev.* **170**, 542 (1968).
20. D. R. Hamann and J. A. Appelbaum, *Phys. Rev.* **180**, 334 (1969).
21. A. J. Heeger and M. A. Jensen, *Phys. Rev. Lett.* **18**, 488 (1967).
22. J. Kondo, *Phys. Rev.* **154**, 644 (1967).
23. A. A. Abrikosov, *Physics (Long Island City, N.Y.)* **2**, 5 (1965).
24. S. Nakajima, *Progr. Theor. Phys.* **39**, 1402 (1968).
25. J. Kondo, *Progr. Theor. Phys.* **40**, 683 (1968).
26. K. Yosida and H. Miwa, *Progr. Theor. Phys.* **41**, 1416 (1969).
27. P. W. Anderson, G. Yuval, and D. R. Hamann, *Phys. Rev. B* **1**, 4464 (1970).

28. H. Shiba, *Progr. Theor. Phys.* **43**, 601 (1970).
29. E. T. Whittaker and G. N. Watson, "Modern Analysis." Cambridge Univ. Press, London and New York, 1927.
30. H. Suhl and D. Wong, *Physics (Long Island City, N.Y.)* **3**, 17 (1967).
31. Y. Nagaoka, *Phys. Rev. A* **138**, 1112 (1965).
32. D. R. Hamann, *Phys. Rev.* **154**, 596 (1967).
33. H. Ishii and K. Yosida, *Progr. Theor. Phys.* **38**, 61 (1967).
34. S. B. Nam and J. W. F. Woo, *Phys. Rev. Lett.* **19**, 649 (1967).
35. H. Ishii, *Progr. Theor. Phys.* **40**, 201 (1968).
36. A. Okiji, A. Kato and H. Shiba, *Progr. Theor. Phys. Suppl.* **46**, 182 (1970).
37. J. Zittartz, *Z. Phys.* **217**, 155 (1968).
38. F. Takano and T. Ogawa, *Progr. Theor. Phys.* **35**, 343 (1966).
39. H. Ishii, *Progr. Theor. Phys.* **43**, 578 (1970).
40. M. A. Ruderman and C. Kittel, *Phys. Rev.* **96**, 99 (1954).
41. K. Yosida, *Phys. Rev.* **106**, 893 (1957).
42. D. R. Hamann, *Phys. Rev. B* **2**, 1373 (1970).
43. Y. Kurata, *Progr. Theor. Phys.* **43**, 621 (1970).
44. H. Ishii, private communication, 1970.

Part III
MAGNETIC MOMENT EFFECTS IN SUPERCONDUCTORS

10. Paramagnetic Impurities in Superconductors

M. Brian Maple

Institute for Pure and Applied Physical Sciences
University of California, San Diego
La Jolla, California

I. Introduction 289
II. Long-Lived Local Moments in Superconductors 291
   1. Experimental Verification of the Abrikosov–Gor'kov Theory 291
   2. Multiple Pair Breaking Effects in Superconductors 295
   3. Superconductors Containing Impurities with Crystal-Field Split Energy Levels 297
   4. The Kondo Effect in Superconductors 299
III. The Effect of Nonmagnetic Resonant States on Superconductivity 308
IV. The Effect of Localized Spin Fluctuations on Superconductivity 313
V. Magnetic–Nonmagnetic Transitions of Impurities in Superconductors 318
   References 323

I. Introduction

The effect of magnetic impurities on superconductivity has received a great deal of attention ever since the discovery that they produce a precipitous drop in the superconducting transition temperature T_c [1, 2]. That the conduction-electron–impurity-spin exchange interaction could account for this strong depression of T_c was first noted by Herring [3] and by Suhl and Matthias [4]. Assuming that exchange scattering of conduction electrons by the impurity spins may be adequately described within the first Born approximation (to second order in the exchange interaction parameter $\mathscr{I}$), Abrikosov and Gor'kov (hereafter AG) developed in 1960 [5] what is now a classic theory for superconductors with paramagnetic impurities. Their theory successfully explained the basic features of these early experiments and further predicted the striking phenomenon of gapless superconductivity. This prediction was subsequently verified by experiment.

In recent years, however, normal state studies of local moments in metals have shown that the assumptions upon which the AG theory is founded are not applicable to many matrix–impurity systems which nonetheless exhibit a strong depression of T_c. First, the supposition that the solute spins are well defined does not apply to weakly magnetic systems (such as **Al**Mn) in which the localized spins apparently fluctuate with a finite frequency τ_{sf}^{-1} (τ_{sf} is the localized spin fluctuation lifetime). Secondly, even when the impurity spins are well defined (when τ_{sf} may be regarded as essentially infinite), the effect of exchange scattering of conduction electrons to higher order than $\mathscr{J}^2$ (i.e., the Kondo effect) can be very significant. These important developments have stimulated the current interest in the effects of impurities on superconductivity which are outside the scope of the basic AG theory.

The purpose of this chapter is therefore to review with emphasis on experiment the main aspects of the modification of superconductivity by the presence of paramagnetic impurities. In classifying the local moments associated with paramagnetic solutes, we adopt the notion that the moments fluctuate in time and may be characterized phenomenologically by an appropriate lifetime τ_{sf}. A brief topical breakdown of the remaining four sections of the chapter is outlined below.

The long-lived local moment limit is discussed in Section II which is divided further into four subsections. The first three deal with circumstances where the scattering of conduction electrons by the impurity spins may be treated adequately without going beyond the first Born approximation. Section II, 1 is concerned with experimental verification of the AG pair breaking theory, while Section II, 2 considers multiple pair breaking effects due to the penetration of an external magnetic field into a type II superconducting matrix in the vortex state. In Section II, 3 we discuss superconductors containing rare earth (hereafter RE) impurities whose Hund's rule ground states are split by crystal fields.

The scattering of conduction electrons by paramagnetic impurity spins to higher order than the first Born approximation is considered in Section II, 4. Since this problem has elicted considerable theoretical interest (reviewed in Chapter 12 by Müller-Hartmann) we contend here with the rather limited amount of experimental work contributed to the subject.

Localized spin fluctuations with finite τ_{sf} have been studied experimentally in systems with characteristic temperatures $T_0 = h/k_B \tau_{sf}$ appreciably greater than superconducting temperatures. The resultant pair weakening (rather than pair breaking) effect on superconductivity can be quite pronounced, and is experimentally very similar to that expected from nonmagnetic resonant states in the sense of the Friedel–Anderson model. For this reason we consider nonmagnetic resonant states in Section III and localized spin fluctuations in Section IV. As in Section II, 4 some discussion is devoted to the normal state properties since these indicate the appropriate lifetime τ_{sf} of the impurity spin.

Finally, in Section V we review an interesting class of experiments associated with magnetic–nonmagnetic transitions induced by pressure or by changing the properties of the matrix by alloying. In these experiments, the magnetic character of the impurities appears to pass smoothly from long-lived local moment to nonmagnetic resonant state regimes. Once we have an understanding of how magnetic and nonmagnetic impurities affect superconductivity, we will have a sensitive probe for studying local moment formation in metals—one of the most fundamental problems in magnetism today.

II. Long-Lived Local Moments in Superconductors

1. Experimental Verification of the Abrikosov–Gor'kov Theory

The classic theory of superconductivity in the presence of paramagnetic impurities, formulated by Abrikosov and Gor'kov [5], has been experimentally verified for a number of matrix–impurity systems where the first Born approximation apparently provides an adequate representation of exchange scattering of conduction electrons by long-lived solute spins. We describe here briefly some basic results of the theory and mention recent experiments which verify certain aspects of it. The reader is referred to several excellent theoretical expositions and detailed calculations pertaining to the AG theory (for example, deGennes and Sarma [6], Phillips [7], Skalski et al. [8]), as well as an earlier review of the theory and related experiments in a previous volume of this treatise [9].

Abrikosov and Gor'kov assumed that a paramagnetic impurity spin **S** interacts with the conduction electron spin density **s** at the impurity site via an exchange interaction

$$\mathscr{H}_{int} = -2\mathscr{J}\mathbf{S}\cdot\mathbf{s}, \tag{2.1}$$

where the exchange interaction parameter is denoted by $\mathscr{J}$. At the outset it was supposed that the impurity spins were randomly distributed in space and uncorrelated with one another. If one regards the impurity spin direction as fixed, the exchange interaction is not time-reversal invariant. As a result, the lifetime τ of the time-reversed single-particle paired states of which the superconducting wave function is comprised is no longer infinite.

Within the first Born approximation, the inverse lifetime τ^{-1} (or pair breaking parameter α) which characterizes the superconducting–paramagnetic impurity system is given by

$$\alpha \equiv \tau^{-1} = \hbar^{-1} n N(E_F) \mathcal{I}^2 S(S+1), \tag{2.2}$$

where n is the paramagnetic impurity concentration and $N(E_F)$ is the density of states at the Fermi level (for one spin direction). For RE impurities where the orbital angular momentum $\mathbf{L}$ is not quenched, $\mathbf{S}$ is replaced by its projection on the total angular momentum vector $\mathbf{J} = \mathbf{L} + \mathbf{S}$ of the Hund's rule ground state. Then the interaction Hamiltonian becomes

$$\mathcal{H}_{\text{int}} = -2\mathcal{I}(g-1)\mathbf{J}\cdot\mathbf{s} \tag{2.3}$$

since $\mathbf{S} \to \langle \mathbf{S} \cdot \mathbf{J} \rangle \mathbf{J} / J(J+1)$; g is the Landé g factor. The corresponding expression for α is

$$\alpha \equiv \tau^{-1} = \hbar^{-1} n N(E_F) \mathcal{I}^2 (g-1)^2 J(J+1). \tag{2.4}$$

The theory predicts a second-order transition to the superconducting state and a rapid decrease of the transition temperature with α given by the universal relation

$$\ln\left(\frac{T_c}{T_{c_0}}\right) = \Psi\left(\frac{1}{2}\right) - \Psi\left(\frac{1}{2} + \frac{0.14\alpha T_{c_0}}{\alpha_{\text{cr}} T_c}\right). \tag{2.5}$$

T_{c_0} corresponds to $\alpha = 0$, $\alpha_{\text{cr}} \equiv k_B T_{c_0}/4\hbar\gamma$ (ln γ is Euler's constant) corresponds to $T_c = 0$ (complete destruction of superconductivity), and Ψ is the digamma function. This expression has the *linear* asymptotic form

$$T_c/T_{c_0} = 1 - 0.691(\alpha/\alpha_{\text{cr}}) \tag{2.6}$$

as $\alpha \to 0$.

When α is varied by changing the solute concentration n, $\alpha/\alpha_{\text{cr}}$ may be replaced by n/n_{cr} in expressions (2.5) and (2.6), where n_{cr} is the critical concentration for the complete suppression of superconductivity. Thus the AG theory correctly describes the initial *linear* depression of T_c of La by RE impurities which is approximately proportional to the deGennes factor $(g-1)^2 J(J+1)$ of the RE solutes [2, 10]. It should be noted that, according to the AG theory, the depression of T_c depends on the magnitude of $\mathcal{I}$ but not on its sign. However, neglected in the AG theory are contributions to α of *higher order* than $\mathscr{I}^2$ which must be taken into account when $\mathscr{I}$ is *negative*. These contributions give rise to the very interesting *temperature-dependent* pair breaking effects which are considered in Section II, 4. Here, we confine our attention to superconducting matrices containing Gd impurities for which $\mathscr{I}$ is expected to be *positive* and the first Born approximation embodied in the AG theory can be employed to describe conduction-electron–impurity-spin exchange scattering.

For most superconducting–magnetic impurity systems studied, the experimental $T_c(n)$ curves are in reasonable agreement with the AG theory up to concentrations $n/n_{cr} \sim 0.7$ (corresponding to $T_c/T_{c_0} \sim 0.45$). However, at higher concentrations, striking departures of $T_c(n)$ from Eq. (2.5) are often observed [11–13].[†] These effects have been attributed to the onset of magnetic order among the impurity spins [15], suggested by the fact that the curve of normal state magnetic ordering temperatures $\theta(n)$ intersects the $T_c(n)$ curve at appreciable temperatures compared to T_{c_0}.[‡] The AG theoretical T_c vs. n curve has only recently been verified to concentrations near n_{cr} (to $T_c/T_{c_0} \sim 0.1$) in the systems (**La**, Gd)Al$_2$ {where $T_c(n)$ and $\theta(n)$ do not appear to intersect; Fig. 1 (Maple [16])} and **Th**Gd [17].

One of the most striking predictions of the AG theory is the phenomenon of gapless superconductivity. This results from the fact that the finite lifetime τ corresponds to an energy-broadening $\Gamma \sim h/\tau$ which introduces states into the gap and spreads out the BCS peak in the density of states. As a consequence, the energy gap $\Omega_G(\alpha)$ no longer corresponds to the order parameter $\Delta(\alpha)$ and with increasing α (or n), $\Omega_G(\alpha)$ goes to zero faster than $T_c(\alpha)$. For all concentrations the superconductor is gapless at temperatures sufficiently near T_c; while for $n > 0.91 n_{cr}$, the superconductor is gapless at all temperatures. For example, the theory predicts an attenuated specific heat jump at the transition (compared to the BCS law of corresponding states; see Fig. 9), and a linearly temperature-dependent term in the specific heat below T_c in the gapless region $n > 0.91\ n_{cr}$.

[†] The anomalous departure of $T_c(n)$ from the AG theory reported by Kuwasawa et al. [13] for the (La, Gd)$_3$Al system was not observed in a recent specific heat study on this system by Mamiya et al. [14]. This discrepancy may be associated with the different manner in which the alloys were prepared in the two investigations.

[‡] The effect on T_c of impurity spin ordering in the superconducting state has recently been investigated theoretically by J. Keller and R. Benda [*J. Low Temp. Phys.* 2, 141 (1970)]. They conclude that the anomalous dependences of T_c on the concentration of magnetic impurities observed in some matrix-impurity systems [11–13] cannot be ascribed to spin-dynamic effects [15] unless the impurity spin polarization energy is *inordinately large* (of the order of the Debye energy).

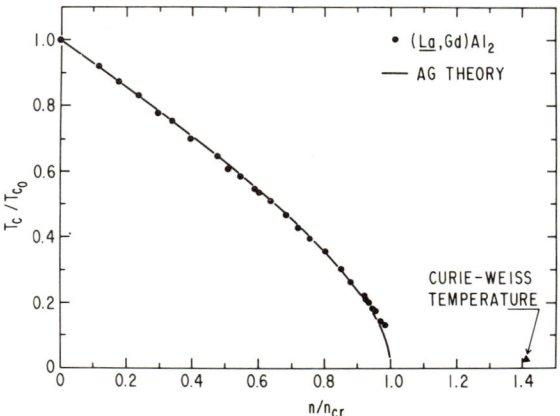

FIG. 1. Reduced transition temperature T_c/T_{c_0} versus reduced concentration n/n_{cr} for (**La**, Gd)Al$_2$ compared to the AG theory [solid curve; Eq. (2.5)]. $T_{c_0} = 3.24°K$ and $n_{cr} = 0.590$ at. % Gd substitution in La. The reduced Curie–Weiss temperature θ_p/T_{c_0} measured at $n/n_{cr} = 1.41$ is denoted by the solid triangle. Underlining in figures corresponds to boldface letters elsewhere (after Maple [16]).

The basic phenomenon of gapless superconductivity has been verified by tunneling experiments on **Pb**Gd quenched films by Woolf and Reif [18] and by specific heat measurements on bulk **La**Gd alloys by Finnemore et al. [19]. Recently, the AG theory has been tested by critical field measurements on type I **Th**Gd alloys [20, 21]. The results agree with the theory to within an accuracy of better than 0.5 % (Fig. 2). Thermal conductivity measurements on **Th**Gd alloys in the superconducting state [22] are also in good agreement (to within a few percent) with calculations based on the AG theory [23]. In a far-infrared study of **Pb**Gd quenched films, Dick and Reif [24] found that the frequency dependence of the real part of the conductivity (which reflects the characteristics of the energy gap) is in good quantitative agreement with predictions [8] of the AG theory.[†]

[†] Several recent experiments have provided additional verification of the AG theory. The specific heat jumps ΔC at T_c of (**La**, Gd)Al$_2$ alloys are in excellent agreement with the AG theory [C. A. Luengo and M. B. Maple, Solid State Commun. 12, 757 (1973)], in accordance with the close correspondence of the T_c vs. n curve with the theory [16]. Moreover, an electron paramagnetic resonance (EPR) study [D. Davidov, A. Chelkowski, C. Rettori, R. Orbach, and M. B. Maple, Phys. Rev. B 7, 1029 (1973)] shows that when the "bottleneck" in the EPR of the (**La**, Gd)Al$_2$ system is broken by reducing the Gd concentration, or by adding a second "chemical" impurity to increase spin-lattice relaxation, there is a positive Gd g-shift which gives a value for the s-f exchange interaction parameter $\mathscr{J}$ of $+0.1$ eV. This is in very good agreement with the absolute value of $\mathscr{J}$

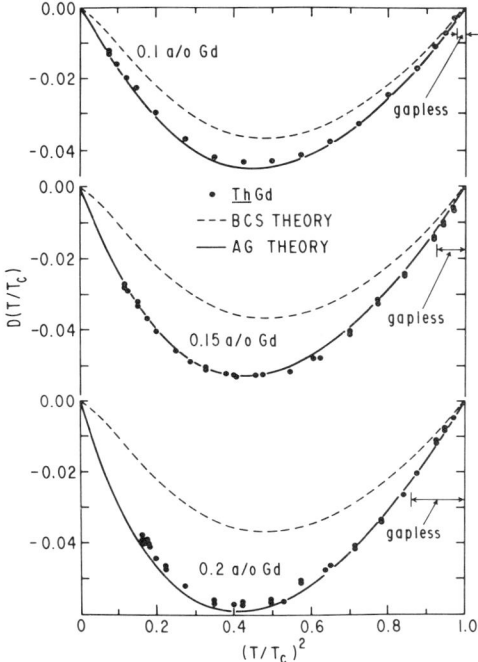

FIG. 2. Critical field deviation functions $D(T/T_c) \equiv [H_c(T/T_c)/H_c(0)] - [1 - (T/T_c)^2]$ of ThGd alloys compared to the BCS theory (dashed curves) and the AG theory (solid curves) (after Decker and Finnemore [21]).

2. Multiple Pair Breaking Effects in Superconductors

The AG expression [Eq. (2.5)] can be generalized to include other pair breaking situations. According to the theoretical work of Fulde and Maki [25], the reduced pair breaking parameters are additive in certain circumstances, and may be expressed as

$$\alpha/\alpha_{\mathrm{cr}} = \sum_i \alpha_i/\alpha_{\mathrm{cr}i}. \tag{2.7}$$

---

deduced from the measured depression of T_c [16] and expression (2.5). For the system ThGd, EPR measurements yield the value $\mathscr{J} = +0.042$ eV which compares favorably with the value derived from the depression of T_c [D. Davidov, R. Orbach, C. Rettori, D. Shaltiel, L. J. Tao, and B. Ricks, *Phys. Rev. B* **5**, 1711 (1972)]. Electron thermal conductivity measurements on superconducting InGd and PbGd quench-condensed films are in good agreement with calculations based on the AG theory [B. J. Mrstik and D. M. Ginsberg, to appear in *Phys. Rev. B*, June (1973)].

If the matrix for paramagnetic impurities is a bulk superconductor of the second kind, there are three important pair breaking effects: (1) spin depairing caused by exchange scattering of conduction electrons by paramagnetic impurities; (2) momentum depairing resulting from the penetration of an external magnetic field into the superconducting matrix in the vortex state; and (3) Pauli or exchange-field depairing due to spin polarization of the conduction band by the exchange field of the impurity spins which polarize according to a Brillouin function in the penetrating external field.

These three depairing mechanisms are characterized by the reduced pair breaking parameters α_1/α_{cr1}, α_2/α_{cr2}, and α_3/α_{cr3}, respectively. In the Fulde–Maki theory, pair breaking interaction (2) is subject to the constraint $l_{tr} \ll \xi_0$ where l_{tr} is the transport mean free path and ξ_0 is the superconducting coherence length and (3) to the conditions $l_{tr} \ll \xi_0$ and $l_{so} \ll \xi_0$ where l_{so} is the spin–orbit mean free path. Discussed in detail elsewhere [25, 26], the three depairing terms are given by

$$\sum_{i=1}^{3} \frac{\alpha_i}{\alpha_{cri}} = \frac{n}{n_{cr}} + \frac{H_{c2}(n, T)}{H_{c2}(0, 0)} + \frac{P}{P_{cr}}. \qquad (2.8)$$

$H_{c2}(n, T)$ is the upper critical field for concentration n and temperature T; while the Pauli polarization term is

$$P = \tau_{so}[n\mathscr{I}\langle S_z \rangle]^2, \qquad (2.9)$$

where τ_{so} is the spin–orbit scattering time, and $\langle S_z \rangle$ is the average z component of the impurity spin. The upper critical field is then

$$H_{c2}(n, T) = H_{c2}(0, T) - H_{c2}(0, 0) \left[\frac{n}{n_{cr}} + \frac{P}{P_{cr}} \right]. \qquad (2.10)$$

Multiple pair breaking situations have been studied experimentally in several superconducting matrices of the second kind with paramagnetic impurities. We cite two examples, (**Th**, Gd)$_{0.95}$La$_{0.05}$ and (**La**, Gd)$_3$In. Other examples are discussed in Chapter 11 by Fischer and Peter. In the (**Th**, Gd)$_{0.95}$La$_{0.05}$ system, the third depairing mechanism is unimportant, and the theory [Eq. (2.10) with $P = 0$] is in good agreement with the measurements of $[dH_{c2}(n, T)/dT]_{T=T_c}$ as a function of T_c/T_{c_0} [27] shown in Fig. 3. The third depairing mechanism is quite appreciable in the (**La**, Gd)$_3$In system and gives rise to nonmonotonic or reentrant critical field curves [28] due to the fact that $\langle S_z \rangle$ responds to the penetrating external magnetic field according to the Brillouin function $\langle S_z \rangle = SB_S(\mu H/kT)$. Nonmonotonic critical field curves were in fact predicted by deGennes and Sarma [29].

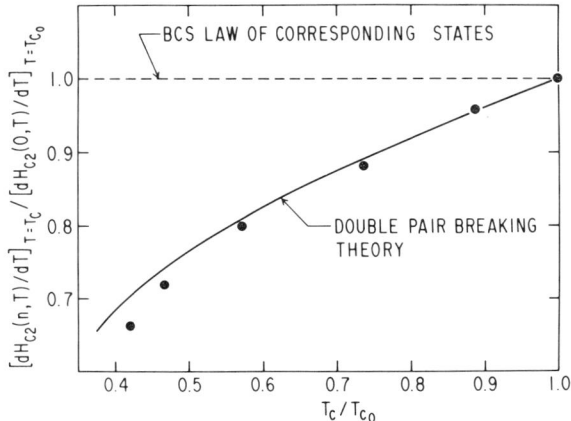

FIG. 3. Slope of the upper critical field curve at T_c of $(\mathbf{Th}, \mathrm{Gd})_{0.95}\mathrm{La}_{0.05}$ alloys versus reduced transition temperature compared to the double pair breaking theory [Eq. (2.10) with $P = 0$] and the BCS law of corresponding states (after Guertin et al. [27]).

3. Superconductors Containing Impurities with Crystal-Field Split Energy Levels

Crystal fields often lift the Zeeman degeneracy of the Hund's rule ground states of RE ions in metals. Since spin-dependent exchange scattering will depend on the relative population of the crystal-field split levels at a given temperature, the depression of T_c of superconductors containing RE impurities with partially filled $4f$ shells should reflect these splittings, especially if they are of the order of T_{c_0}. From the theoretical work of Fulde et al. [30], it has been found that the transition temperature is modified by two competing effects: (1) inelastic charge scattering of conduction electrons from the aspherical part of the $4f$ shell which *enhances* T_c, and (2) the usual exchange interaction which *depresses* T_c. The latter effect can be important via off-diagonal matrix elements even when the relevant levels are nonmagnetic.

Experiments on superconductors containing impurities with crystal-field split energy levels have been conducted by Bucher et al. [31] and Cooper [32, 33]. In the former experiments, the ground state was a nonmagnetic singlet, and it was observed that T_c as a function of paramagnetic impurity concentration deviated markedly from the AG theory. Fulde and Hoenig [34] analyzed these experiments in terms of the theory of Fulde et al. [30] considering only transitions between the ground state and the first excited state, and neglecting all higher levels. In this situation pair breaking is inelastic and depends on the ratio

$x \equiv \delta/2T_c$, where δ is the splitting in degrees Kelvin. It was also assumed that the first excited state was a nonmagnetic singlet or that δ was so much larger than T_{c_0} that thermal population of the first excited level at superconducting temperatures was negligible. The latter criterion was satisfied in the systems investigated by Bucher et al. [31]. Theory and experiment agree qualitatively, as shown in Fig. 4. Crystal-

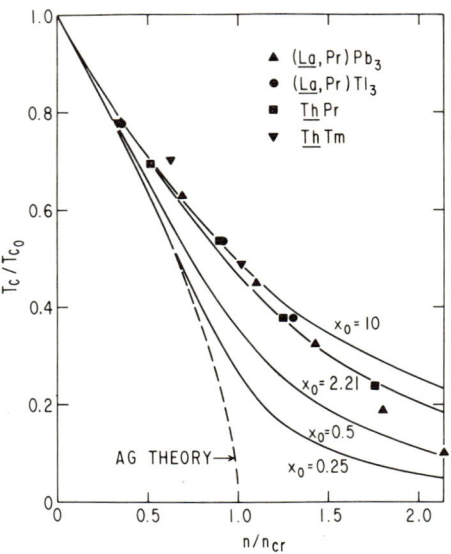

FIG. 4. Reduced transition temperature versus impurity concentration [in reduced units relative to the AG theory (dashed curve) by normalizing the initial slopes] for superconductors containing RE impurities with nonmagnetic singlet ground states [31]. The data are compared to the theoretical curves of Fulde and Hoenig [34] for various values of the parameter $x_0 \equiv \delta/2T_{c_0}$ where δ is the splitting (in degrees Kelvin) between the ground state and the first excited state. Values of x_0 from the specific heat measurements of Bucher et al. [31] are 2.21 for (**La**, Pr) Pb$_3$ and 12.3 for (**La**, Pr) Tl$_3$ (after Fulde and Hoenig [34]).

field splitting of paramagnetic impurity levels should also affect the upper critical field of type II superconductors [34] and electron tunneling characteristics [30]. Curves of T_c versus paramagnetic impurity concentration for both Kramer's and non-Kramer's ions have recently been calculated numerically by Keller and Fulde [35]. The depression of T_c due to interband mixing of 4f and conduction electrons for systems with nonmagnetic gound states has been considered theoretically by Capellmann [36].

4. The Kondo Effect in Superconductors

As discussed previously, the AG theory considers the scattering of conduction electrons by impurity spins only to order $\mathscr{I}^2$. However, the Kondo effect arises from higher order terms in $\mathscr{I}$ and may be expected to affect superconducting properties as profoundly as it does normal state properties. The Kondo effect in superconductors has recently received considerable attention from theorists, and some rather unusual properties have been predicted such as the appearance of a bound state in the gap. Experimentally, however, the effect has not been studied extensively, probably because the classic Kondo systems consist of first-row transition elements dissolved in noble metal hosts which, unfortunately, are not superconducting to the lowest temperatures presently accessible (~ 1 m°K). Most of the experimental work has therefore been concentrated on systems with relatively complex matrices and solutes that contribute spins from unfilled f shells.

The most interesting situation arises for antiferromagnetic exchange coupling between the conduction electron and magnetic impurity spins, that is, for $\mathscr{I} < 0$. By considering conduction-electron–impurity-spin exchange scattering to third order in $\mathscr{I}$, Kondo [37] explained the normal state resistivity minimum phenomenon in metals containing magnetic impurities. The resultant perturbation theory expression for the magnetic contribution to the resistivity (ρ_m) varied as ($-\ln T$) in agreement with experiment. However, the expression diverged for negative $\mathscr{I}$ at a characteristic temperature T_K given by

$$T_K \sim T_F \exp(-1/N(E_F)|\mathscr{I}|), \tag{2.11}$$

where T_F is the Fermi temperature. Since Kondo's original work, much theoretical effort has been devoted to removing the divergence at the Kondo temperature T_K and to understanding the various normal state properties for all temperatures $T \gtrsim T_K$. A physical interpretation which has emerged from these theories is that T_K is a characteristic temperature below which the impurity spins tend to be compensated by the conduction-electron spins, the degree of compensation increasing smoothly with decreasing temperature. Since the relevant energy of this so-called quasibound state is of order $k_B T_K$, whereas the pertinent energy for pairing of time-reversed states in a superconductor is of order $k_B T_{c_0}$, the superconducting properties are expected to be most strongly affected when $T_K \sim T_{c_0}$. For example, recent theories show that the initial depression of T_c with impurity concentration exhibits a maximum as a function of T_K/T_{c_0} near $T_K/T_{c_0} \sim 10$ [38, 39].

Generally, $\mathscr{I}$ is approximated as

$$\mathscr{I} \sim \mathscr{I}_0 + \mathscr{I}_1, \tag{2.12}$$

where $\mathscr{I}_0(>0)$ is the Heisenberg exchange term, while $\mathscr{I}_1(<0)$ arises from covalent mixing between the local impurity state and conduction-electron states. In terms of the Schrieffer–Wolff transformation [40], $\mathscr{I}_1$ is given by

$$\mathscr{I}_1 \sim \frac{\langle V_{kd}^2 \rangle U}{E_d(E_d + U)} \tag{2.13}$$

when $|E_d| \gg \varDelta$. Here V_{kd} is a matrix element mixing local states with itinerant states, $\varDelta = \pi \langle V_{kd}^2 \rangle N(E_F)$ is the Hartree–Fock (HF) half-width of the resonant state, E_d is the energy separating the local state and the Fermi level E_F, and U is the intra-atomic Coulomb repulsion which splits the spin-up and spin-down states in the Friedel–Anderson model [41, 42].[†]

For transition metal solutes in noble metal matrices, the dominant contribution to $\mathscr{I}$ is in nearly all cases $\mathscr{I}_1$, whereas for RE solutes it is usually $\mathscr{I}_0$. An important exception is the RE solute Ce where in most instances $\mathscr{I}$ is characterized by $\mathscr{I}_1$ since the Ce 4f level lies very close to E_F ($E_f \sim -0.1$ eV). We consider first the systems most extensively studied—La and La compounds containing Ce solutes.

a. La and La compounds containing Ce solutes. La and certain La compounds with Ce impurities simultaneously exhibit both superconductivity and a Kondo effect. Although the systems discussed here [**La**Ce and (**La**, Ce)Al$_2$] are somewhat more complex than desired (inasmuch as there is evidence that the Ce $J = \frac{5}{2}$ Hund's rule ground state is split by crystal fields), they provide an opportunity to study the influence of the Kondo effect on superconductivity in the regime $T_K/T_{c_0} \lesssim 1$. Before considering the superconducting properties of these systems, we briefly summarize various normal state measurements used to estimate T_K and thereby establish the ratio T_K/T_{c_0}.

Anomalies associated with the Kondo effect for a RE solute were first observed in the **La**Ce system by Sugawara and Eguchi [43]. The low-temperature normal state resistivity in this system exhibits a minimum near 6°K and the incremental resistivity varies as $(-\ln T)$ between 0.4 and 6°K for sufficiently dilute Ce concentrations [44]. These measurements have recently been extended to lower temperatures

---

[†] The subscript d (e.g., E_d and V_{kd}) which denotes a local d state will sometimes be replaced by the subscript f for a local f state, since we refer to both.

where the resistivity shows a concentration-dependent maximum which extrapolates to $T_\mathrm{m} = 0.17°\mathrm{K}$ as $n \to 0$ [45]. In the dilute limit where the solute ions no longer interact with one another, the resistivity should saturate to the unitarity limit for $T \lesssim T_\mathrm{K}$; therefore it seems plausible to identify T_m (in the limit $n \to 0$) with the Kondo temperature (i.e., $T_\mathrm{K} \sim 0.2°\mathrm{K}$). This value is very close to other estimates of the Kondo temperature of **La**Ce alloys which range from about 0.1 to about 0.6°K [44, 46–48]. The absence of a peak in the thermoelectric power above 7°K also suggests a low Kondo temperature [44]. This is in contrast to the **Y**Ce system for which $T_\mathrm{K} \sim 20\text{--}40°\mathrm{K}$; for example, the low-temperature electrical resistivity of **Y**Ce resembles that of **Cu**Fe and the thermoelectric power shows a large peak near 20°K [49, 50].

The susceptibility of **La**Ce alloys has also been cited as evidence for a Kondo effect. Edelstein [51] described his measurements on **La**Ce below about 50°K by a $T^{-1/2}$ dependence proposed by Anderson [52] for Kondo systems with $T \ll T_\mathrm{K}$. However, this analysis has been criticized [53] because the weakly temperature-dependent susceptibility is observed at temperatures much greater than the inferred value of T_K (16°K). An alternative interpretation [54] is that the temperature dependence of the susceptibility is due to a crystal-field splitting of the Ce $J = \frac{5}{2}$ multiplet. This interpretation has also been applied to the susceptibility of dilute (**La**, Ce)Al$_2$ alloys [55, 56] which resembles the dependence expected for a crystal-field splitting of the Ce $J = \frac{5}{2}$ state into a quartet and a ground-state doublet with a splitting $\delta \sim 100°\mathrm{K}$. However, the effective moment as $T \to 0$ is about 30% smaller than the theoretical value for a doublet ground state and this has been attributed to the Kondo effect [56]. Finally, the susceptibility of (**La**, Ce)Al$_2$ does not fit a $T^{-1/2}$ dependence, although the low-temperature electrical resistivity is much like that of the **La**Ce system. The incremental resistivity varies as $(-\ln T)$ between 1 and 10°K, indicating a relatively low Kondo temperature. Hence the normal state properties suggest that T_K ($\lesssim 0.5°\mathrm{K}$) is significantly less than T_{c_0} in the **La**Ce and (**La**, Ce)Al$_2$ systems ($T_{c_0} = 6.0$ and 3.24°K, respectively) and that conduction electrons at superconducting temperatures exchange scatter from a Ce doublet ground state characterized by an effective spin of $\frac{1}{2}$.

A striking manifestation of the large negative $\mathscr{I}$ for Ce impurities is the anomalous depression of T_c. This is shown in Fig. 5 where $-(dT_\mathrm{c}/dn)_{n=0}$ is plotted versus RE impurity dissolved in the superconductors La [2] and LaAl$_2$ [57]. The anomalous depression arises in part from the large magnitude of $\mathscr{I}$, but also from the Kondo effect, since when $\mathscr{I}$ is *negative*, the depression of T_c can be enhanced over the Born approximation value (see discussion by Müller-Hartmann in

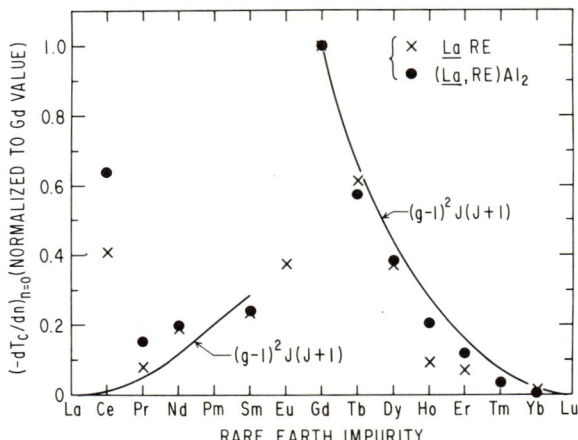

FIG. 5. Initial rate of depression of the superconducting transition temperature of **LaRE** alloys [2] and (**La**, RE) Al_2 alloys [57] versus RE impurity (normalized to the Gd value). The solid line is the deGennes factor $(g - 1)^2 J(J + 1)$ [Eq. (2.4)], normalized to the Gd value. The values of $-(dT_c/dn)_{n=0}$ for Gd impurities are 5.3 and 3.79°K/at. % Gd substitution in La for **LaRE** and (**La**, RE) Al_2, respectively.

Chapter 12). The other RE impurities depress T_c in the manner expected for a nearly constant $\mathscr{I}$ across the RE series (i.e., as the deGennes factor).

Because of the temperature dependence of the spin-flip scattering amplitude, some rather unusual variations of the transition temperature with paramagnetic impurity concentration have been predicted for superconducting Kondo systems [58, 59]. For example, in the theory of Müller-Hartmann and Zittartz, the pair breaking parameter α varies with n and T as

$$\alpha(n, T) = nf(T) \sim n \left[\frac{\pi^2 S(S + 1)}{\ln^2(T/T_K) + \pi^2 S(S + 1)} \right]. \quad (2.14)$$

This function has a maximum at $T = T_K$ and must be solved simultaneously with the AG expression (2.5) for $T_c(n)$. An interesting dependence of T_c on n is predicted when $T_K \ll T_{c_0}$. Expressions (2.5) and (2.14) then yield three solutions for $T_c(n)$ in a particular range of n which depends on the ratio T_K/T_{c_0}. As the temperature is lowered, an alloy of the appropriate composition should first become superconducting at a temperature $T_{c1}(n)$, remain superconducting to a lower temperature $T_{c2}(n)$ at which it becomes normal, and then remain normal to yet a lower temperature $T_{c3}(n)$ at which it again becomes superconducting.

At both boundaries of this composition range there are two solutions for $T_c(n)$ (one is a turning point), whereas outside this range T_c is a single-valued function of n.

A maximum in the ac diamagnetic susceptibility of certain arc-melted (**La**, Ce)Al$_2$ alloys as a function of temperature, recently reported by Riblet and Winzer [60] and confirmed by Maple and Huber [61], is apparently associated with a second transition temperature $T_{c2}(n)$ below $T_{c1}(n)$. However, the transitions in the arc-melted specimens were so broad that the magnitude of the susceptibility maximum failed to attain the fully superconducting value. The transitions can be sharpened considerably by suitable heat treatment, and heat-treated alloys with Ce concentrations not too close to the turning point of the T_c vs. n curve experience a full transition to the superconducting state before their return to the normal state at lower temperatures. The normalized transition signal versus temperature for several heat-treated (**La**, Ce)Al$_2$ alloys [62], measured by a standard ac mutual inductance method (the superconducting state corresponds to a transition signal of unity), is shown in Fig. 6a. Transition curves denoted a through k exhibit complete superconductivity, whereas l, m, and n show only partial superconductivity. Curves f through n display a second transition at lower temperatures back to the normal state which never quite reaches completion.

The T_c vs. n curve for the (**La**, Ce)Al$_2$ system is presented and compared to the AG curve in Fig. 6b. The reentrant behavior is clearly evident. Transition temperatures are defined from the transition curve midpoints, and transition widths (vertical bars in the figure) from the 10 and 90% values. The turning point of the T_c vs. n curve has been estimated from the temperature (0.63°K) of the peak in the partially superconducting transition curves l, m, and n and the interpolated concentration (0.67 ± 0.01 at.% Ce) which would give a normalized transition signal peak amplitude of 0.5. The broad transitions near the turning point are presumably due to concentration gradients and strains. Concentration gradients cause larger transition widths near the turning point where $|dT_c/dn|$ is very large ($|dT_c/dn|$ is infinite at the turning point), while strains contribute to the widths because the magnetic state of Ce ions is extremely sensitive to pressure. A strained sample is likely to have a distribution of Kondo temperatures which results in a distribution of transition temperatures (see Section V). Selected samples with two transitions were measured down to 6 m°K ($T_c/T_{c_0} \sim 2 \times 10^{-3}$), but there was no evidence for a third transition back to the superconducting state to this temperature.

The temperature dependence (below T_{c_0}) of the spin-flip scattering

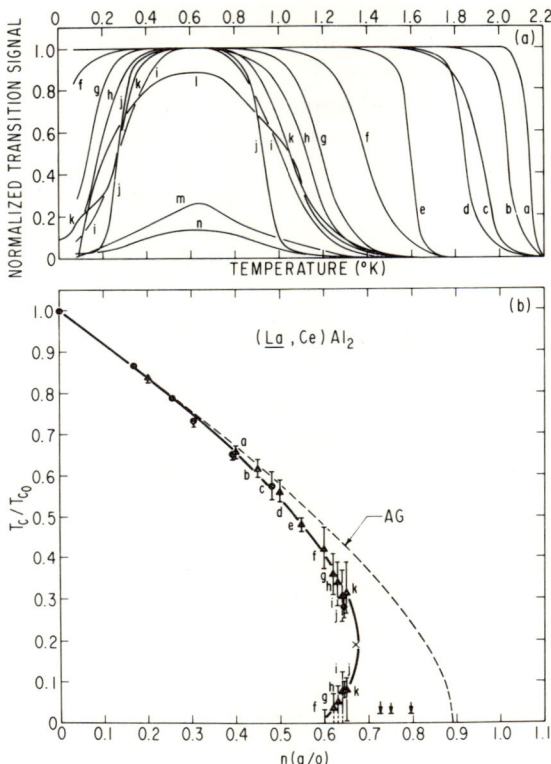

FIG. 6. (a) Normalized transition signal versus temperature for several (**La**, Ce) Al$_2$ alloys. The superconducting state corresponds to a transition signal of unity. (b) Reduced transition temperature versus impurity concentration for the (**La**, Ce) Al$_2$ system. Transition temperatures corresponding to the transition curves in Fig. 6a are identified. The symbol (×) denotes the estimated turning point of the T_c vs. n curve (see text), while the solid circles and triangles distinguish two separately prepared sets of alloys. The AG curve (dashed) is shown for comparison (after Maple et al. [62]).

amplitude of a superconducting Kondo system may be extracted from measurements of $T_c(n)$. This is accomplished by solving expression (2.5) for $f(T_c) = n^{-1}\alpha(n, T_c)$. Within experimental accuracy, the $f(T)$ so deduced from the (**La**, Ce)Al$_2$ data (Fig. 6b) increases with decreasing temperature as $(-\ln T)$ by about 50% between about 0.1 and 2.9°K, suggesting a Kondo temperature $T_K \lesssim 0.1°$K [62]. Another method for determining $f(T)$ which has recently been exploited in purposely prepared "dirty" **La**Ce alloys employs double pair breaking [63]. Here the effect on T_c of the pair breaking parameter due to spin-flip scattering of conduction electrons by paramagnetic Ce ions is studied by varying the pair breaking parameter arising from momentum depairing in an externally applied magnetic field (see Section II, 2). From their $H_{c2}(n, T)$ measurements, Chaikin and Mihalisin [63] derived for the **LaCe** system a 300% increase of $\alpha(n, T)$ with decreasing temperature between 1 and 4°K. However, $\alpha(n, T)$ does not scale with n, which the authors attribute to interactions between Ce ions. Such interactions may also account for the fact that the **LaCe** T_c vs. n curve, measured to temperatures as low as 0.05°K ($T_c/T_{c_0} \sim 10^{-2}$) [64], remains a single-valued function of n even though the ratio $T_K/T_{c_0} \lesssim 1/15$ for this system should correspond to the reentrant case. Previous **LaCe** critical field studies [54, 64] reveal a depression of $H_{c2}(n, T)$ relative to the values calculated to second order in $\mathscr{I}$ by Fulde and Maki [25], in qualitative accord with the behavior expected of a superconducting Kondo system with $T_K \ll T_{c_0}$. The T_c vs. n curve of another system, **(La, Ce)$_3$In** [65], is described well by the theory of Müller-Hartmann and Zittartz if T_K/T_{c_0} is assumed equal to 0.125 (although T_K for this system has not been estimated from normal state properties, the value $T_K \sim 1.2°K$ does not seem unreasonable for a matrix compound where Ce ions replace La ions).

The superconducting properties of the **LaCe** system have also been investigated by means of electron tunneling and specific heat measurements. Tunneling measurements on **LaCe** alloys [66–68] indicate that there are more states in the gap than expected from the AG theory. Conductance curves for **LaCe** alloys [66] are compared to the theoretical AG curves in Fig. 7. For all concentrations studied (0.2–1.0 at.% Ce), the alloys are apparently gapless and the number of states in the gap is proportional to the Ce concentration between 0.2 and 0.5 at.% Ce. This result is in qualitative agreement with recent theoretical predictions. Culbert and Edelstein [69] measured the specific heat of **LaCe** alloys (0.10–0.75 at.% Ce) in the superconducting state and found a term linear in temperature which they identify with the zero bias density of states observed in the tunneling measurements. This term vanishes at 0.10 at.% Ce, implying that the energy gap is finite below this concentration. In the concentration range where tunneling and specific heat measurements overlap, the values obtained for the zero-energy density of states are in reasonable agreement.

b. *Other systems.* Experiments on a number of dilute alloy systems other than those discussed above reveal modifications of superconducting properties which may be traced to the Kondo effect. For example, the early tunneling studies of Woolf and Reif [18] on **PbMn** and **InFe** quenched films yield a density of states at low energies much greater

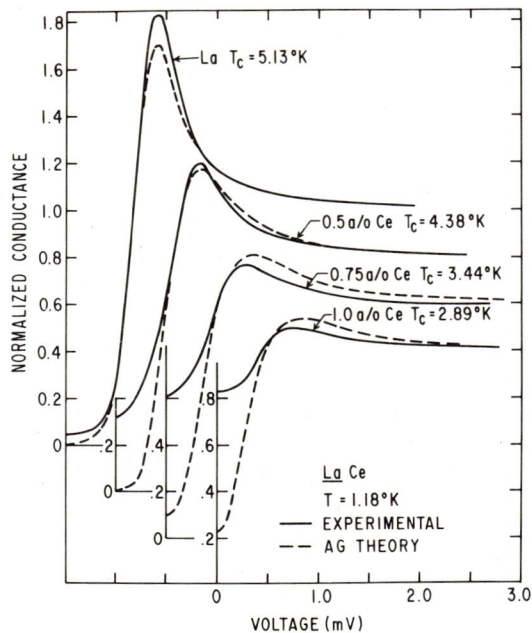

FIG. 7. Electron tunneling conductance curves for **LaCe** films at 1.18°K compared to the AG theory (dashed curves). Successive sets of curves are displaced by 0.5 mV in voltage and 0.2 in normalized conductance (after Edelstein [66]).

than that predicted by the AG theory. Far-infrared absorption measurements on **PbMn** quenched films [24] also provide evidence for additional states in the energy gap. Since the exchange interaction for transition metal solutes is almost invariably large and negative, these deviations from the AG theory may be manifestations of the Kondo effect in the superconducting state. As noted in Section II, 1 both tunneling [18] and far-infrared absorption [24] data are in good agreement with the AG theory for **PbGd** quenched films where the exchange interaction is expected to be relatively weak and positive.

The temperature dependence of the critical magnetic fields of a series of **ZnMn** alloys ($T_K \sim 0.2°$K) has recently been measured by Smith [70]. The calculated entropy difference between the normal and superconducting states is found to be proportional to T as $T \to 0$, indicating a finite density of states at the Fermi level. This has been taken as evidence for a bound state in the energy gap [70]. Specific heat jumps at T_c were also calculated from the critical field data and found to be significantly decreased in magnitude relative to the corresponding AG values.

10. PARAMAGNETIC IMPURITIES IN SUPERCONDUCTORS

Another interesting superconducting Kondo system is $(\text{Zr, Pr})\text{B}_{12}$ [71]. For this system, $(dT_c/dn)_{n=0} \sim -13°\text{K/at.}\%$ Pr [whereas for $(\text{Zr, Gd})\text{B}_{12}$, $(dT_c/dn)_{n=0} \sim -0.6°\text{K/at.}\%$ Gd]. In the normal state there is a pronounced resistivity minimum and the susceptibility follows a Curie–Weiss law with an effective moment of about 3.6 μ_B per Pr ion, the free ion value.

In contrast to the reentrant behavior of $T_c(n)$ in the limit $T_K/T_{c_0} \ll 1$, the theories of Müller-Hartmann and Zittartz [58] and of Ludwig and Zuckermann [59] predict that T_c is a single-valued function of n with positive curvature in the limit $T_K/T_{c_0} \gg 1$. An excellent fit of theory to the experimental $T_c(n)$ curve of the **Th**U system [72] is obtained by Müller-Hartmann and Zittartz for $T_K/T_{c_0} = 32$ and $S = \frac{1}{2}$. The system has been discussed in terms of localized spin fluctuations [72], but if one identifies T_K with the characteristic spin fluctuation temperature $T_0 \sim 100°\text{K}$ inferred from the normal state properties, one obtains $T_K/T_{c_0} \sim 70$, which can be regarded as reasonable agreement. However, it should be pointed out that $S = \frac{1}{2}$ is not really appropriate for this system since the **Th**U normal state magnetic susceptibility at high temperatures is consistent with an effective moment close to the 3.58 μ_B Hund's rule value corresponding to two electrons in the U 5f shell ($^3H_4$ configuration). Moreover, the **Th**U specific heat jumps at T_c [73] follow the BCS law of corresponding states, which indicates that the interaction responsible for the depression of T_c is of *pair weakening* rather than *pair breaking* nature (see Section IV). Finally, a calculation which considers the pair breaking energy dependence [74] does not give a satisfactory fit to $T_c(n)$ for **Th**U and, if correct, would imply that localized spin fluctuations provide the more appropriate description of this system.†

---

† There have been a number of recent experimental investigations of superconducting Kondo systems. Specific heat measurements have been reported for the reentrant system (**La**, Ce)Al$_2$ [C. A. Luengo, M. B. Maple, and W. A. Fertig, *Solid State Commun.* **11**, 1445 (1972); G. von Minnigerode, H. Armbrüster, G. Riblet, F. Steglich, and K. Winzer, *Proc. Int. Conf. Low Temp. Phys., 13th, Boulder, Colorado, 1972* (to be published)] and the system (**La**, Ce)$_3$Al [T. Aoi and Y. Masuda, *Proc. Int. Conf. Low Temp. Phys., 13th, Boulder, Colorado, 1972* (to be published)]. The specific heat jumps ΔC at T_c of both systems are markedly depressed relative to the AG theory, but in accord with a recent calculation by E. Müller-Hartmann and J. Zittartz [*Solid State Commun.* **11**, 401 (1972)] for values of T_K/T_{c_0} close to those estimated from other properties. Similar to the behavior of the initial depression of T_c with n, this theory predicts that the initial depression of $\Delta C/\Delta C_0$ with T_c/T_{c_0} is also a nonmonotonic function of T_K/T_{c_0} with a maximum near $T_K/T_{c_0} \sim 1$. At the maximum, the initial depression of $\Delta C/\Delta C_0$ is largest for $S = \frac{1}{2}$ where it is approximately 2.5 times the AG value. A significant feature of the calculation is that the initial depression of $\Delta C/\Delta C_0$ converges to the AG, rather than the BCS, value in the limit $T_K/T_{c_0} \to \infty$. This strengthens the point of view taken in Section IV; namely, the nearly exponential variation of T_c with n in the systems **Al**Mn and **Th**U should be

III. The Effect of Nonmagnetic Resonant States on Superconductivity

Prior to discussing the effect of localized spin fluctuations on superconductivity (Section IV), we consider in this section the effect on superconductivity of nonmagnetic resonant impurity d or f states, viewed in the sense of the Friedel–Anderson model. Theoretical studies of this somewhat idealized case have been carried out in detail, although recent developments in the local moment problem (for a review, see Heeger [53]) suggest that even in the extreme nonmagnetic limit, $\pi\varDelta \gg U$, the localized impurity spins fluctuate with a finite frequency $\tau_{\rm sf}^{-1} \sim \varDelta/h$. Nonetheless, when the characteristic spin fluctuation temperature $T_0 \equiv h/k_B\tau_{\rm sf}$ is sufficiently greater than T_{c_0}, the impurities behave nonmagnetically at superconducting temperatures and the basic features of superconductivity in a number of matrix–impurity systems with $T_0 \gg T_{c_0}$ are accounted for by these simple theories.

Impurities which form nonmagnetic resonant d or f states depress the transition temperature of a superconductor at a rate intermediate between the strong depression due to magnetic impurities and the rather weak depression due to simple nonmagnetic (i.e., nontransition metal) impurities. This was first pointed out by Boato et al. [75, 76] with regard to their measurements of the depression of T_c of Al by small additions of first-row transition element (Fe group) solutes. Since the work of Friedel [41] and Anderson [42], the apparent nonmagnetic nature of $3d$ transition element impurities in certain simple metals has been regarded as due to the fact that the d-electron states, degenerate in energy with states in the conduction band, become broadened resonances. For Fe group additions to Al, the resonant state widths are so great ($\varDelta \gtrsim 1$ eV) that intra-atomic exchange and Coulomb correlations cannot support a local moment. A more recent systematic study of the

---

attributed to localized spin fluctuations rather than the Kondo effect with $T_K \gg T_{c_0}$ since for both systems $\varDelta C/\varDelta C_0$ vs. T_c/T_{c_0} follows the BCS law of corresponding states. Critical field measurements reveal reentrant behavior of $H_{c2}(T)$ for the system (**La**, **Ce**)Al$_2$ (even for low Ce concentration samples which do not exhibit a second transition in zero field [G. Riblet and K. Winzer, *Solid State Commun.* **11**, 175 (1972)]), but nonreentrant behavior of $H_{c2}(T)$ for the systems (**La**, **Ce**)$_3$Al [T. Aoi and Y. Masuda, *Proc. Int. Conf. Low Temp. Phys., 13th, Boulder, Colorado, 1972* (to be published). and (**La**, **Ce**)$_3$In [H. Jones, Ø. Fischer, G. Bongi, and A. Treyvaud, *Solid State Commun.* **10**, 927 (1972)]. Cerium interaction effects appear to be important in (**La**, **Ce**)$_3$In, as they are in the system **La**Ce. Electron thermal conductivity measurements on superconducting **In**Mn quench-condensed films show deviations from calculations based on the AG theory [A. W. Bjerkaas, D. M. Ginsberg, and B. J. Mrstik, *Phys. Rev. B* **5**, 854 (1972)] which seem most likely due to the Kondo effect.

10. PARAMAGNETIC IMPURITIES IN SUPERCONDUCTORS

superconducting as well as normal state properties of Fe group solutes dissolved in Al has been carried out by Aoki and Ohtsuka [77, 78]. Their results for the reduced initial depression of T_c, $T_{c_0}^{-1}(\Delta T_c/n)$, after correcting for the effect of gap anisotropy, are shown in Fig. 8a.

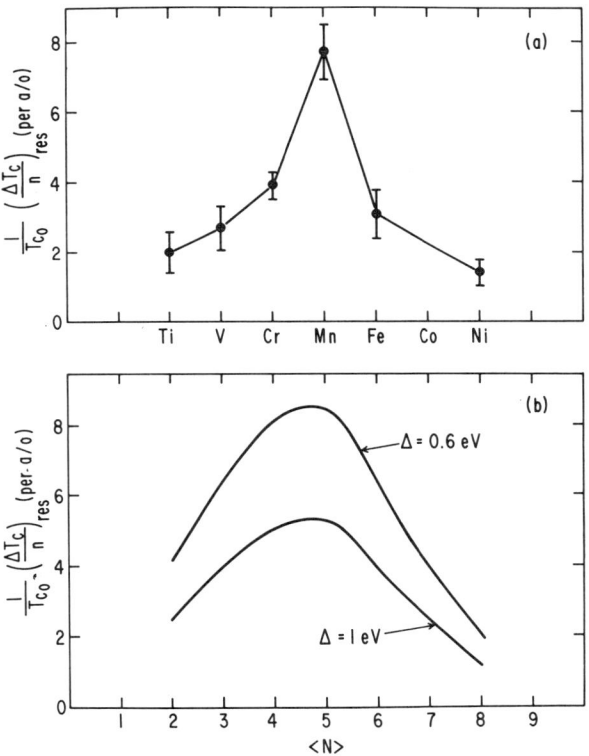

FIG. 8. (a) Initial depression of the transition temperature due to localized d states for $3d$ transition metal impurities in Al (after Aoki and Ohtsuka [77, 78]). (b) Calculated depression of the transition temperature of Al-based alloys due to localized d states versus d-state occupation number $\langle N \rangle$ for $U = 10$ eV (after Ratto and Blandin [81]).

The effect of nonmagnetic resonant states on superconductivity was first treated theoretically by Zuckermann [79]. According to Zuckermann's model the scattering of conduction electrons into the large local density of states at each impurity site gives rise to an initially linear depression of T_c which is inversely proportional to the width Δ of the resonant state. In order to explain the observed depressions of T_c in the Al-based Fe group alloys (Fig. 8a), Δ had to vary rapidly with transition element solute with a minimum value of about 0.1 eV for Mn. Such small level widths are clearly at variance with the nonmagnetic nature of 3d solutes in Al.

This discrepancy was resolved, however, by subsequent theories which included intra-atomic Coulomb repulsion between d electrons with opposite spin at an impurity site [80, 81]. Physically, when a Cooper pair scatters into a local impurity state the two paired electrons with opposite spin are strongly repelled by the intra-atomic Coulomb interaction which leads to a weakened net attraction and a depressed transition temperature. With this modification, Ratto and Blandin showed that $T_{c_0}^{-1}(\Delta T_c/n)$ varies as indicated in Fig. 8b with Δ constant across the first-row transition series. The theory provides a good qualitative description of the data in Fig. 8a for $\Delta \sim 1$ eV and $U = 10$ eV (the nonmagnetic limit still holds since $\pi\Delta > U_{\text{eff}}$ where U_{eff} is defined below); the origin of the quantitative discrepancy at the middle of the transition series (near Mn) is discussed in Section IV.

The most recent theoretical development has been the work of Kaiser [82] who derived from the Ratto–Blandin Hamiltonian the concentration dependence of T_c. According to Kaiser's theory, $T_c(n)$ is given by a modified exponential of the form

$$\frac{T_c}{T_{c_0}} = \exp\left[\frac{-(A+B)n}{\lambda(1-Bn)}\right], \tag{3.1}$$

where

$$A \equiv \frac{N_d(E_F)}{N(E_F)}, \tag{3.2}$$

and

$$B \equiv \frac{N_d(E_F)}{N(E_F)} \frac{N_d(E_F) U_{\text{eff}}}{(2L+1)\lambda}. \tag{3.3}$$

$N_d(E_F)$ is the local density of states at an impurity site (for one spin direction), λ is the BCS coupling constant $[N(E_F)V]$, and U_{eff} is the intra-atomic Coulomb repulsion reduced by correlations [83]. In terms of the HF parameters—Δ, E_d, and U,

$$N_d(E_F) = \frac{2L+1}{\pi} \frac{\Delta}{\Delta^2 + E_d^2} \quad \text{(for one spin direction)}, \tag{3.4}$$

where

$$E_d = \Delta \cot\left[\frac{\pi\langle N\rangle}{2(2L+1)}\right] \tag{3.5}$$

and $\langle N \rangle$ is the d-state occupation number, while

$$U_{\text{eff}} = \frac{U}{1 + (U/\pi E_d)\tan^{-1}(E_d/\varDelta)}. \tag{3.6}$$

The coefficient A comes from the dilution term first studied by Zuckermann [79], whereas B represents the pair weakening term due to the Coulomb repulsion considered by Ratto and Blandin [81]. The theory also predicts a critical concentration given by

$$n_{\text{cr}} = \frac{N(E_{\text{F}})(2L+1)\lambda}{N_d{}^2(E_{\text{F}})\, U_{\text{eff}}}, \tag{3.7}$$

and it should be noted, a law of corresponding states as in the BCS theory.

The modified exponential proposed by Kaiser [82] does indeed describe the T_c vs. n curves of a number of systems including the Al-transition element alloys first studied by Boato et al [75, 76].[†] However, in the **AlMn** and **ThU** alloys to be described in Section IV, spin fluctuation effects are expected to complicate the situation.

The system which appears to satisfy best the conditions of the Kaiser theory is **ThCe** [84, 85]. Shown in Fig. 9 is the specific heat jump $\varDelta C/\varDelta C_0$ versus reduced transition temperature T_c/T_{c_0} for **ThGd** [21] and **ThCe** [86]. The **ThGd** data are in good agreement with the AG pair breaking theory whereas the **ThCe** data follow the BCS theory, indicating that the latter system is nonmagnetic at superconducting temperatures. Accordingly, an excellent fit of Kaiser's modified exponential [Eq. (3.1)] to the experimental T_c vs. n curve for **ThCe** is obtained for various pressures between 0 and 18 kbar as shown in Fig. 10. This is in marked contrast to the **ThGd** system for which the T_c vs. n curve follows the AG prediction [Eq. (2.5)] as noted earlier in Section II, 1. Other evidence for the nonmagnetic nature of **ThCe** alloys at low temperatures is seen in the normal state properties: (a) The electrical resistivity does not show a minimum [87] which indicates that the spin fluctuation temperature is large (presumably $\tau_{\text{sf}}^{-1} \sim \varDelta/h$); (b) the magnetic susceptibility exhibits a maximum and then decreases at low tempera-

---

[†] Superconductivity measurements have recently been extended to concentrated Al-based Fe-group alloys prepared by ultrarapid quenching techniques (E. Babić, P. J. Ford, C. Rizzuto, and E. Salamoni, *J. Low Temp. Phys.* **8**, 219 (1972)]. The dependence of T_c on n remains nearly exponential to the highest concentrations studied (for some impurities, one or two orders of magnitude higher in concentration than previous measurements [75–78]). Exponential depressions of T_c with n have also been observed in films, sputtered at 77°K, of Al-10 wt. % Al_2O_3 with Cr, Mn and Fe impurities [J. J. Hauser, *Phys. Rev. B* **5**, 1830 (1972)].

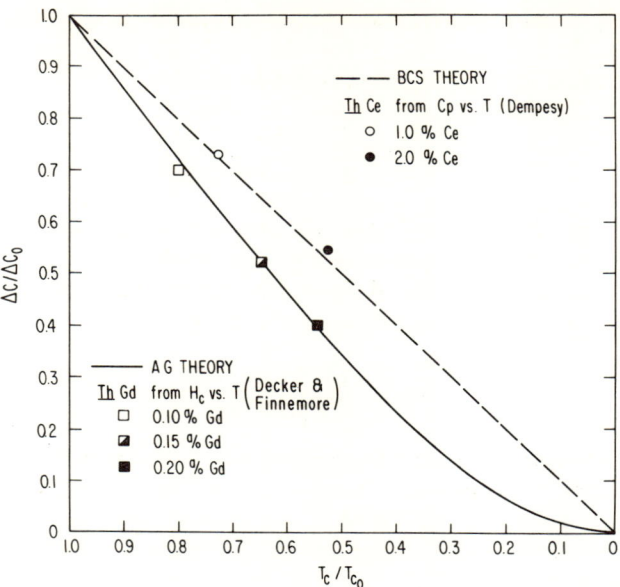

FIG. 9. Reduced specific heat jump ($\Delta C/\Delta C_0$) at T_c versus the reduced transition temperature for **ThCe** alloys [86] and **ThGd** alloys (Decker and Finnemore [21]; measured from critical magnetic fields). The dashed line represents the BCS law of corresponding states, whereas the solid line is the result of the AG theory as calculated by Skalski *et al.* [8] (after Huber and Maple [85]).

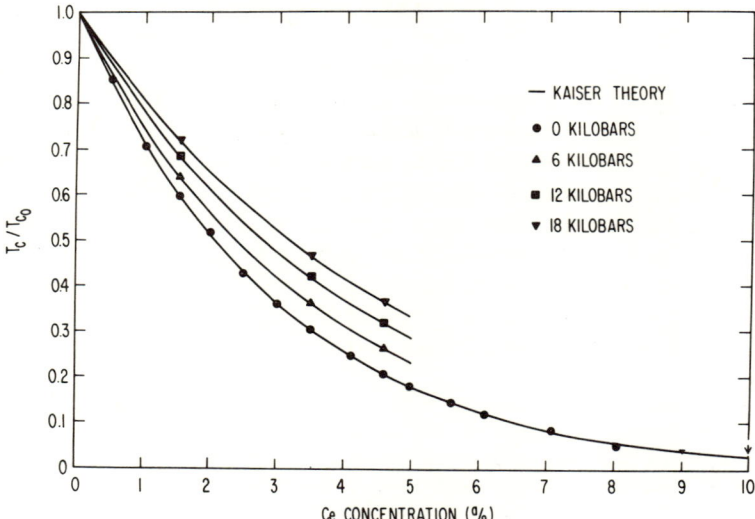

FIG. 10. Reduced transition temperature of **ThCe** alloys versus Ce concentration at various pressures. The curves represent the result of Kaiser's theory [Eq. (3.1)] fitted to the data by the method of least squares (after Huber and Maple [85]).

tures [88]; and (c) the lattice parameter versus composition deviates from Vegard's linear approximation, suggesting a low-temperature valence somewhat greater than 3.25 for Ce between 0 and 20 at.% Ce [89]. From Eq. (3.1) and experiment at zero pressure, the following values for $N_f(E_F)$ and U_{eff} have been determined [85]:

$$N_f(E_F) = 4.86 \text{ states/eV-atom}$$

(in reasonable agreement with the 7.6 states/eV-atom deduced from the low-temperature normal state specific heat [86]), and

$$U_{eff} = 7.1 \times 10^{-2} \text{ eV}.$$

These yield estimates for the parameters of the Friedel–Anderson model:

$$\Delta \gtrsim 1.3 \times 10^{-2} \text{ eV}, \quad E_f \gtrsim 7.6 \times 10^{-2} \text{ eV}, \quad \text{and} \quad U \gtrsim 1.2 \times 10^{-1} \text{ eV}.$$

The value for Δ is in good agreement with previous estimates (see Coqblin and Blandin [90]) but U seems rather small.

IV. The Effect of Localized Spin Fluctuations on Superconductivity

Recent experimental evidence indicates that certain matrix–impurity systems, thought to be nonmagnetic, are actually in the transition region between magnetic and nonmagnetic behavior (i.e., $\pi\Delta \sim U$). The systems best documented in this regard which are also superconducting are **AlMn** and **ThU**. Anomalies in their normal state properties suggest that the localized spins fluctuate with characteristic temperatures T_0 of 530 and 100°K, respectively. In this section we summarize the superconducting and normal state properties of these two systems and consider briefly a number of theoretical attempts at describing the effect of localized spin fluctuations on superconductivity.

The significance of localized spin fluctuations in **AlMn** was first brought to light by Caplin and Rizzuto [91] who measured the low-temperature normal state resistivity of **AlMn** and **AlCr** alloys to very high precision. They found that the impurity contribution to the resistivity varied as

$$\Delta\rho = \rho_0[1 - (T/T_0)^2], \quad (4.1)$$

with $T_0 = 530°K$ for **AlMn** and $1200°K$ for **AlCr**. This led them to suggest that the localized impurity spins fluctuate with a frequency $\tau_{sf}^{-1} \sim k_B T_0/h$. At temperatures much lower than T_0, the spin fluctuations have short lifetimes relative to thermal fluctuation lifetimes, and the spin polarization averages over time which results in a decreased static spin susceptibility relative to a Curie law.

Although the static magnetic susceptibility of **AlMn** alloys is enhanced relative to pure Al, it exhibits essentially the same temperature dependence below 300°K (see, for example, Hedgcock and Li [92] and references to earlier work cited therein). On the other hand, the NMR Knight shift at Al sites varies as $(T - \theta)^{-1}$ with $\theta = -2000 \pm 300$°K (i.e., $|\theta| \sim 4T_0$) which has been attributed to localized spin fluctuations [93, 94].

Additional evidence for the importance of localized spin fluctuations in **AlMn** alloys comes from an enhanced electronic specific heat coefficient [78] and an enhanced thermoelectric power [95]. Within the HF approximation, these measurements give a small virtual bound-state level width of the order of several tenths of an electron volt. Using a localized spin fluctuation approach, Hargitai and Corradi [96] have recently derived from the temperature dependence of the low-temperature normal state electrical resistivity and the enhanced specific heat coefficient of **AlMn** alloys, a level width of order 1.6 eV, and a value for the effective Coulomb interaction $U_{\text{eff}} \sim 4.8$ eV, consistent with the basic nonmagnetic nature of $3d$ solutes in Al.

As indicated in Section III, the depression of T_c for **AlMn** alloys is greater than expected from the Ratto–Blandin theory. Galleani d'Agliano and Ratto [97] proposed that this may be accounted for by a renormalization of the effective Coulomb repulsion due to spin fluctuations; that is, in the Ratto–Blandin theory U_{eff} should be replaced by a parameter U^* to be deduced by comparison with experiment. Except for Mn, U^* equals U_{eff} in the Al-based Fe group alloys for $\Delta = 1$ eV and $U = 5$–10 eV. The value of U^* for Mn impurities is much greater than the largest possible value of U_{eff} corresponding to $U = \infty$ [98].

Of particular interest is the detailed dependence of T_c on n. Shown in Fig. 11a is the measured T_c vs. n curve of the **AlMn** system [99] which exhibits the same general positive curvature as the **ThCe** system. It should be remarked that this curve is described well by the Kaiser theory after applying an energy gap anisotropy correction. Kaiser [100] has suggested that his theory may be applicable to the **AlMn** system when U_{eff} is replaced by U^* since $T_0 \gg T_{c_0}$ ($T_{c_0} = 1.2$°K for Al). A good fit of Eq. (3.1) with the experimental T_c vs. n curve of **AlMn** is obtained for $\Delta = 1.2$ eV and $U^* = 7.3$ eV.

Another system which resembles **AlMn** is **ThU** whose T_c vs. n curve is shown in Fig. 11b [72]. On the basis of the similarity between the T_c vs. n curves for the three systems **ThCe**, **ThU**, and **AlMn**, Huber and Maple [99] proposed that this may be a general result for the superconducting–normal phase boundary when $T_0 \gg T_{c_0}$. In the normal state, the resistivity contributed by the U impurities also varies like

10. PARAMAGNETIC IMPURITIES IN SUPERCONDUCTORS

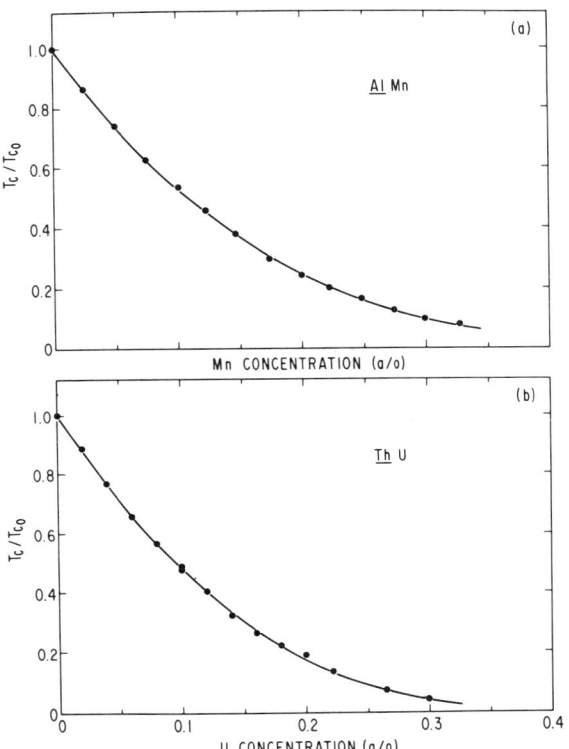

FIG. 11. (a) Reduced transition temperature of AlMn alloys versus Mn concentration; T_{c_0} (1.17°K for Al) has been corrected for gap anisotropy as a function of concentration (after Huber and Maple [99]). (b) Reduced transition temperature of ThU alloys versus U concentration; $T_{c_0} = 1.36°$K for Th (after Maple et al. [72]). The curves are from Kaiser's theory [Eq. (3.1)] fitted to the data by the method of least squares.

expression (4.1) with $T_0 \sim 100°$K, a temperature much larger than $T_{c_0} = 1.4°$K for the Th matrix [72]. The T_c vs. n curve of **ThU** is also described well by the Kaiser theory with U_{eff} presumably replaced by U^*. However, the Kaiser model yields a value for $N_f(E_F)$ which is negligibly small compared to the very large 103 states/eV-atom recently determined from the low-temperature normal state specific heat [73].[†]

[†] A revised value for $N_f(E_F)$ of 57 states/eV-atom has recently been determined from a more extensive specific heat investigation of the **ThU** system [C. A. Luengo, J. M. Cotignola, J. Sereni, A. R. Sweedler, and M. B. Maple, *Proc. Int. Conf. Low Temp. Phys.*, *13th, Boulder, Colorado, 1972* (to be published)]. This value is still very much larger than that deduced from the Kaiser theory [82] and the **ThU** experimental T_c vs. n curve [72].

Thus even though the form of the Kaiser expression for T_c vs. n is correct when localized spin fluctuations are present and $T_0 \gg T_{c_0}$, the parameters $N_f(E_F)$ and U_{eff} both apparently require renormalization. It would be highly desirable to have a theory that begins with the Kaiser model and allows U to increase continuously until a local moment can be said to exist, tracing the behavior of $T_c(n)$. Such a theory exists for the normal state [101, 102], giving a continuous transition from nonmagnetic to magnetic behavior. Although the Curie constant predicted by the theory is too small, it may describe a weakly magnetic situation quite well.

Other normal state properties of **Th**U alloys indicate a characteristic temperature $T_0 \sim 100°K$ [72]. There is a "giant" positive thermoelectric power peak near 80°K. The magnetic susceptibility follows a Curie–Weiss law at temperatures above about 100°K with a Curie–Weiss temperature $\theta_p \sim -309 \pm 50°K$. This implies, in analogy with transition element impurities in noble metal hosts, a characteristic temperature $T_0 \sim |\theta|/4 \sim 80°K$. The effective moment $\mu_{\text{eff}} = 3.4 \pm 0.2\,\mu_B$ is close to the Hund's rule configuration value ($3.58\,\mu_B$) expected for two f electrons per U atom. At temperatures below 100°K, the susceptibility deviates somewhat from the Curie–Weiss law and approaches a finite value as $T \to 0$.

Riblet [103] has recently measured the depression of T_c of Ir by Fe, Co, and Ni impurities. In order to account for the approximately exponential decrease of T_c with concentration in these alloys and in the **Th**U system as well, Riblet inserted an electron–paramagnon coupling constant λ_s into the McMillan [103a] formula. The modified expression takes the form

$$T_c = \frac{\theta_D}{1.45} \exp\left[\frac{-1.04(1 + \lambda + \lambda_s)}{\lambda - \lambda_s - \mu^*(1 + 0.62\lambda)}\right], \qquad (4.2)$$

where the parameters have their usual meanings; the Coulomb pseudopotential μ^*, in terms of the Coulomb repulsion μ, is given by

$$\mu^* = \frac{\mu}{1 + \mu \ln(E_F/k\theta_D)}. \qquad (4.3)$$

Assuming λ_s varies linearly with concentration, this expression is similar to the modified exponential relation between T_c and n derived by Kaiser. Riblet found that the values of $d\lambda_s/dn$ from Eq. (4.2) for Fe, Co, and Ni solutes dissolved in Ir were in reasonable agreement with values deduced from the enhancement of the magnetic susceptibility and the electronic specific heat coefficient.

Morandi [104] developed a theory for the effect of localized spin fluctuations on superconductivity in Ir-based alloys using a local exchange enhancement model [105]. In this theory T_c curves positively with increasing concentration and approaches zero with vanishing slope at a critical concentration n_{cr} about one-half of the concentration which leads to the magnetic instability.

A different theoretical approach due to Bennemann [106] assumes that localized spin fluctuations break pairs and that the universal relation between T_c and α of the AG theory is applicable. The characteristic positive curvature or "slowing down" of T_c vs. n for systems such as **AlMn** can then be obtained by assuming α is temperature as well as concentration dependent; that is, $\alpha = nf(T)$. Since scattering of conduction electrons by spin fluctuations varies as T^2 at low temperatures, Bennemann suggested $\alpha \sim nT^2$ which gives a T_c vs. n curve with positive curvature in qualitative agreement with experiment. It was also argued that *interatomic* exchange and Coulomb coupling damped the spin fluctuations. However, the critical field curves of **AlMn** [32, 107] and the specific jumps at T_c of **AlMn** [108] and **ThU** [73] follow the BCS law of corresponding states.[†] This indicates that, to the extent that localized spin fluctuations are responsible for the depression of T_c in **AlMn** alloys, they do *not* break pairs. Moreover, when the T_c vs. n curves of **AlMn** [99] and **ThU** [72] are analyzed within the context of the Bennemann theory, it is found that α does not vary as T^2, but rather is given by

$$\alpha/\alpha_{cr} = n[1.2 + 7.0(T/T_{c_0})^{1/2}]$$

for **AlMn** and

$$\alpha/\alpha_{cr} = n[1.9 + 6.9(T/T_{c_0})^{1/2}]$$

for **ThU** where n is expressed in atomic percent.

A model based on an instantaneous exchange interaction between conduction-electron spins and fluctuating impurity spins has recently been developed by Rivier and MacLaughlin [109]. A single phenomenological parameter, the spin fluctuation lifetime τ_{sf} (or temperature $T_0 \equiv h/k_B\tau_{sf}$), characterizes the dependence of T_c on n. The AG result for $T_c(n)$ is obtained for temperatures $T \gtrsim T_0$ in the limit $T_0/T_{c_0} \ll 1$, whereas an exponential dependence of T_c on n, similar to the Kaiser result, is found for $T_0/T_{c_0} \gg 1$. Thus the superconducting behavior is

[†] Recent critical field measurements on superconducting **ThU** alloys indicate that $H_c(n, T)$ follows the BCS law of corresponding states [D. K. Finnemore, private communication (1973); H. L. Watson, D. T. Peterson, and D. K. Finnemore, *Proc. Int. Conf. Low Temp. Phys., 13th, Boulder, Colorado, 1972* (to be published)].

implicitly related to the parameters of the Friedel–Anderson model via τ_{sf}.

The effect of localized spin fluctuations on the critical temperature of dilute superconducting alloys in both the weakly magnetic and strongly magnetic limits has also been studied theoretically by Zuckermann [110].

V. Magnetic–Nonmagnetic Transitions of Impurities in Superconductors

In the preceding sections we have considered separately long-lived local moments, nonmagnetic resonant states, and localized spin fluctuations. We now discuss a class of experiments in which smooth transitions between these regimes are induced by applying pressure or by varying the composition of a binary alloy matrix. The former method is more desirable because a pure metal or intermetallic compound may be used as the host, whereas the latter involves a statistical distribution of two types of matrix metal atoms and the complications incurred thereof [111].

Pressure-induced magnetic–nonmagnetic transitions have been studied in La, La compounds, and Y containing Ce impurities. Due to the inherent instability of the Ce $4f$ shell, the magnetic character of Ce solutes is very sensitive to pressure. This is reflected dramatically in the superconducting properties of the matrix–Ce impurity system. A large increase of the depression of T_c in **La**Ce alloys with pressure to 10 kbar was first observed by Smith [112]. This experiment was interpreted by Coqblin and Ratto [113] in terms of the Schrieffer–Wolff transformation [Eq. (2.13)]; assuming $\langle V_{kf}^2 \rangle$ was approximately independent of pressure, the inferred increase of $|\mathscr{J}|$ with pressure was attributed to a corresponding decrease of $|E_f|$ with pressure. The effect was later observed in (**La**, Ce)Al$_2$ alloys by Maple and Smith [114].

If the energy separating the Ce $4f$ level and Fermi level continues to decrease with pressure, one would ultimately expect a transition from a magnetic to a nonmagnetic state. In the context of the Friedel–Anderson model, this transition is shown schematically in Fig. 12. As the spin-up state begins to overlap the Fermi level, the net Coulomb repulsion which splits the spin-up and spin-down states decreases. At a critical value of E the spin-up and spin-down states collapse onto the Fermi level, overlapping it to the extent that a fraction of an f electron is contained in the spin degenerate local state. Such a transition has been inferred from the nonmonotonic depression of T_c with pressure observed in the systems (**La**, Ce)$_3$In [65] and **La**Ce [116]. Figure 13 depicts T_c as a function of pressure for the **La**Ce system where the measurements extend to 140 kbar.

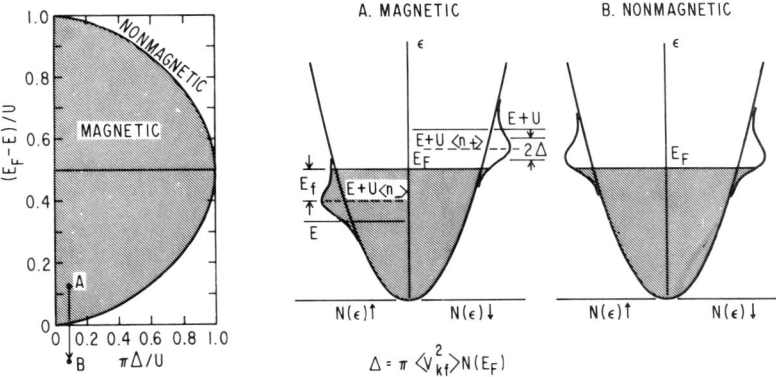

FIG. 12. Schematic diagram of the magnetic–nonmagnetic transition of Ce impurities in the Friedel–Anderson model. The transition is assumed to be driven by a decrease of the energy separating the Ce $4f$ level and the Fermi level with pressure (after Maple [115]).

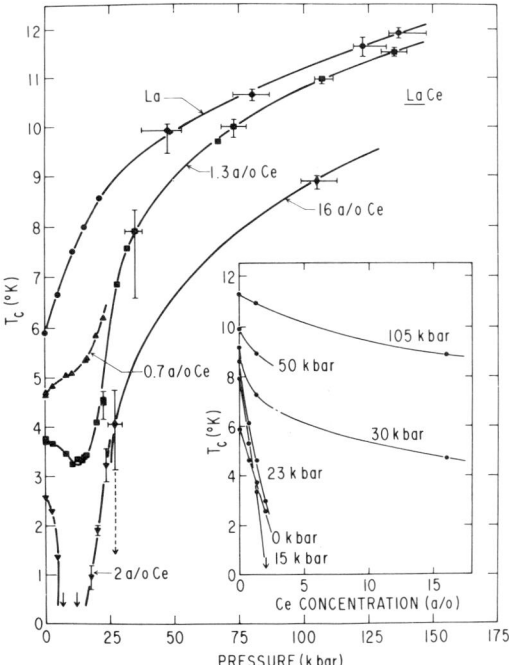

FIG. 13. Pressure dependence of the superconducting transition temperature of as-cast predominantly fcc La and **La**Ce alloys to very high pressure. The vertical bars represent the transition widths and the horizontal bars the pressure inhomogeneity in the high-pressure cell. Isobars of T_c versus Ce concentration are shown in the inset (After Maple et al. [116]).

The depression of T_c first increases with pressure, reaches a maximum near 15 kbar, and then decreases from the maximum to a value more than an order of magnitude smaller above about 100 kbar. It has been suggested [116] that the appreciable depression of T_c ($\sim$0.24°K/at.% Ce) at high pressure ($\gtrsim$100 kbar) results from scattering of conduction electrons into the nonmagnetic Ce resonant states as occurs, for example, in the **Th**Ce system at normal pressure.

An alternative explanation for the nonmonotonic dependence of T_c on pressure based on the exchange model has been pointed out by Maple and Kim [65], Müller–Hartmann and Zittartz [39], and Gey and Umlauf [47]. From the theoretical work of Zuckermann [38] and Müller-Hartmann and Zittartz [39], the depression of T_c goes through a maximum as a function of T_K/T_{c_0}. If $|\mathscr{I}|$ were to continue to increase with pressure, one would expect T_K to increase exponentially with pressure [Eq. (2.11)]. This is certainly expected for the **La**Ce system at low pressures where the exchange Hamiltonian provides an appropriate description. If this remained so throughout the entire pressure range (to 140 kbar), it would provide an explanation for the maximum in the depression of T_c as a function of pressure. The essential difference between the two models is therefore the following. In the first case, the increase and subsequent decrease in the depression of T_c with pressure are associated with demagnetization of the Ce 4f level as it crosses the Fermi level—the depression of T_c at low pressure is due to *pair breaking*, whereas the depression at high pressure is due to *pair weakening*. In the second case, the Ce 4f level never crosses the Fermi level—the impurity always possesses a well-defined spin (although the *matrix-impurity* system may be nonmagnetic at temperatures much less than T_K) and the nonmonotonic depression of T_c is due to the competition between superconducting electron singlet spin pairing and antiferromagnetic conduction-electron–impurity-spin pairing. Here the depression of T_c is always due to *pair breaking* interactions.

The first mechanism seems more likely. Kim and Maple [46] noted that if one takes the argument based on the exchange model to its logical conclusion T_K would be about 10^6 °K at 140 kbar (assuming ΔT_c and T_K are related through the theory of Müller-Hartmann and Zittartz). This value for T_K is unphysically large and implies that the exchange model is an unrealistic description of **La**Ce alloys at high pressure. Moreover, the low-temperature normal state resistivity of **La**Ce under pressure [46, 47] shows that the slope $|d\rho_m/d\ln T|$ in the temperature range where ρ_m is linear in ($-\ln T$) initially increases with pressure and then exhibits a maximum near 15 kbar, the pressure at which the depression of T_c attains a maximum. If T_K were to increase continuously with pressure, one would expect a corresponding increase in the slope $| d\rho_m/d \ln T |$ in the region where ρ_m is linear in $(- \ln T)$, with curvature and saturation of ρ_m at lower temperature (for example, Suhl and Wong [117]). No evidence is seen for this up to 18 kbar, although from the theory of Müller-Hartmann and Zittartz and the observed depression of T_c, T_K would be about 100°K at this pressure. In the YCe system, where $T_K \sim 20\text{-}40°K$ at zero pressure, ρ_m tends toward saturation at about 10°K [49].

A phenomenological description of the depression of T_c as a function of pressure has recently been given by Coqblin et al. [118]. Toulouse and Coqblin [119] have shown that inclusion of potential scattering in the Schrieffer–Wolff transformation leads to an approximate expression for $\mathscr{J}$ given by

$$\mathscr{J} \sim \frac{-2\langle V_{kf}^2 \rangle | E_f |}{E_f^2 + \varDelta^2} \quad \text{when} \quad | E_f | \ll U, \tag{5.1}$$

which implies that $| \mathscr{J} |$ has a maximum at $E_f = -\varDelta$ and equals zero when $E_f = 0$. Assuming $| E_f |$ decreases linearly with pressure, and the AG expression for the depression of T_c (Kondo effects are neglected here since T_K is assumed much less than T_{c_0}), the theoretical expression describes the data rather well (Fig. 14). Above the magnetic–nonmagnetic transition the Ratto–Blandin theory has been employed, again assuming that E_f varies linearly with pressure. Both theories break down in the region of the transition (near 30 kbar) where localized spin fluctuations

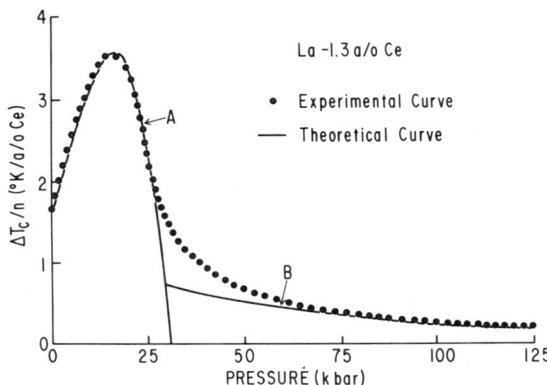

FIG. 14. Depression of the transition temperature of a LaCe (1.3 at. % Ce) alloy versus pressure compared to the AG theory with $\mathscr{J}$ given by Eq. (5.1) (curve A) and the Ratto–Blandin theory (curve B) (after Coqblin et al. [118]).

are also presumed to be important. Since the slope $|d\rho_m/d \ln T|$ depends on $\mathscr{I}$, the maximum in $|d\rho_m/d \ln T|$ may also be explained by the maximum in $|\mathscr{I}|$. It should be emphasized, however, that a general theory of magnetic–nonmagnetic transitions in superconductors must go *smoothly* from *pair breaking* to *pair weakening* regimes. Theories employing the functional integral method appear to be very promising in this respect and a recent calculation of the critical temperature through the magnetic–nonmagnetic transition [120] qualitatively describes the non-monotonic variation of the depression of T_c of the **La**Ce system with pressure.

Several other superconducting–magnetic impurity systems appear to exhibit magnetic–nonmagnetic transitions. Maple and Wittig [121] have measured the low-temperature electrical resistivity of **Y**Ce under pressure to 140 kbar. The normal state resistivity anomaly associated with the Kondo effect is suppressed by pressure and disappears completely by about 80 kbar, while the depression of T_c of Y by Ce impurities above 100 kbar is consistent with that expected for scattering of conduction electrons into nonmagnetic Ce 4f resonant states. Superconductivity experiments in which the composition of a binary matrix is varied to induce changes in the magnetic character of the impurities have been conducted in the systems (**Zn, Al**)Mn [122] and (**Y, Th**)Ce [123].[†]

ACKNOWLEDGMENTS

The author wishes to thank Dr. J. G. Huber, Dr. E. Müller-Hartmann, Dr. C. F. Ratto and Dr. D. Wohlleben for comments regarding this manuscript and useful discussions, and Professor P. Fulde for illuminating correspondence. The support of the Air Force Office of Scientific Research under Grant No. AF-AFOSR-631-67-A during the preparation of this article is gratefully acknowledged.

[†] Superconductivity has recently been studied in fcc (**La, Th**)Ce alloys at zero pressure [S. Ortega, M. Roth, C. Rizzuto, and M. B. Maple *Solid State Commun.* (in press)]. The depression of T_c due to the Ce impurities exhibits a maximum as a function of Th concentration at about 55 at. % Th substitution for La. The height of the maximum and the decrease of the lattice parameter (relative to La) at which it occurs have values approximately equal to those encountered in pressurized **La**Ce alloys at 15 kbar. Apparently, the decrease of the lattice parameter caused by the addition of Th drives the magnetic–nonmagnetic transition of the Ce impurities in the (**La, Th**)Ce system as it does through the application of pressure to the **La**Ce system [116]. This suggests that the *nonmagnetic* **Th**Ce system at zero pressure is indeed the analog of the **La**Ce system at high pressure. The pressure dependence of the resistance anomaly of **Y**Ce alloys has been reinvestigated [M. Dietrich, W. Gey, and E. Umlauf, *Solid State Commun.* 11, 655 (1972)]. These authors interpret their results in terms of a monotonic increase of T_K with pressure, rather than a pressure-induced magnetic–nonmagnetic transition.

References

1. W. Buckel and R. Hilsch, *Z. Phys.* **128**, 324 (1950).
2. B. T. Matthias, H. Suhl, and E. Corenzwit, *Phys. Rev. Lett.* **1**, 92 (1958).
3. C. Herring, *Physica* **24**, S 184 (1958).
4. H. Suhl and B. T. Matthias, *Phys. Rev.* **114**, 977 (1959).
5. A. A. Abrikosov and L. P. Gor'kov, *Zh. Eksp. Teor. Fiz.* **39**, 1781 (1960); *Sov. Phys. JETP* **12**, 1243 (1961).
6. P. G. deGennes and G. Sarma, *J. Appl. Phys.* **34**, 1380 (1963).
7. J. C. Phillips, *Phys. Rev. Lett.* **10**, 96 (1963).
8. S. Skalski, O. Betbeder-Matibet, and P. R. Weiss, *Phys. Rev. A* **136**, 1500 (1964).
9. M. A. Jensen and H. Suhl, in "Magnetism" (G. T. Rado and H. Suhl, eds.), Vol. IIB, pp. 183–214. Academic Press, New York, 1966.
10. B. T. Matthias, H. Suhl, and E. Corenzwit *J. Phys. Chem. Solids* **13**, 156 (1960).
11. R. A. Hein, R. L. Falge Jr., B. T. Matthias, and E. Corenzwit, *Phys. Rev. Lett.* **2**, 500 (1959).
12. J. E. Crow and R. D. Parks, *Phys. Lett.* **21**, 378 (1966).
13. Y. Kuwasawa, K. Sekizawa, N. Usui, and K. Yasukochi, *J. Phys. Soc. Jap.* **27**, 590 (1969).
14. T. Mamiya, T. Aio, K. Iwahashi, and Y. Masuda, *J. Phys. Soc. Jap.* **31**, 485 (1971).
15. K. H. Bennemann, *Phys. Rev. Lett.* **17**, 438 (1966).
16. M. B. Maple, *Phys. Lett. A* **26**, 513 (1968).
17. R. P. Guertin, Ph. D. Thesis, Univ. of Rochester, Rochester, New York, 1968, unpublished.
18. M. A. Woolf and F. Reif, *Phys. Rev. A* **137**, 557 (1965).
19. D. K. Finnemore, D. L. Johnson, J. E. Ostenson, F. H. Spedding, and B. J. Beaudry, *Phys. Rev. A* **137**, 550 (1965).
20. W. R. Decker, D. T. Peterson, and D. K. Finnemore, *Phys. Rev. Lett.* **18**, 889 (1967).
21. W. R. Decker and D. K. Finnemore, *Phys. Rev.* **172**, 430 (1968).
22. R. L. Cappelletti and D. K. Finnemore, *Phys. Rev.* **188**, 723 (1969).
23. V. Ambegaokar and A. Griffin, *Phys. Rev. A* **137**, 1151 (1965).
24. G. J. Dick and F. Reif, *Phys. Rev.* **181**, 774 (1969).
25. P. Fulde and K. Maki, *Phys. Rev.* **141**, 275 (1966).
26. R. D. Parks, in "Superconductivity" (P. R. Wallace, ed.), Vol. 2, pp. 625–690. Gordon & Breach, New York, 1969.
27. R. P. Guertin, W. E. Masker, T. W. Mihalisin, R. P. Groff, and R. D. Parks, *Phys. Rev. Lett.* **20**, 387 (1968).
28. J. E. Crow, R. P. Guertin, and R. D. Parks, *Phys. Rev. Lett.* **19**, 77 (1967).
29. P. G. deGennes and G. Sarma, *Solid State Commun.* **4**, 449 (1966).
30. P. Fulde, L. L. Hirst, and A. Luther, *Z. Phys.* **230**, 155 (1970).
31. E. Bucher, K. Andres, J. P. Maita, and G. W. Hull, *Helv. Phys. Acta* **41**, 723 (1968).
32. J. R. Cooper, Ph.D. Thesis, Univ. of London, London, 1969, unpublished.
33. J. R. Cooper, *Solid State Commun.* **9**, 1429 (1971).
34. P. Fulde and H. E. Hoenig, *Solid State Commun.* **8**, 341 (1970).
35. J. Keller and P. Fulde, *J. Low Temp. Phys.* **4**, 289 (1971).
36. H. Capellmann, *J. Phys. Chem. Solids* **32**, 2439 (1971).
37. J. Kondo, *Progr. Theor. Phys.* **32**, 37 (1964).
38. M. J. Zuckermann, *Phys. Rev.* **168**, 390 (1968).
39. E. Müller-Hartmann and J. Zittartz, *Z. Phys.* **234**, 58 (1970).

40. J. R. Schrieffer and P. A. Wolff, *Phys. Rev.* **149**, 491 (1966).
41. J. Friedel, *Nuovo Cimento Suppl.* **12**, 1861 (1958).
42. P. W. Anderson, *Phys. Rev.* **124**, 41 (1961).
43. T. Sugawara and H. Eguchi, *J. Phys. Soc. Jap.* **21**, 725 (1966).
44. T. Sugawara and H. Eguchi, *J. Phys. Soc. Jap.* **26**, 1322 (1969).
45. J. J. Wollan and D. K. Finnemore, *Phys. Lett. A* **33**, 299 (1970).
46. K. S. Kim and M. B. Maple, *Phys. Rev. B* **2**, 4696 (1970).
47. W. Gey and E. Umlauf, *Z. Phys.* **242**, 241 (1971).
48. J. Flouquet, *Phys. Rev. Lett.* **27**, 515 (1971).
49. T. Sugawara and S. Yoshida, *J. Phys. Soc. Jap.* **24**, 1399 (1968).
50. H. Nagasawa, S. Yoshida, and T. Sugawara, *Phys. Lett. A* **26**, 561 (1968).
51. A. S. Edelstein, *Phys. Rev. Lett.* **20**, 1348 (1968).
52. P. W. Anderson, *Phys. Rev.* **164**, 352 (1967).
53. A. J. Heeger, *Solid State Phys.* **23**, 283–411 (1969).
54. T. Sugawara and H. Eguchi, *J. Phys. Soc. Jap.* **23**, 965 (1967).
55. J. A. White, H. J. Williams, J. H. Wernick, and R. C. Sherwood, *Phys. Rev.* **131**, 1039 (1963).
56. M. B. Maple and Z. Fisk, *Proc. Int. Conf. Low Temp. Phys., 11th, St. Andrews, Scotland, 1968* (J. F. Allen, D. M. Findlayson, and D. M. McCall, eds.), Vol. 2, pp. 1288–1292. St. Andrews, Scotland, 1968.
57. M. B. Maple, *Solid State Commun.* **8**, 1915 (1970).
58. E. Müller-Hartmann and J. Zittartz, *Phys. Rev. Lett.* **26**, 428 (1971).
59. A. Ludwig and M. J. Zuckermann, *J. Phys. F* **1**, 516 (1971).
60. G. Riblet and K. Winzer, *Solid State Commun.* **9**, 1663 (1971).
61. M. B. Maple and J. G. Huber, unpublished data, 1971.
62. M. B. Maple, W. A. Fertig, A. C. Mota, L. E. DeLong, D. Wohlleben, and R. Fitzgerald, *Solid State Commun.* **11**, 829 (1972).
63. P. M. Chaikin and T. W. Mihalisin, *Solid State Commun.* **10**, 465 (1972).
64. E. Umlauf, J. Schneider, R. Meier, and H. Kreuzer, *J. Low Temp. Phys.* **5**, 191 (1971).
65. M. B. Maple and K. S. Kim, *Phys. Rev. Lett.* **23**, 118 (1969).
66. A. S. Edelstein, *Phys. Rev. Lett.* **19**, 1184 (1967).
67. A. S. Edelstein, *Phys. Rev.* **180**, 505 (1969).
68. N. Tsuda, *J. Phys. Soc. Jap.* **27**, 1025 (1969).
69. H. V. Culbert and A. S. Edelstein, *Solid State Commun.* **8**, 445 (1970).
70. F. W. Smith, *J. Low Temp. Phys.* **5**, 683 (1971).
71. Z. Fisk and B. T. Matthias, *Science* **165**, 279 (1969).
72. M. B. Maple, J. G. Huber, B. R. Coles, and A. C. Lawson, *J. Low Temp. Phys.* **3**, 137 (1970).
73. C. A. Luengo, J. M. Cotignola, J. Sereni, A. R. Sweedler, M. B. Maple, and J. G. Huber, *Solid State Commun.* **10**, 459 (1972).
74. E. Müller-Hartmann, private communication, 1971.
75. G. Boato, G. Gallinaro, and C. Rizzuto, *Phys. Lett.* **5**, 20 (1963).
76. G. Boato, G. Gallinaro, and C. Rizzuto, *Phys. Rev.* **148**, 353 (1966).
77. R. Aoki and T. Ohtsuka, *J. Phys. Soc. Jap.* **23**, 955 (1967).
78. R. Aoki and T. Ohtsuka, *J. Phys. Soc. Jap.* **26**, 651 (1969).
79. M. J. Zuckermann, *Phys. Rev. A* **140**, 899 (1965).
80. K. Takanaka and F. Takano, *Progr. Theor. Phys.* **36**, 1080 (1966).
81. C. F. Ratto and A. Blandin, *Phys. Rev.* **156**, 513 (1967).
82. A. B. Kaiser, *J. Phys. C* **3**, 409 (1970).

10. PARAMAGNETIC IMPURITIES IN SUPERCONDUCTORS 325

83. J. R. Schrieffer and D. C. Mattis, *Phys. Rev. A* **140**, 1412 (1965).
84. M. B. Maple, J. G. Huber, and K. S. Kim, *Solid State Commun.* **8**, 981 (1970).
85. J. G. Huber and M. B. Maple, *J. Low Temp. Phys.* **3**, 537 (1970).
86. C. W. Dempesey, private communication of unpublished data, 1970.
87. D. T. Peterson, D. F. Page, R. B. Rump, and D. K. Finnemore, *Phys. Rev.* **153**, 701 (1967).
88. L. F. Bates and M. M. Newmann, *Proc. Phys. Soc. London* **72**, 345 (1958).
89. I. R. Harris and G. J. Raynor, *J. Less-Common Metals* **6**, 70 (1964).
90. B. Coqblin and A. Blandin, *Advan. Phys.* **17**, 281 (1968).
91. A. D. Caplin and C. Rizzuto, *Phys. Rev. Lett.* **21**, 746 (1968).
92. F. T. Hedgcock and P. L. Li, *Phys. Rev. B* **2**, 1342 (1970).
93. H. Launois and H. Alloul, *Solid State Commun.* **7**, 525 (1969).
94. H. Alloul and H. Launois, *J. Appl. Phys.* **41**, 923 (1970).
95. G. Boato and J. Vig, *Solid State Commun.* **5**, 649 (1967).
96. C. Hargitai and G. Corradi, *Solid State Commun.* **7**, 1535 (1969).
97. E. Galleani d'Agliano and C. F. Ratto, *Nuovo Cimento B* **60**, 121 (1969).
98. C. F. Ratto, Ph. D. Thesis, Univ. of Paris, Paris, 1969, unpublished.
99. J. G. Huber and M. B. Maple, *Solid State Commun.* **8**, 1987 (1970).
100. A. B. Kaiser, Ph.D. Thesis, Univ. of London, London, 1970, unpublished.
101. H. Suhl, *Phys. Rev. Lett.* **19**, 442 (1967).
102. M. Levine and H. Suhl, *Phys. Rev.* **171**, 567 (1968).
103. G. Riblet, *Phys. Rev. B* **3**, 91 (1971).
103a. W. L. McMillan, *Phys. Rev.* **167**, 10 (1968).
104. G. Morandi, *Solid State Commun.* **6**, 561 (1968).
105. P. Lederer and D. L. Mills, *Phys. Rev. Lett.* **20**, 1036 (1968).
106. K. H. Bennemann, *Phys. Rev.* **183**, 492 (1969).
107. F. W. Smith, *J. Low Temp. Phys.* **6**, 435 (1972).
108. D. L. Martin, *Proc. Phys. Soc. London* **78**, 1489 (1961).
109. N. Y. Rivier and D. E. MacLaughlin, *J. Phys. F* **1**, L48 (1971).
110. M. J. Zuckermann, *J. Phys. C* **3**, 2130 (1970).
111. V. Jaccarino and L. R. Walker, *Phys. Rev. Lett.* **15**, 258 (1965).
112. T. F. Smith, *Phys. Rev. Lett.* **17**, 386 (1966).
113. B. Coqblin and C. F. Ratto, *Phys. Rev. Lett.* **21**, 1065 (1968).
114. M. B. Maple and T. F. Smith, *Solid State Commun.* **7**, 515 (1969).
115. M. B. Maple, *Proc. Conf. Superconductivity in d- and f-Band Metals, Univ. of Rochester, 1971* (D. H. Douglass, ed.), AIP Conf. Proc. No. 4, pp. 175–203. Amer. Inst. Phys., New York, 1972.
116. M. B. Maple, J. Wittig, and K. S. Kim, *Phys. Rev. Lett.* **23**, 1375 (1969).
117. H. Suhl and D. Wong, *Physics (Long Island City, N.Y.)* **3**, 17 (1967).
118. B. Coqblin, M. B. Maple, and G. Toulouse, *Int. J. Magn.* **1**, 333 (1971).
119. G. Toulouse and B. Coqblin, *Solid State Commun.* **7**, 853 (1969).
120. A. Theumann, *Phys. Rev. B* **5**, 4382 (1972).
121. M. B. Maple and J. Wittig, *Solid State Commun.* **9**, 1611 (1971).
122. G. Boato and C. Rizzuto, *Proc. Int. Conf. Low Temp. Phys., 11th, St Andrews, Scotland, 1968* (J. F. Allen, D. M. Findlayson, and D. M. McCall, eds.), Vol. 2, pp. 1062–1065. St. Andrews, Scotland, 1968.
123. J. G. Huber and M. B. Maple, *Proc. Int. Conf. Low Temp. Phys., 13th, Boulder, Colorado, 1972*, to be published.

11. Recent Work on Ferromagnetic Superconductors

Ø. Fischer and M. Peter

Départment de Physique de la Matière Condensée
Université de Genève
Geneva, Switzerland

I. Introduction 327
II. Magnetic Ordering in a Superconductor. The Low-Concentration Limit . . 328
III. Coexistence in More Concentrated Systems. The Case of $Ce_{1-x}Gd_xRu_2$ . . 336
IV. Superconductor in a Molecular Field. Compensation of the Exchange Field by an External Field 343
V. Final Remarks 350
References 351

I. Introduction

The BCS theory, with its singlet pairing of the conduction electrons, apparently excludes the coexistence of ferromagnetism and superconductivity. The experiments of Matthias *et al.* [1] on $La_{1-x}Gd_x$ clearly showed the destructive effect of magnetic ions on superconductivity. However, further measurements [2, 3] showed that superconductivity and ferromagnetism could coexist in the same sample. The phase diagrams obtained did not correspond to what one expected from a simple BCS superconductor and a long-range ferromagnetic polarization. Different explanations were suggested, essentially supposing that the coexistence of a ferromagnetic alignment and superconductivity occurred in domains smaller than the coherence length ξ [4, 5]. Subsequent theoretical work has extended the BCS theory to include different scattering effects, thus reopening the question of coexistence. These, and related theoretical and experimental works, are briefly reviewed in Section II.

The approach generally taken in this problem is the one of a very dilute system where the ions only interact indirectly via the superconducting electrons. However, there exist many systems [3] where superconductivity remains up to relatively high impurity concentrations, and where the above-mentioned interaction is certainly not the only one possible. In Section III we discuss some of these systems, particularly $Ce_{1-x}Gd_xRu_2$, which was recently reinvestigated in detail by Wilhelm and Hillenbrand [5] and by Peter et al. [6].

One important experimental fact is that when the ions start to align, the exchange field acting on the conduction electrons is the interaction mainly responsible for the suppression of superconductivity in most cases. Therefore one may look at a ferromagnetic superconductor as a superconductor with a molecular field acting on the conduction-electron spins. In Section IV we discuss the coexistence problem from this point of view.

II. Magnetic Ordering in a Superconductor. The Low-Concentration Limit

The experiments of Matthias and co-workers showed two interesting features. One was the apparent coexistence, and the other a strange anomaly in the T_C versus concentration curve for $La_{1-x}Gd_x$ [7]. Since this anomaly occurred just before the point where curves T_c and θ_C (θ_C = ferromagnetic Curie temperature) versus concentration met, it was reasonable to think that it was connected with ordering phenomena. The works that we discuss are essentially concerned with these two features.

The first extension of the theory of Abrikosov and Gorkov [8] (AG) to include magnetic ordering was done by Gorkov and Rusinov [9] (GR). Although this work has been reviewed elsewhere in this treatise [10], we will give a short description of the physical ideas here. As in AG, they (GR) considered localized magnetic ions in a superconducting electron gas. In the absence of superconductivity this system can become ferromagnetic due to the indirect RKKY interaction between the ions via the conduction electrons. The polarizing field will be proportional to $\chi(0)$, the Pauli susceptibility of the conduction electrons. If the electron gas is superconducting, one has in the simplest BCS case $\chi_s(0) = 0$, and coexistence of ferromagnetism and superconductivity turns out to be impossible in this model. However, the scattering of the electrons by the impurities which produce the decrease of T_c will also produce a finite susceptibility since the spin is not conserved in this interaction.

Thus at high impurity concentrations it is possible to get magnetic ordering, even in the presence of superconductivity. Now, as the spins order, there will be a mean exchange field H_s acting on the superconducting electrons. As first pointed out by Clogston [11] and Chandrasekhar [12], this mean exchange field will produce a first-order transition from a superconducting to a normal state if

$$\tfrac{1}{2}(\chi_n - \chi_s) H_s^2 = \tfrac{1}{2}\Delta^2 N(0), \qquad (2.1)$$

where Δ is the superconducting energy gap and $N(0)$ is the density of states at the Fermi energy. For $\chi_s(0) = 0$ this defines the Clogston limit

$$H_{p0} = \sqrt{2}\,\Delta/g\mu_B. \qquad (2.2)$$

Gorkov and Rusinov found that, as one decreased the temperature, in a certain range of concentration the system would first become superconducting, and then by a second-order phase transition go over to a ferromagnetic superconducting state. Finally, due to the effect of the mean exchange field on the Cooper pair the system would go over to a pure ferromagnet. This transition may be first or second order, depending on the strength of the exchange scattering. In Fig. 1 we show the phase diagram that may result from this model. This is very different from the ones found by Matthias et al. [2, 3] for $Ce_{1-x}Gd_xRu_2$ and

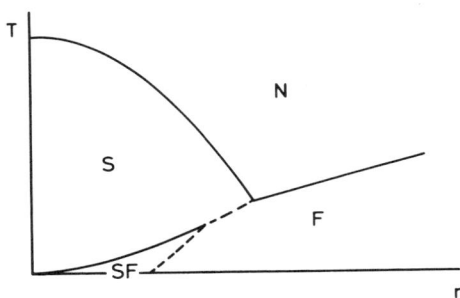

FIG. 1. Schematic phase diagram with account of exchange interaction, but neglecting spin–orbit scattering (after Gorkov and Rusinov [9]).

$Y_{1-x}Gd_xOs_2$, which are shown in Figs. 2 and 3. Gorkov and Rusinov pointed out that the spin–orbit scattering on nonmagnetic impurities could be a very effective way to increase $\chi_s(0)$ and thus also increase the maximum exchange field that a superconductor could support. In fact, Abrikosov and Gorkov [13] find that in the limit of strong spin–orbit

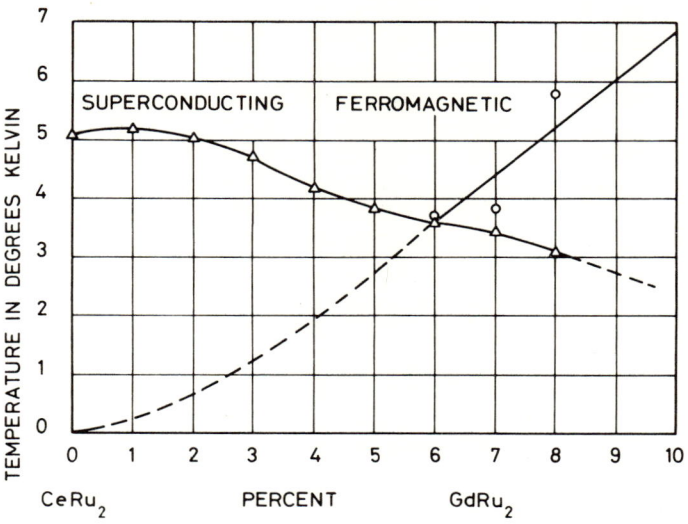

FIG. 2. Phase diagram for $Ce_{1-x}Gd_xRu_2$ obtained by Matthias et al. [2].

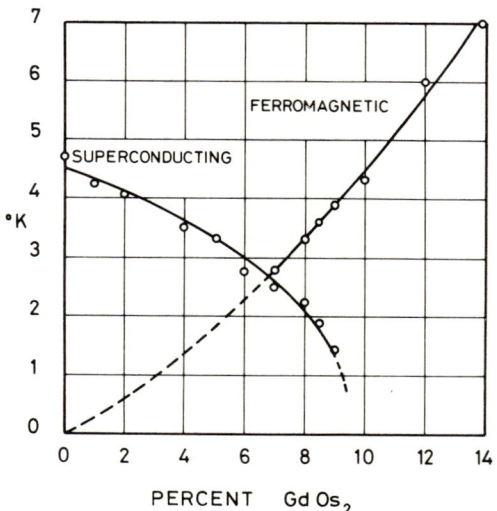

FIG. 3. Phase diagram for $Y_{1-x}Gd_xOs_2$ obtained by Suhl et al. [3].

scattering the susceptibility of the superconducting electrons is given by

$$\chi_s = \chi_n \left(1 - \frac{3\pi\tau_{so}}{4\hbar} \Delta(T) \tanh\left(\frac{\Delta(T)}{2T}\right)\right) \qquad (2.3)$$

The spin–orbit scattering time τ_{so} is defined by

$$\hbar/\tau_{so} = \tfrac{4}{3}\pi n N(0) U_{so}^2, \qquad (2.4)$$

where n is the concentration of impurities and U_{so} the Fourier transform of the spin–orbit potential. Fulde and Maki [14] have discussed in detail the effect of spin–orbit scattering on superconductors with magnetic impurities. In the case that the spin–orbit interaction is strong enough to produce a second-order transition, and in the absence of external fields and currents, their result for T_c can be written

$$\ln\left(\frac{T_c}{T_{c0}}\right) + \frac{1}{2}\left\{1 + \frac{1}{[1-(\hbar^{-1}\tau_{so}I)^2]^{1/2}}\right\}$$

$$\times \Psi\left\{\frac{1}{2} + \frac{\hbar}{2\pi k_B T_c}\left(\frac{1}{\tau_s} + \frac{1}{\tau_{so}}[1-(\hbar^{-1}\tau_{so}I)^2]^{1/2}\right)\right\} + \frac{1}{2}\left\{1 - \frac{1}{[1-(\hbar^{-1}\tau_{so}I)^2]^{1/2}}\right\}$$

$$\times \Psi\left\{\frac{1}{2} + \frac{\hbar}{2\pi k_B T_c}\left(\frac{1}{\tau_s} - \frac{1}{\tau_{so}}[1-(\hbar^{-1}\tau_{so}I)^2]^{1/2}\right)\right\} - \Psi\left(\frac{1}{2}\right) = 0. \qquad (2.5)$$

τ_s is the exchange scattering time, defined by

$$\hbar/\tau_s = \pi c S(S+1) J^2 N(0) \qquad (2.6)$$

and I is the mean exchange field, defined by

$$I = cJ(q=0)\langle S_z\rangle; \qquad (2.7)$$

c is the concentration of magnetic impurities, J is the Fourier transform of the exchange interaction, S is the spin of the ions and ψ is the digamma function.

In the limit $(\hbar^{-1}\tau_{so}I) \ll 1$ this reduces to

$$\ln\left(\frac{T_c}{T_{c0}}\right) + \Psi\left(\frac{1}{2} + \frac{\rho}{2\pi k_B T_c}\right) - \Psi\left(\frac{1}{2}\right) = 0. \qquad (2.8)$$

This is formally the same expression as obtained by AG, but with a new pair breaking parameter.

$$\rho = \left(\frac{\hbar}{\tau_s} + \frac{I^2\tau_{so}}{2\hbar}\right). \qquad (2.9)$$

Thus in the limit $\tau_{so} \to 0$ there will be no change in the T_c versus concentration curve when the spins start to order (i.e., $I \neq 0$). In this limit one will also have $\chi_s \approx \chi_n$ and the indirect RKKY interaction will not be changed when the system goes into the superconducting state. This picture is consistent with the experiments of Suhl et al. [3]. For this to be true one needs $\xi \gg l_{so} \equiv v_F \tau_{so}$ where ξ is coherence length. However, recent work on high-field superconductors [15, 16], especially the work by Crow et al. [17] on the system $La_{3-x}Gd_xIn$, shows that this is generally not the case. If we introduce reasonable values for I and τ_{so}, we find that $\hbar/\tau_s$ and $\frac{1}{2}I^2\tau_{so}\hbar^{-1}$ are of the same order of magnitude in the ferromagnetic case. This can be seen by calculating the ratio between the two. Assuming $\langle S_z \rangle = S$, one finds from Eqs. (2.6) and (2.7)

$$\frac{1}{2\hbar^2} I^2 \tau_{so} \tau_s = \frac{cS}{3\pi^2 N(0) k_B T_{c0}(S+1) \lambda_{so}}, \qquad (2.10)$$

where λ_{so} is the spin–orbit parameter introduced by Werthamer et al. [15]

$$\lambda_{so} = \frac{2\hbar}{3\pi \tau_{so} k_B T_{c0}}. \qquad (2.11)$$

Taking typical values as $c = 10^{-2}$, $N(0) = 1$ state/eV-atom, $S = \frac{7}{2}$, and $T_c = 5°K$ one finds

$$\frac{1}{2\hbar^2} I^2 \tau_{so} \tau_s = \frac{0.6}{\lambda_{so}}. \qquad (2.12)$$

Since λ_{so} is normally found to be of the order unity, the two pair-breaking mechanisms are indeed of the same order of magnitude. We note that this statement is essentially independent of the exchange potential J and the spin S, but that the ratio increases with the concentration of impurities, making the effect of the exchange field more important at high concentrations. Thus from Eqs. (2.8) and (2.9) one expects a sharp drop in T_c when the spins start to order ferromagnetically. This is also what is generally seen for different systems; that is, the T_c versus concentration curve falls sharply to zero at high concentrations, indicating increasing magnetic order. The reason why this is not the case in the two systems reported by Matthias is not fully clarified. For the case of $Ce_{1-x}Gd_xRu_2$ it has recently been shown [5] that samples leading to a diagram similar to the one shown in Fig. 2 are not single-phase systems. This is discussed in more detail in the next section.

In the preceding discussion we have left out that τ_s is sensitive to magnetic ordering. Bennemann [18] pointed out that when the spin starts to align, the spin-flip scattering becomes inelastic, and finally, in the limit of very high exchange fields acting between the ions, the spin-flip scattering processes will be frozen in. He finds that

$$\frac{\hbar}{\tau_s} \sim c \left(S(S+1) - S \left\langle B_s \left(\frac{\omega_z}{k_B T} \right) \tanh \left(\frac{\omega_z}{2 k_B T} \right) \right\rangle_{av} \right), \qquad (2.13)$$

where ω_z is the local exchange field acting on the ions. Between the randomly oriented spins and the fully aligned spins there will be a reduction in the spin scattering by a factor $S/(S+1)$. Thus when the spins start to order there will be two competing mechanisms determining T_c, and under special conditions an increase in T_c may result, explaining the anomaly first seen by Hein et al. [7] in $La_{1-x}Gd_x$. The effect can be seen qualitatively by inserting Eq. (2.13) into Eq. (2.9). If the decrease in $\hbar/\tau_s$ is stronger than the increase in $\frac{1}{2}I^2\tau_{so}\hbar^{-1}$, T_c will increase. The actual value of I will of course depend strongly on the nature of the magnetic ordering. For ferromagnetic alignment I will rapidly become large; for antiferromagnetic alignment I will essentially be zero. Therefore, as noted by Bennemann, the form of the T_c versus concentration curve should reflect the nature of magnetic ordering. This result is in qualitative agreement with the result of Finnemore et al. [19] who found that Gd in La orders antiferromagnetically. This suggests T_c values which are higher than predicted by AG, near the critical concentration, and this is also what is observed. A similar situation was observed by Guertin and Parks [20] in $Th_{1-x}Er_x$. Their results are shown in Fig. 4. According to Andres and Bucher [21] Er orders antiferromagnetically

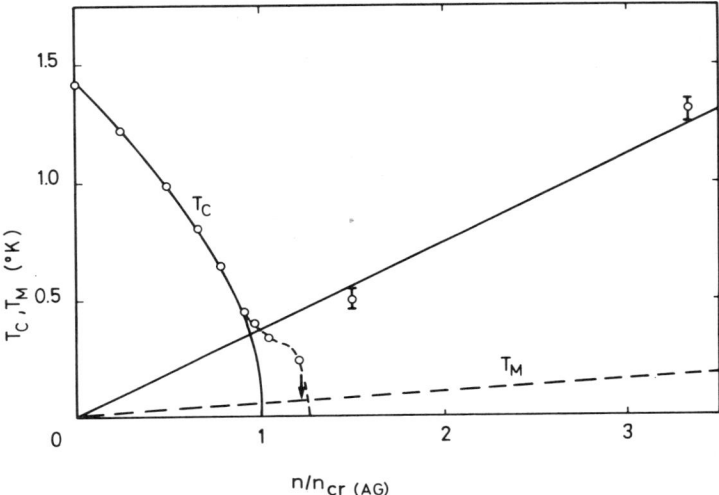

FIG. 4. Superconducting transition temperature T_c and magnetic ordering temperature T_M versus concentration of magnetic impurities for $Th_{1-x}Er_x$. The points with error bars denote the temperatures at which the magnetic susceptibility departs from the Curie law (after Guertin and Parks [20]).

in Th at higher concentration. Thus one expects again that $I \approx 0$ and that T_c would lie above the AG curve, as seen in the experimental results. We note, however, that the ratio of the critical concentration c_2 to the one predicted by AG, c_{AG}, is definitely smaller than $(S+1)/S$ expected from the theory by Bennemann.

In the case of $La_{3-x}Gd_xIn$ studied by Crow and Parks [22] the T_c values fall below the ones expected from AG, showing that a net exchange field I is acting on the conduction electrons, and thus indicating ferro- or ferrimagnetic ordering.

In all cases cited above the deviations from AG are observed above the temperatures where long-range magnetic ordering is expected to occur. The ordering effects producing the deviations are therefore of short range. Parks [23] noted that if the correlation distance between the magnetic ions is much smaller than the coherence distance ξ, there will be no pair breaking effect due to the mean exchange field. In La the coherence distance ξ is relatively large ($\xi \approx 300$ Å). Thus small ferromagnetic clusters of, say, 50 to 100 Å in diameter will produce essentially the same effect as a long-range antiferromagnetic order. In $La_{3-x}Gd_xIn$ the coherence distance is much smaller ($\xi \approx 70$ Å) and one expects the exchange field due to clusters of the same size to be pair breaking in agreement with experiments.

Thus the only conclusion that one can draw from T_c measurements is on the presence or absence of ferro- or ferrimagnetic ordering over distances of the same size or larger than the coherence length of the system under consideration.

The reduction of the spin scattering will also be reflected in other properties of the system. Bennemann and co-workers have calculated the effect on different properties, such as the density of states [24], the electronic specific heat [25], and the thermal conductivity [26]. Bennemann et al. [27] also discussed the critical field of a superconductor with magnetic impurities, with the aim of explaining the curious behavior of $H_{c_2}(T)$ in $La_{3-x}Gd_xIn$ [17]. This system shows different interesting features:

1. A sharp drop of $H_{c_2}(T=0, x)$ at small concentrations. This clearly indicates the effect of the exchange field.

2. An anomalous dependence of H_{c_2} on T for intermediate concentrations. This effect, which was suggested by the Gennes and Sarma [28] and later discussed in more detail by Cyrot [29], supports the idea of a strong exchange field.

3. A sharp knee in $H_{c_2}(T=0, x)$ and a slow decrease of $H_{c_2}(T=0, x)$ at higher concentrations. Bennemann et al. [27] find that

this shows short-range ferrimagnetic order rather than long-range ferromagnetic order. They also exclude antiferromagnetic order since any long-range order should show itself in a reduction $S/(S+1)$ in the spin scattering amplitude.

However, the reduction in the spin scattering amplitude has recently been reinvestigated by Keller and Benda [30]. They discussed in detail the effect of spin ordering on the critical concentration c_2 and compared it with c_{AG}. They found that in some cases $c_2/c_{AG} > 1$, but that one needs anomalously high exchange fields K acting on the ions to get $c_2/c_{AG} = (S+1)/S$. This also remains true in the limit $\tau_{so} \to 0$ (which means that the pair breaking effect of I will disappear). Only when K reaches the Debye frequency ω_0 are all dynamical processes eliminated. This is not the case for most experimental situations. In Fig. 5 we show

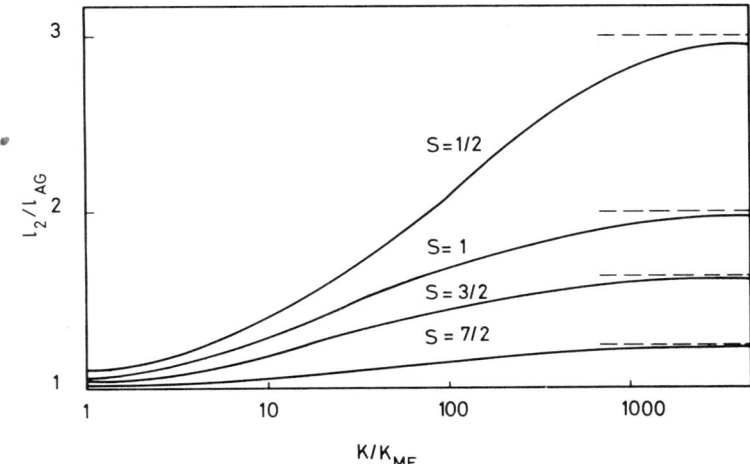

FIG. 5. Critical impurity concentration c_2 relative to the value predicted by AG, c_{OU}, versus polarization field K relative to the value given by the mean field approximation. $K_{MF} = S/\pi\tau_s$. The dashed lines indicate the upper limits given by Bennemann $c_2/c_{AG} = (S+1)/S$ (after Keller and Benda [30]).

their results for c_2/c_{AG} as a function of K in the limit $\tau_{so} \to 0$. They conclude that the anomaly in the T_c versus concentration curve cannot be explained only by a reduction in the spin scattering amplitude. We note in this connection that two of the most prominent examples of magnetic ordering in a superconductor ($Y_{1-x}Gd_xOs_2$ and $Ce_{1-x}Gd_xRu_2$) do not show any sign of this effect.

III. Coexistence in More Concentrated Systems. The Case of $Ce_{1-x}Gd_xRu_2$

A detailed investigation has been initiated by Wilhelm and Hillenbrand [5] on the system $Ce_{1-x}Gd_xRu_2$. They also extended their investigation to include other systems of the type $Ce_{1-x}(RE)_xRu_2$ (RE = rare earth metals) and $Ce(Ru_{1-x}TM_x)_2$ (TM = transition metals). Their results concerning the coexistence problem are twofold. They found that samples leading to a phase diagram as shown in Fig. 2 contained several phases with different Gd content. After an appropriate thermal treatment they obtained single-phase and homogeneous samples and the phase diagram changed markedly. The point $T_c = \theta_C$ shifted to higher concentrations ($\sim$4.5 at.% Gd) and the superconducting transitions to the right of this point disappeared. They also found three new systems showing the same behavior as $Ce_{1-x}Gd_xRu_2$ ($Ce_{1-x}Tb_xRu_2$, $Ce_{1-x}Dy_xRu_2$, $Ce_{1-x}Ho_xRu_2$). The points $T_c = \theta_c$ are located at 6, 7, and 9 at.%, respectively. It is clear that we cannot treat these systems solely as "dilute systems." The phase diagrams obtained are shown in Fig. 6. We noted in the previous section that it seems difficult to explain the phase diagrams in Figs. 2 and 3 solely on the basis of spin–orbit scattering. In the new phase diagram for $Ce_{1-x}Gd_xRu_2$ the superconducting transitions in the ferromagnetic region are suppressed, supporting the idea of a relatively weak spin–orbit coupling. However, in the $Ce_{1-x}Tb_xRu_2$ system Hillenbrand and Wilhelm found one sample to the right of $T_c = \theta_C$ showing first a ferromagnetic transition and then a superconducting one [32]. (See Fig. 6.) A similar behavior has been reported by Willens and Buehler [33] for MoC with 2 at.% Fe.

These measurements leave us with a whole series of systems with a region of possible coexistence to the left of the point $T_c = \theta_C$. This region is perhaps the most interesting one. If we keep the concentration x fixed and vary the temperature, we have three possibilities:

1. Superconductivity remains down to 0°K and the sample does not become ferromagnetic. This means that ferromagnetism is suppressed by superconductivity. In this case we expect to see a change in the coupling between the ions at $T = T_c$, and it would support the idea of indirect RKKY coupling between the ions via the superconducting electrons.

2. The system becomes ferromagnetic, and superconductivity is destroyed. This would be essentially the Gorkov–Rusinov picture.

3. The system becomes ferromagnetic, and superconductivity is not destroyed. In this case the line between the coexistence phase and the

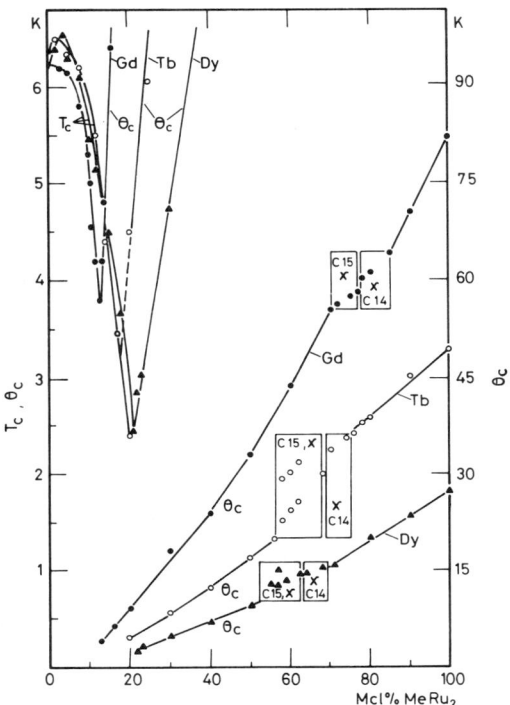

FIG. 6. Phase diagrams obtained by Wilhelm and Hillenbrand [31] for the systems $Ce_{1-x}Gd_xRu_2$, $Ce_{1-x}Tb_xRu_2$, and $Ce_{1-x}Dy_xRu_2$.

ferromagnetic phase can give information on the strength of the spin–orbit coupling.

A study of the region to the left of the point $T_c = \theta_C$ for the system $Ce_{1-x}Gd_xRu_2$ has recently been carried out by Peter et al. [6]. It revealed some anomalous magnetic properties of the system which might also be characteristic for several other systems. The problem is that to detect a ferromagnetic transition by magnetization measurements one has to apply a magnetic field strong enough to suppress superconductivity partly or fully. Thus such measurements do not give conclusive answers to the question of coexistence. In the work of Peter et al. [6] specific heat and thermal expansion measurements were made to detect the magnetic properties of the system without destroying superconductivity. In principle speed of sound measurements may also be effective to detect ferromagnetic transitions [34]. However, the $Ce_{1-x}Gd_xRu_2$ samples were not suitable for such measurements. The specific heat results are shown in Fig. 7. The superconducting transitions are easily seen and they agree well with the inductive and resistive transitions. A magnetic contribution to C/T is also seen but there is absolutely no anomaly at the points where the magnetic measurements show a Curie point. This could lead one to think that superconductivity destroys ferromagnetism. However, there is no change in the magnetic contribution to C/T at $T = T_c$. This unexpected behavior of C/T can be understood by looking at the magnetization curves at different temperatures in the paramagnetic region above T_c. Doing so, Hillenbrand and Wilhelm [35] found that in spite of the relative dilute concentrations the magnetization curves can

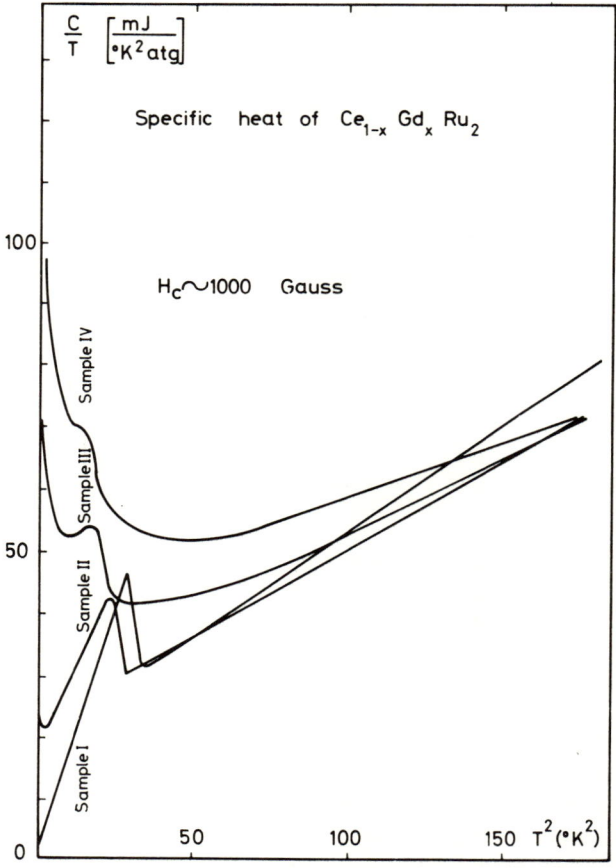

FIG. 7. Specific heat versus temperature for $Ce_{1-x}Gd_xRu_2$. Sample I, $x = 0$; sample II, $x = 0.05$; sample III, $x = 0.09$; sample IV, $x = 0.11$ (after Peter et al. [6]).

be very well reproduced by Brillouin curves if one chooses different molecular field constants λ for different temperatures. The function $\lambda(T)$ determined in this way is constant above a certain temperature T_λ and decreases linearly with temperature below T_λ. The value of T_λ varies between 10 and 20°K, depending on the concentration. This means that we can define a temperature-dependent Curie temperature $\theta(T)$,

$$\theta(T) = \mu_{\text{eff}}^2 N \lambda(T)/3k_B,$$

where N is the number of magnetic ions and μ_{eff} is their effective magnetic moment. The actual transition temperature θ_C is given by

$$\theta(\theta_C) = \theta_C. \qquad (3.1)$$

The experimental points are shown in Fig. 8. Using the experimental data, the solution of Eq. (3.1) agrees within experimental error with the θ_C determined by other methods.

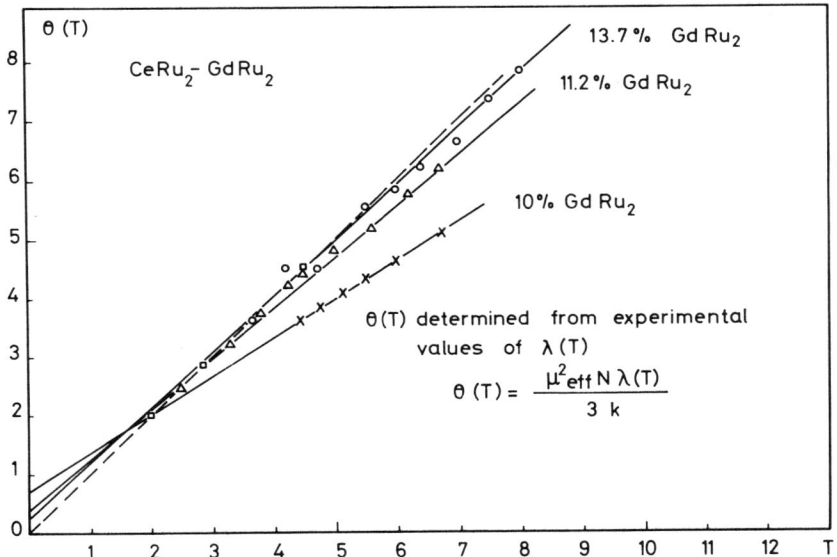

FIG. 8. $\theta(T)$ determined from experimental values of $\lambda(T)$ for $Ce_{1-x}Gd_xRu_2$; ($\times$) $x = 0.10$, ($\triangle$) $x = 0.112$, ($\circ$) $x = 0.137$ (after Peter et al. [6]).

This unusual behavior of $\lambda(T)$ is not yet understood. It is quite different from the behavior of FeCr and CrMn alloys, measured by Wei et al. [36], and the behavior of Cu–Mn alloys measured by

Zimmermann and Hoare [37]. In these systems the impurities, at low temperature, add a constant to C/T. This constant is independent of concentration and the temperature where it appears is proportional to the concentration of impurities. This anomalous behavior has been explained by Marshall [38] and by Klein and Brout [39] on the basis of randomly oriented spins which form clusters of different sizes. In the case of $CeRu_2$ we note also that since the Gd atoms will probably go only on the places of Ce atoms, at 5 at.% Gd we have already substituted 15% of the total number of Gd ions possible. Thus we are considering concentrations much higher than the ones considered by them. Liu [40], generalizing the ideas of Marshall, calculated the specific heat and the susceptibility of more concentrated alloys. He found a behavior for C/T that below about 1% impurity concentration is qualitatively similar to the one measured on $Ce_{1-x}Gd_xRu_2$. For higher concentration it can be seen from his results that C/T as a function of T should pass through a maximum located near the maximum of the initial susceptibility. Furthermore one expects from this theory that the specific heat approaches the molecular field specific heat as the concentration of impurities is increased. Both these results are in disagreement with the experimental results on $Ce_{1-x}Gd_xRu_2$.

Although we do not understand the behavior of $\lambda(T)$, we may take it as an experimental fact and try to determine the magnetization and the specific heat anomaly for such a temperature-dependent interaction. In view of the very good agreement of the Brillouin curves and the magnetization curves for the paramagnetic samples, it seems reasonable to use a molecular field picture, although it will not be valid near the transition temperature [41].

We derive the specific heat from the free energy

$$G(H, T) = -k_B NT \log Z \left(\frac{g\mu_B(H + \lambda m)S}{k_B T} \right) + \frac{\lambda(T)}{2} m^2. \qquad (3.2)$$

This is the form of the free energy for the molecular field model if λ = constant. When λ depends on T the form of the free energy will depend on the model chosen to explain the T dependence of λ. However, Eq. (3.2) is consistent with the experimental magnetization results where it was found that λ did depend on T. The entropy of the system is now given by

$$S = -\left(\frac{\partial G}{\partial T} \right)_H = k_B N \log Z \left(\frac{g\mu_B(H + \lambda m)S}{k_B T} \right) - \frac{(H + \lambda m)m}{T} + \frac{d\lambda}{dT} \frac{m^2}{2}.$$

$$(3.3)$$

In the limit of $T \to \infty$ this gives as expected $S = \log(2S + 1)$, the value for free spins. To satisfy Nernst's theorem, $S = 0$ for $T = 0$, we must have $(d\lambda/dT)_{T=0} = 0$. However, by introducing the experimental values for λ into Eq. (3.3) one finds that S remains large as long as $(T/\theta)(d\theta/dT) \approx 1$. Only when λ approaches a constant (this happens below the experimental range) does S fall to zero. This is also what is found experimentally. By integrating the experiment C/T curves one finds that less than 50% of the entropy difference is accounted for in the experimental range [42].

The magnetization and the specific heat become

$$m = \left(\frac{\partial G}{\partial H}\right)_T = g_B N S B_S \left(\frac{g\mu_B(H + \lambda m)S}{k_B T}\right) \tag{3.4a}$$

$$C = -T \frac{\partial^2 G}{\partial T^2}$$

$$= -H \frac{\partial m}{\partial T} - \lambda m \frac{\partial m}{\partial T} + Tm \frac{\partial m}{\partial T} \frac{\partial \lambda}{\partial T} + \frac{1}{2} m^2 \frac{\partial^2 \lambda}{\partial T^2}. \tag{3.4b}$$

By knowing $\lambda(T)$ one can calculate from Eqs. (3.4a) and (3.4b) $m(T)$ and $C(T)$. Of special interest is the jump in the specific heat at $T = \theta_C$. Expressing $\lambda(T)$ in terms of $\theta(T)$ and assuming $\theta(T) = A + BT$, the specific heat jump becomes

$$C = \left(\frac{A}{\theta}\right)^2 (\Delta C)_0 = \left(\frac{A}{\theta_C}\right)^2 \frac{5S(S+1)}{2S^2 + 2S + 1} k_B N, \tag{3.5}$$

where $(\Delta C)_0$ is the jump if λ were independent of temperature and corresponded to θ_C. The jump is strongly reduced and for the interesting region around $T_c = \theta_C$ we have from Fig. 8

$$(A/\theta_C)^2 \lesssim 10^{-2}.$$

Thus we cannot expect to see any jump in the experimental curves. However, since the entropy decreases to zero at $T = 0$ we now expect C/T to increase as the temperature decreases. The result of a calculation using Eqs. (3.4a) and (3.4b) is shown in Fig. 9, assuming $\theta(T)$ corresponding to a sample with 13.7% $GdRu_2$. The result agrees qualitatively with the experiments. Quantitative agreement cannot be expected due to the short-range ordering that occurs near θ_C. For $T < A$ the ratio C/T decreases, and at very low temperatures it falls exponentially to zero. Experimentally it is found that $A < 1°$; thus this part of the curve was not within the experimental range.

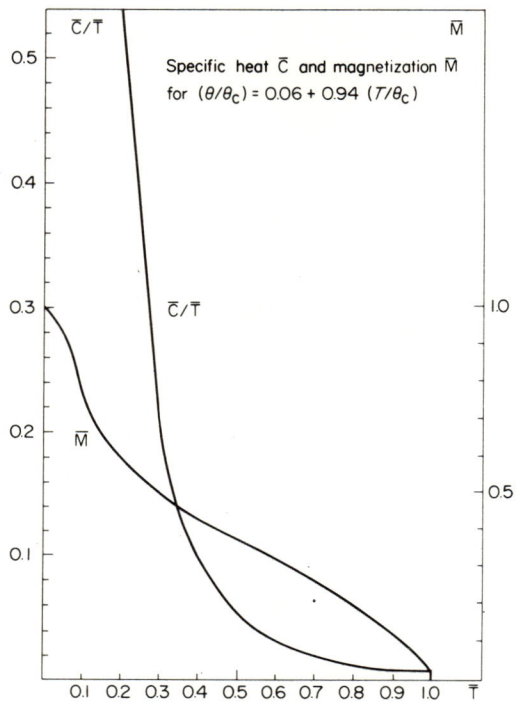

FIG. 9. Specific heat $\bar{C} = C/k_B N$ and magnetization $M = M/g\mu_B NS$ in the molecular field approximation for $\theta(T)/\theta_C = 0.06 + 0.94(T/\theta_C)$ (after Peter *et al.* [6]).

In Fig. 9 the spontaneous magnetization $m(T)$ is shown as a function of temperature; $m(T)$ is strongly reduced compared to the case $\lambda =$ constant and the saturation magnetization is only reached at very low temperatures. The thermal expansion results for a sample in the region $T_c = \theta_C$ are shown in Fig. 10. Again there is no anomaly corresponding to θ_C, but T_c is easily seen. The upper curve shows the thermal expansion in an applied field of 1.6 kG. The result is similar to that of the specific heat and is in agreement with the expected behavior of $m(T)$. The critical field of this sample may be lower than 1.6 kG; thus these magnetostriction measurements are not conclusive in the coexistence question. We conclude that the anomalous behavior of the specific heat and the magnetostriction simply reflects the slow buildup of the spontaneous magnetization. Since the superconducting transition does not influence the anomalous behavior of C/T, we further conclude that the interaction between the ions is not influenced in any significant manner by the transition into the superconducting state. Thus these samples become magnetic at the temperature θ_C determined by extrapolating the magnetic measurements in the region $\theta_C > T_c$ to the region $\theta_C < T_c$. However, the samples remain superconducting down to $T = 1.2°$. This coexistence is probably favored by the anomalously slow increase of the spontaneous magnetization below θ_C. It is not excluded, however, that a sample near the concentration where $\theta_C = T_c$ will become normal at very low temperatures when all spins are fully aligned.

We now go back to the phase diagrams in Figs. 2 and 3, and the one for $Ce_{1-x}Tb_xRu_2$ shown in Fig. 6. By admitting the possibility of a temperature-dependent interaction between the ions, one must admit that such phase diagrams are in fact possible even in the case of a weak spin–orbit scattering, and it may be an accident that in the correct phase diagram for $Ce_{1-x}Gd_xRu_2$ the superconductivity stops at the "critical concentration" ($\theta_C = T_c$).

Concerning the interaction itself, different mechanisms should be considered: statistical distribution of spins, temperature-dependent effective density of states of the conduction electrons providing the coupling, may be phase transitions. Experimental information on the mechanism actually at work may be obtained by extending the measurements to lower temperature and higher fields. At high fields ($H \to \infty$) the entropy takes the form

$$S = \frac{1}{2}\frac{d\lambda}{dT} \cdot m_s^2$$

If the T dependence of λ is due to a distribution of molecular fields alone, we expect $S \to 0$ for $H \to \infty$. Thus at high fields the T dependence of λ should disappear. However, if it is due to some anomalous behavior of the density of states of the conduction electrons, the T dependence may remain also in high fields.

IV. Superconductor in a Molecular Field. Compensation of the Exchange Field by an External Field

Soon after the BCS theory Baltensperger and Strässler [44] generalized the argument of Anderson [45] for dirty superconductors and showed that a well-ordered antiferromagnet may become a superconductor. They also showed that the indirect interaction between the two electrons in a Cooper pair via spin waves is repulsive, but that it is less strong than the attractive phonon interaction. This point has been discussed more in detail by Klose et al. [46] for the ferromagnetic case. They found that the effective interaction is given by

$$V^M \sim \frac{J^2 S}{E_m} \log\left(\frac{E_m}{2\mu H_0}\right)$$

where E_m is the spin wave energy for short wavelengths, $H_0 \mu$ is the long-wavelength limit of the spin wave spectrum, S is the spin, and J is the exchange constant. Thus by choosing H_0 large enough, V^M will be smaller than the phonon-induced interaction. Therefore it seems that for the superconductor the essential difference between a ferromagnet and an antiferromagnet lies in the mean exchange field acting on the conduction electrons in the ferromagnet.

From this point of view one may treat a ferromagnetic superconductor as a superconductor in a molecular field H_s acting only on the conduction electron spins. The first approach to this problem is to look at the behavior of a BCS superconductor as a function of field H_s and temperature T. This has been done by several authors. Sarma [47] found that for $0.58 T_{c_0} < T < T_{c_0}$ the s–n transition is second order. Below $0.58 T_{c_0}$ he found a first-order transition which at $T = 0$ becomes the transition predicted by Chandrasekhar and Clogston [Eq. (2.2)]. If one assumes Δ to be space independent, this is the highest field for which superconductivity will persist. However, Fulde and Ferrel [48] (FF) pointed out that at low temperatures and near $H_s = H_{p0}$ space-dependent solutions of the BCS gap equation exist, having a lower energy than the space-independent solution. The general idea is that in order to break one pair for $T = 0$ in the BCS solution one needs a field higher than H_{p0} by a factor $\sqrt{2}$; that is, the collective state collapses before the field is strong enough to break one pair. Now at $H_s \approx H_{p0}$ if one allows for a certain number, but not all, of the pairs to be broken, the gap will decrease and so will the field necessary to destroy one pair. This allows a second-order phase transition from this, the FF state, to the normal state. Fulde and Ferrell found that this transition takes place at the field $H_s = 1.06 H_{p_0}$. Thus the critical field enhancement produced by this state is relatively small. However, in discussing systems where the exchange field H_s is of the order H_{p0} one has to take into account the possibility of this state occurring, since it has some peculiar properties such as a nearly normal metal specific heat and an anisotropic Meissner effect. For a detailed calculation of these properties we refer the reader to the paper by Takada and Izuyama [49].

The space dependence of Δ, coming from the asymmetric distribution of normal and superconducting regions in the Fermi sphere, is periodic:

$$\Delta(r) = \Delta e^{i\mathbf{q}\mathbf{r}},$$

meaning that the Cooper pair has a finite momentum **q**. The states used to build up Cooper pairs are $\langle k + q/2, \uparrow |$ and $\langle -k + q/2, \uparrow |$. However, these states are not time-reversed states. Therefore the Anderson argument for nonmagnetic impurities is not valid and one must expect nonmagnetic impurities to produce pair breaking [50, 51].

In a recent paper Takada [52] calculated in detail the influence of scattering on the FF state. He found that when $\hbar/\tau \sim \Delta_0$ the FF state disappears (τ is the lifetime of the conduction electrons). This excludes the observation of this state in the case of high-field superconductors, but Takada concluded that it may be observed in connection with ordering of magnetic impurities in clean superconductors. This is a very optimistic conclusion because the magnetic ordering in such cases will always occur near the critical concentration, where we have $\hbar/\tau_s \sim \Delta_0$ (τ_s is the spin scattering time). Now, since the nonmagnetic scatterings produced by the magnetic ions themselves are at least several times stronger than the magnetic ones, it follows from the arguments above that a FF state is unlikely under these conditions.

The effect of the spin–orbit scattering has been discussed in the first section. Since it increases H_p it will reduce the possibility of observing the FF state. Takada finds that when $1/\tau_{so} \sim \Delta_0$ the FF state disappears. This limit is easily reached and the paramagnetic limit of the BCS state will tend to be higher than the critical field for the FF state.

When discussing these effects, one has to bear in mind that as seen from different types of experiments (superconductivity, Kondo effect, resonance experiments, etc.), the exchange field of aligned magnetic impurities often reaches the level of the Clogston limit of the pure superconductor at a concentration of a few atomic percent. Thus when one goes to higher concentrations one must expect to meet exchange fields several times larger than the Clogston limit. It is therefore very important to know *how* effectively the spin–orbit scattering raises the paramagnetic limit. This can be estimated from Eqs. (2.8) and (2.9). Putting $1/\tau_s = 0$ and taking the limit $T_c \to 0$ we get

$$H_p(\tau_{so}) \equiv \frac{I_{\text{crit}}}{g\mu_B} = \left(\frac{\pi \hbar k_B T_{c0}}{\gamma \tau_{so}}\right)^{1/2},$$

where γ is Eulers constant. Again introducing λ_{so} [15] and using the BCS relation between Δ and T_{c0} we get

$$H_p(\lambda_{so}) \approx 1.2 \lambda_{so}^{1/2} H_{p0}. \tag{4.1}$$

As noted in Section II λ_{so} is usually of the order of unity. In some cases it is found to be about 2 [53]. For these cases the paramagnetic limit may be about twice the Clogston limit. Note that the formula (4.1) is only valid in the limit of strong spin–orbit coupling. In the weak limit we have in any case $H_p \approx H_{p0}$ (first-order transition). However, since H_p goes with the square root of λ_{so}, one would need anomalously strong spin–orbit scattering to increase the paramagnetic limit by the order of magnitude necessary for all dense ferromagnetic systems, except those with exceptionally low exchange fields.

Jaccarino and Peter [54] suggested that in cases where the exchange field H_s is opposite to the magnetization of the ferromagnet the action of an external field on the conduction-electron spins will counteract the exchange field. At $H = -H_s$ the mean spin polarization will vanish, thus allowing for superconductivity in a field interval $H_s - H_p < H < H_s + H_p$. This will only be true if one neglects the orbital effect of the external field, that is, $H_{c2}^* \gg H_p$ (where H_{c2}^* is the critical field from purely orbital effects [55]. Experimentally it is found that for most high-field superconductors one has $H_{c2}^* \approx H_p$, and since we suppose $H_s > H_p$ there is a serious limitation to the compensation effect because of the orbital effects of the external field.

Avenhaus et al. [56] have discussed the influence of the magnetic ions on the orbital motion of the conduction electrons. Since the only long-range interaction involved is the dipolar field of the ions, the only contribution similar to the external field will be the one of the mean magnetization $4\pi M$. It is therefore not possible to compensate the effect of the external field on the orbits of the conduction electrons by an effective field, analogous to the exchange field.

Note that the argument of Baltensperger and Strässler [44] is only valid for a pure and well-ordered system. However, this means a low H_{c2}^*. Starting from such an ordered system we may introduce disorder in the system to increase H_{c2}^*. Such a disorder will involve magnetic scatterings and thus reduce T_c and consequently H_{c2}^*. We therefore get a maximum exchange field that can be compensated. Estimates of this field for different cases have been worked out by Fischer and Peter [57], using data representative of known superconducting and magnetic systems. It was found that the compensation should be possible in favorable but realistic cases. Similar estimates can be made for thin films [58].

One difficulty in observing this effect is that one has to start with a ferromagnetic material of which one does not know if it will become conducting when the exchange field is removed. Schwartz and Gruenberg [59] suggested that the effect could also be seen in the case of paramagnetic impurities in a known superconductor, provided that the temperature is low enough so that the spins align in the external field at

$H = H_{c2}^*$. They predicted that the superconductivity should disappear as soon as the spins align and then reappear at the field $H = H_s - H_p$ and finally disappear at $H = H_s + H_p$. Parks reported an attempt to find this effect in $\text{La}_{3-x}\text{Ce}_x\text{In}$ [23], but he did not observe the predicted behavior. However, to observe this effect one needs $H_{c2}^* \gg H_p$, and this is certainly not the case for the La_3In system.

Even in cases where $H_{c2}^* < H_p$ it is possible to observe the effect of exchange-field compensation, provided that the temperature is low enough to align the spins at $H = H_{c2}^*$. To discuss this effect quantitatively one needs to know the equation for the s–n phase transition in the presence of nonmagnetic, magnetic, and spin–orbit scattering, a mean exchange field and an external magnetic field. To simplify the discussion we shall assume that we are in the dirty limit with relatively strong spin–orbit scattering. In this case the pair breaking effects turn out to be additive [23], and from Fulde and Maki [14] it follows that the phase transition is given by Eq. (2.8) where the pair breaking parameter ρ is now given by

$$\rho = \frac{\hbar}{\tau_s} + \frac{\tau_{tr} v_F^2 e}{3c} H_{c_2} + \frac{(g\mu_B)^2 \tau'_{so}}{2\hbar}(H_{c_2} + H_s)^2. \qquad (4.2)$$

The first term comes from the exchange scatterings, the second term comes from the orbital effect of the external magnetic field, and the last term describes the paramagnetic effect of the external field and the exchange field; τ_{ir} is the transport scattering time and τ_{so} is a generalized spin–orbit scattering time defined by

$$\frac{1}{\tau'_{so}} = \frac{1}{\tau_{so}} - \frac{3}{2\tau_s}\left(\frac{S(S+1) - \langle S_z^2 \rangle}{S(S+1)}\right).$$

Equation (2.8) may be viewed as a universal pair breaking equation having as a solution a universal function $\rho(T)$. If we insert this $\rho(T)$ into Eq. (4.2), we get an explicit equation for $H_{c2}(c, T)$ (where c is the concentration of magnetic impurities). In the limit $T \to 0$, Eq. (4.2) may be rewritten [43]

$$H_{c2}(c, 0) = H_{c2}^*(0, 0)\left(1 - \frac{\tau_{s\,crit}}{\tau_s}\right) - a\frac{\tau'_{so}}{\tau_{tr}}(H_{c_2}(c, 0) + H_s(c, 0))^2. \qquad (4.3)$$

$H_{c2}^*(0, 0)$ is the orbital critical field of the superconductor without magnetic impurities. It is obtained from Eq. (4.2) by neglecting the spin scattering term and the paramagnetic term. $\tau_{s\,crit}$ is the spin scattering time corresponding to the critical concentration of magnetic impurities in absence of magnetic fields and a is an unimportant constant. One notes that $\tau_{\text{scrit}}/\tau_{\text{s}}$ can be determined by T_c measurements. We now define a reduced critical field $H_{\text{red}}(c, 0)$

$$H_{\text{red}}(c, 0) \equiv \frac{H_{c2}(c, 0)}{(1 - \tau_{\text{s crit}}/\tau_{\text{s}})}$$

$$= H_{c2}^*(0, 0) - \frac{a(\tau'_{\text{so}}/\tau_{\text{tr}})(H_{c2}(c, 0) + H_s(c, 0))^2}{(1 - \tau_{\text{s crit}}/\tau_{\text{s}})}. \quad (4.4)$$

If H_s has the same sign as H, H_{red} will decrease as we add magnetic impurities. We assume here that τ_{tr} does not change when we add the magnetic impurities. This can be checked by adding nonmagnetic impurities. On the contrary, if H_s has the opposite sign to H, H_{red} will increase with increasing c. A maximum will occur when

$$Hc_2(c, 0) = H_s(c, 0) = \frac{cJ(q=0)\langle S_z \rangle}{g\mu_B}, \quad (4.5)$$

and after that H_{red} will decrease quickly. The compensation of the paramagnetic effect on H_{c2} should therefore show itself in a maximum of H_{red} ocurring at a certain concentration. The position of the maximum allows one to calculate J, the exchange constant from Eq. (4.5). The height of the maximum gives $H_{c2}^*(0, 0)$. Together with the measured critical field $H_{c2}(0, 0)$ of the pure sample this allows us to determine the Maki parameter α, defined as

$$\alpha = \sqrt{2}\, H_{c2}^*(0, 0)/H_{p0}$$

and the spin–orbit parameter λ_{so}, two central parameters of a high-field superconductor. This effect has recently been seen in the system $Mo_{1-x}Mn_xGa_4$ [43, 60]. In Fig. 10 are shown the results for T_c, H_{c2}, and H_{red} as functions of the concentration of Mn. Also shown is a calculation of H_{red} obtained by generalizing the expression given by Werthamer et al. [15] to include magnetic impurities [43]. The parameters calculated from the position of the maximum are $J = -0.30$ eV, $\alpha = 0.83$, $\lambda_{\text{so}} = 0.5$. A significant increase in the critical field was not achieved. The maximum occurred at low concentration (0.1 at. % Mn) and it must be left for future studies to find the effect in more concentrated alloys. However, it is obvious that in systems with larger α and where $N(0) \cdot J$ is small, one may actually increase the critical field by a large amount by adding an appropriate amount of magnetic impurities. In the case of a thin film (or the surface layer of a bulk sample, where the orbital critical field is given by H_{c3}), the orbital effects are less important than in

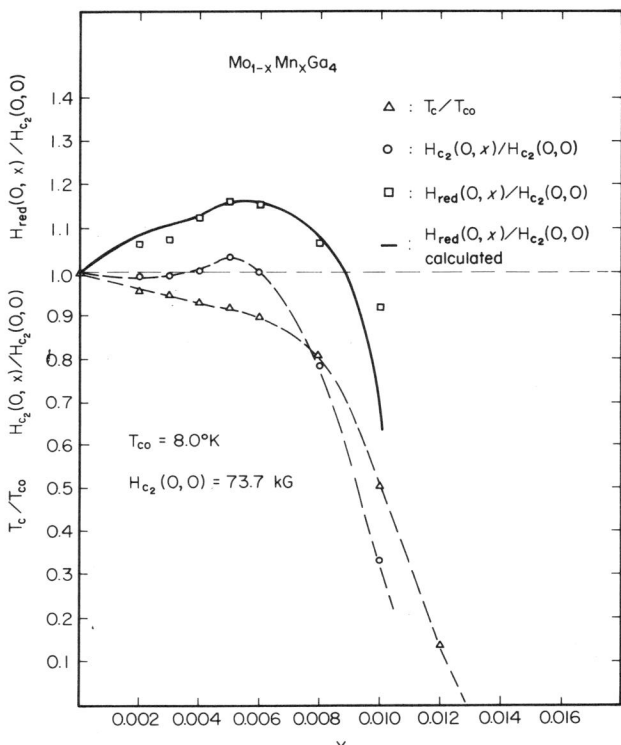

FIG. 10. T_c/T_{c0}, $H_{c2}(T=0, x)/H_{c2}(T=0, x=0)$, and $H_{red}(T=0, x)/H_{c2}(T=0, x=0)$ versus concentration x for $Mo_{1-x}Mn_xGa_4$ (after Fischer [43]).

a bulk sample. We expect the paramagnetic effects to be more easily seen in such cases [43].[†]

In the discussion above we have assumed that the conduction-electron states $\phi_n(r)$ are plane waves. If this is not the case $J(q=0)$ in H_c is to be replaced by $\bar{J}$:

$$\bar{J} = \left\langle \int \phi_n^*(r) J(r) \phi_n(\mathbf{r}) \, d^3r \right\rangle_n ,$$

where the mean value over the states n is to be taken at the Fermi

---

[†] Similar effects have now also been seen in $Ce_{1-x}Gd_xRu_2$ [B. Hillenbrand, K. Schuster, and M. Wilhelm, *Phys. Lett.* **36A**, 383 (1971)] $Mo_3Os_{1-x}Fe_x$ [G. Bongi, Ø. Fischer, H. Jones, R. Flükiger, and A. Treyvaud, *Helv. Phys. Acta* **45**, 13 (1972)] and $Mo_3Ir_{1-x}Fe_x$ [Ø. Fischer, G. Bongi, H. Jones, R. Flükiger, and A. Treyvaud, to be published].

surface. This value $\bar{J}$ might actually be considerably different from J ($q = 0$) since $J(r)$ and $\phi_n(r)$ will tend to have maxima near the impurities.

One may ask how one can expect to realize a dense ferromagnetic superconductor, when, until now, no antiferromagnetic superconductor has been seen. To answer this point, one has to note the crucial importance of the crystallographic order in the argument of Baltensperger and Strässler. In fact a nonmagnetic impurity in a magnetic lattice will produce magnetic as well as nonmagnetic scatterings. Thus from experiments on magnetic impurities in nonmagnetic superconductors one must conclude that about 1% of impurities or disorder in a ferromagnetic or antiferromagnetic superconductor will be sufficient to destroy the possibility of superconductivity. It is interesting to note in this connection that Petalas and Baltensperger [61] found that an Overhauser spin density wave can coexist with superconductivity. To show this they formed their Cooper pairs from states $\langle k, \uparrow |, \langle -k + Q, \downarrow |$. This pairing is analogous to the pairing used in the FF state, and from our previous discussion one might be led to the conclusion that a low concentration of nonmagnetic impurities will also destroy this state. This is, however, not obvious since the Q's involved here are about 1000 times larger than the q's in the FF state.

V. Final Remarks

In view of the results reported above it can be regarded as established that magnetic order and superconductivity may coexist. However, the magnetic properties of the systems which show this coexistence turn out to be anomalous, and conclusions drawn from theories based on the simple RKKY interaction may be wrong. Actually the behavior of these systems can only be determined by looking at many different properties experimentally. Such detailed investigations would certainly be worthwhile in all cases where there is a tendency to coexistence. The most interesting systems to be investigated might be the ones with small negative electron–ion exchange interactions.

Acknowledgments

We thank Dr. M. Wilhelm and Dr. B. Hillenbrand for communicating their results to us prior to publication, and Dr. H. Jones for carefully reading the manuscript and suggesting many changes, as well as Professor R. J. Schrieffer and Professor A. B. Harris for discussions.

References

1. B. T. Matthias, H. Suhl, and E. Corenzwit, *Phys. Rev. Lett.* **1**, 92 (1958).
2. B. T. Matthias, H. Suhl, and E. Corenzwit, *Phys. Rev. Lett.* **1** 449 (1958).
3. H. Suhl, B. T. Matthias, and E. Corenzwit, *J. Phys. Chem. Solids* **19**, 346 (1959).
4. P. W. Anderson and H. Suhl, *Phys. Rev.* **116**, 898 (1959).
5. M. Wilhelm and B. Hillenbrand, *J. Phys. Chem. Solids* **31**, 559 (1970).
6. M. Peter, P. Donzé, Ø. Fischer, A. Junod, J. Ortelli, A. Treyvaud, E. Walker, M. Wilhelm, and B. Hillenbrand, *Helv. Phys. Acta* **44**, 345 (1971).
7. R. A. Hein, R. L. Falge, B. T. Matthias, and E. Corenzwit, *Phys. Rev. Lett.* **2**, 500 (1959).
8. A. A. Abrikosov and L. P. Gorkov, *Sov. Phys. JETP* **12**, 1243 (1961).
9. L. P. Gorkov and A. I. Rusinov, *Sov. Phys. JETP* **19**, 922 (1964).
10. M. A. Jensen and H. Suhl, in "Magnetism" (G. T. Rado and H. Suhl, eds.), Vol. IIB, p. 183. Academic Press, New York, 1966.
11. A. M. Clogston, *Phys. Rev. Lett.* **9**, 266 (1962).
12. B. Chandrasekhar, *Appl. Phys. Lett.* **1**, 7 (1962).
13. A. A. Abrikosov and L. P. Gorkov, *Sov. Phys. JETP* **15**, 752 (1962).
14. P. Fulde and K. Maki, *Phys. Rev.* **141**, 275 (1966).
15. N. R. Werthamer, E. Helfand, and P. C. Hohenberg, *Phys. Rev.* **147**, 295 (1966).
16. R. R. Hake, *Appl. Phys. Lett.* **10**, 186 (1967).
17. J. E. Crow, R. P. Guertin, and R. D. Parks, *Phys. Rev. Lett.* **19**, 77 (1967).
18. K. H. Bennemann, *Phys. Rev. Lett.* **17**, 438 (1966).
19. D. K. Finnemore, D. C. Hopkins, and P. E. Palmer, *Phys. Rev. Lett.* **15**, 891 (1965).
20. R. P. Guertin and R. D. Parks, *Solid State Commun.* **7**, 59 (1969).
21. K. Andres and E. Bucher, *Helv. Phys. Acta* **42**, 590 (1969).
22. J. E. Crow and R. D. Parks, *Phys. Lett.* **21**, 378 (1966).
23. R. D. Parks, in "Superconductivity" (P. R. Wallace, ed.), Vol. 2, p. 625. Gordon & Breach, New York, 1969.
24. K. H. Bennemann and S. Nakajima, *Phys. Rev. Lett.* **16**, 243 (1966).
25. K. H. Bennemann and J. W. Garland, *Phys. Rev.* **159**, 369 (1967).
26. K. H. Bennemann and F. M. Mueller, *Phys. Rev.* **176**, 546 (1968).
27. K. H. Bennemann, J. W. Garland, and F. M. Mueller, *Phys. Rev. Lett.* **23**, 169 (1969).
28. P. G. de Gennes and G. Sarma, *Solid State Commun.* **4**, 449 (1966).
29. M. Cyrot, *J. Phys. Chem. Solid* **29**, 663 (1968).
30. J. Keller and R. Benda, *J. Low Temp. Phys.* **2**, 141 (1970).
31. M. Wilhelm and B. Hillenbrand, *Z. Naturforsch.* **26A**, 141 (1971).
32. B. Hillenbrand and M. Wilhelm, *Phys. Lett. A* **31**, 448 (1970).
33. R. H. Willens and E. Buehler, *J. Appl. Phys.* **38**, 405 (1967).
34. J. Ortelli, C. Susz, E. Walker, and M. Peter, *Helv. Phys. Acta* **42**, 284 (1969).
35. B. Hillenbrand and M. Wilhelm, "Sommerschule für Supraleitung," Pegnitz, Germany, 1970.
36. C. T. Wei, C. H. Cheng, and P. A. Beck, *Phys. Rev. Lett.* **2**, 95 (1959).
37. J. E. Zimmerman and F. E. Hoare, *J. Phys. Chem. Solids* **17**, 52 (1960).
38. W. Marshall, *Phys. Rev.* **118**, 1519 (1960).
39. M. W. Klein and R. Brout, *Phys. Rev.* **132**, 2412 (1963).
40. S. H. Liu, *Phys. Rev.* **157**, 411 (1967).
41. B. M. Boerstoel, Thesis, Univ. of Leiden, Leiden, Netherlands, 1970.
42. B. Hillenbrand, private communication (1970).

43. Ø. Fischer, Thesis, Univ. of Geneva, Geneva, Switzerland, 1971, *Helv. Phys. Acta* **45**, 331 (1972).
44. W. Baltensperger and S. Strassler, *Phys. Kondens. Mat.* **1**, 20 (1963).
45. P. W. Anderson, *J. Phys, Chem. Solids* **11**, 26 (1959).
46. W. Klose and M. Peter, to be published.
47. G. Sarma, *J. Phys. Chem. Solids* **24**, 1029 (1963).
48. P. Fulde and R. A. Ferrell, *Phys. Rev. A* **135**, 550 (1964).
49. S. Takada and T. Izuyama, *Progr. Theor. Phys.* **41**, 635 (1969).
50. L. W. Gruenberg and L. Gunther, *Phys. Rev. Lett.* **16**, 996 (1966).
51. L. G. Aslamazov, *Sov. Phys. JETP* **8**, 773 (1969).
52. S. Takada, *Progr. Theor. Phys.* **43**, 27 (1970).
53. K. Hechler and E. Saur, *Proc. Int. Conf. Low Temp. Phys.*, 11th, *St. Andrews, Scotland*, 1968 (J. F. Allen, D. M. Findlayson, and D. M. McCall, eds.), St. Adrews, Scotland 1968.
54. V. Jaccarino and M. Peter, *Phys. Rev. Lett.* **9**, 280 (1962).
55. E. Helfand and N. R. Werthamer, *Phys. Rev.* **147**, 288 (1966).
56. R. Avenhaus, Ø. Fischer, B. Giovannini, and M. Peter, *Helv. Phys. Acta* **42**, 649 (1969).
57. Ø. Fischer and M. Peter, *Conf. Sci. Superconductivity*, Standford, California, 1969, *Physica* **55**, 597 (1971).
58. Ø. Fischer and M. Peter, *Tagung. Schweiz. Phys. Gesellschaft*, St. Gallen 1969, *Helv. Phys. Acta* **42** (1969).
59. B. B. Schwartz and L. W. Gruenberg, *Phys. Rev.* **177**, 747 (1969).
60. Ø. Fischer, H. Jones, G. Bongi, C. Frei, and A. Treyvaud, *Phys. Rev. Lett.* **26**, 305 (1971).
61. P. Petalas and W. Baltensperger, *Helv. Phys. Acta* **41**, 388 (1968).

12. Recent Theoretical Work on Magnetic Impurities in Superconductors

E. Müller-Hartmann[†]

Department of Physics
University of California, San Diego
La Jolla, California

I. Introduction 353
II. Model and Green's Functions 355
III. Nagaoka Approximation 358
IV. Solution of Hamann's Equation 361
V. Results at Low Impurity Concentration 369
VI. Treatment of Finite Impurity Concentrations 375
VII. Conclusion 380
References 381

I. Introduction

The problem of magnetic impurities in metals has received a great amount of attention from solid-state physicists since Kondo's discovery of a breakdown of perturbative treatments. Several chapters of this volume are devoted to reviewing the resulting experimental and theoretical work on normal-conducting alloys. As it was known that magnetic impurities have a profound effect on superconductors, there arose the question of what influence the Kondo divergence might have on the properties of superconducting alloys. The experimental side of this question is the subject of Chapter 10, by Maple. The theoretical side is discussed here.

Throughout this chapter, we shall regard magnetic impurities in metals as those having a spin S of infinite lifetime which exchange-

[†] Present address: Institut für Festkorperforschung, Jülich, Germany.

interacts with the conduction electrons. Potential scattering from the same impurity is also included in the theory. It should be emphasized, however, that developing a theory in superconductors for finite-lifetime spins, as described by the Anderson [1], Wolff [2], or spin fluctuation [3] models, which would cover the whole range from nonmagnetic to magnetic impurities in a continuous fashion, is highly desirable and strongly recommended by experiments. Only the nonmagnetic limit of such a theory has been explored in detail so far; see Kaiser [4], where earlier references can also be found.

Theoretical work on the Kondo effect in superconductors has actually been reviewed by Griffin [5] at an earlier stage. Since 1968, when Griffin wrote his article, the theoretical picture has been clarified substantially; hence, it seems worthwhile to describe the current state of this subject. Progress that has been made since 1968 is predominantly of a technical, that is, mathematical, nature. That is why this chapter has a strong mathematical emphasis.

We do not attempt to include an explicit review of all existing approaches to the problem. Instead, aiming at a self-contained article within the limited space available, it was decided to present the thermodynamic Green's function approach at some length. The dispersion relation approach discussed in Griffin's article can be shown to lead to identical results [6]. The Yosida theory has the practical disadvantage of yielding only zero-temperature results; its present standing is nicely summarized in a paper by Nagaoka and Matsuura [7].

The thermodynamic Green's function method, based on the Nagaoka decoupling as the crucial approximation, was applied to superconductors by Zuckermann [8], Takano and Matayoshi [9], and Zittartz and Müller-Hartmann [10]. The correct solution of the Hamann equation in superconductors was given in the last reference; we shall therefore follow this reference closely in Section IV. Before this, however, we discuss the model and the relevant Green's functions in Section II, including potential scattering from the magnetic impurities. It is clear that the latter generalization is a trivial one, since the canonical transformation which removes the potential scattering does not change the BCS interaction [11] and only modifies the coupling strength of the spin exchange interaction [12]. Nonetheless, it is instructive to see how potential scattering can be easily incorporated here. In Section III, the Nagaoka approximation [13, 14] is rederived in a slightly different fashion and contact is made with the "classical" theories of Abrikosov and Gor'kov [15] (hereafter denoted as AG) and Shiba [16]. Section V presents theoretical results that are obtained for very low impurity concentration; it is mainly based on work by Müller-Hartmann and Zittartz [17] and

Zittartz [18] and includes some previously unpublished material. The treatment of larger impurity concentrations is discussed in Section VI, and Section VII indicates some possible further work on the subject of this chapter.

II. Model and Green's Functions

Throughout this chapter, our superconducting metal with magnetic impurities will be described by the Hamiltonian

$$H = H_0 + H_{\text{imp}}. \quad (2.1)$$

Here the Hamiltonian of the pure superconductor

$$H_0 = \sum_{k\mu} \varepsilon_k c_{k\mu}^+ c_{k\mu} - \Delta \sum_k [c_{k\uparrow}^+ c_{-k\downarrow}^+ + c_{-k\downarrow} c_{k\uparrow}] \quad (2.2)$$

is chosen in the spirit of Gor'kov [19], and the order parameter Δ is determined by the self-consistency condition

$$\Delta = \frac{\lambda}{N} \sum_k^{(\omega_D)} \langle c_{-k\downarrow} c_{k\uparrow} \rangle. \quad (2.3)$$

The operators $c_{k\mu}^+$ and $c_{k\mu}$ create and destroy, respectively, a conduction electron of momentum k and spin direction $\mu = \pm 1$; ε_k is the energy of the corresponding conduction electron measured from the Fermi level. The symbol $\lambda(>0)$ represents the superconducting coupling energy, N is the number of atomic cells, and the momentum sum in Eq. (2.3) is restricted to $|\varepsilon_k| \lesssim \omega_D$, where ω_D is the Debye energy. The thermal average in Eq. (2.3) has to be performed with the total Hamiltonian H; therefore, the order parameter will change in the presence of impurities. The model ansatz excludes the possibility of a spatial variation of the order parameter in the vicinity of an impurity. This spatial variation was investigated to some extent by Tsuzuki and Tsuneto [20], Heinrichs [21], Kitamura [22], Soda [23], and Kuroda [24]. It seems to be very weak. We contend that it does not have any major influence on the observable properties of the system. This would be in line with the fact that in a normal metal, the Kondo effect does not create any large local disturbance in the conduction band [25, 26].

The interaction of conduction electrons with impurities is given by

$$H_{\text{imp}} = \sum_j \frac{1}{N} \sum_{kk'} \exp[i(k-k') r_j] \sum_{\mu\mu'} c_{k\mu}^+ c_{k'\mu'} (V\delta_{\mu\mu'} - J\mathbf{S}_j \cdot \mathbf{s}_{\mu\mu'}). \quad (2.4)$$

The jth impurity is assumed to have a random position r_j and the spin operator $\mathbf{S}_j$ operates on the $(2S + 1)$-dimensional space of states of the jth impurity; $s_{\mu\mu'}$ are the components of the electron spin matrices which are one-half times the Pauli matrices. The interaction (2.4) includes potential scattering and spin exchange scattering from each impurity and V and $-J$ are the corresponding interaction energies. A high-energy cutoff of the density of k states will be used to make the interactions nonsingular.

Our theoretical efforts aim at the one-electron thermodynamic Green's functions of the model just described which are conveniently arranged in a Nambu matrix

$$\hat{G}_{kk'} = \tfrac{1}{2}\sum_\mu \begin{pmatrix} \langle c_{k\mu} \mid c^+_{k'\mu} \rangle & \mu \langle c_{k\mu} \mid c_{-k'\,-\mu} \rangle \\ \mu \langle c^+_{-k\,-\mu} \mid c^+_{k'\mu} \rangle & \langle c^+_{-k\,-\mu} c_{-k'\,-\mu} \rangle \end{pmatrix}. \qquad (2.5)$$

By summing over spin directions $\mu = \pm 1$, we do not lose any information, since the system is rotationally invariant. The corresponding Green's function of the pure superconductor is diagonal in momentum and the diagonal elements are given by

$$\hat{G}_k^0(z) = \begin{pmatrix} z - \varepsilon_k & \varDelta \\ \varDelta & z + \varepsilon_k \end{pmatrix}^{-1}, \qquad (2.6)$$

where z denotes the complex energy variable.

As the Kondo effect arises from multiple scattering of conduction electrons from one impurity, we will in the following concentrate our considerations on this by reducing the Hamiltonian to interaction with just one impurity. The implications of many impurities are considered in Section VI.

The remainder of this section is devoted to establishing a relationship between the Green's function $\hat{G}$ and the non-spin-flip scattering amplitudes for the scattering of superconducting conduction electrons from one magnetic impurity. We will do so by making use of the equations of motion

$$zG_{A,B}(z) = \langle\{A, B\}\rangle + G_{[A,H],B}(z) \qquad (2.7a)$$
$$= \langle\{A, B\}\rangle + G_{A,[H,B]}(z) \qquad (2.7b)$$

for a Green's function $G_{A,B}$, where the subscripts denote the operators from which the Green's function is formed. Applying first (2.7b) to $\hat{G}_{kk'}$ we obtain

$$\hat{G}_{kk'}\hat{G}_k^{0-1} = \delta_{kk'} + (V/N)\sum_l \hat{G}_{kl}\hat{\tau}_3 - (J/2N)\sum_l \hat{X}_{kl}, \qquad (2.8)$$

12. MAGNETIC IMPURITIES IN SUPERCONDUCTORS

where

$$\hat{\tau}_3 = \begin{pmatrix} 1 & 0 \\ 0 & -1 \end{pmatrix}$$

is a Nambu space matrix and where the new Green's function

$$\hat{X}_{kk'} = \sum_{\mu\mu'} \begin{pmatrix} \langle c_{k\mu} | \mathbf{S} \cdot \mathbf{s}_{\mu'\mu} c^+_{k'\mu'} \rangle & -\mu \langle c_{k\mu} | \mathbf{S} \cdot \mathbf{s}_{-\mu\mu'} c_{-k'\mu'} \rangle \\ \mu \langle c^+_{-k-\mu} | \mathbf{S} \cdot \mathbf{s}_{\mu'\mu} c^+_{k'\mu'} \rangle & -\langle c^+_{-k-\mu} | \mathbf{S} \cdot \mathbf{s}_{-\mu\mu'} c_{-k'\mu'} \rangle \end{pmatrix} \quad (2.9)$$

appears. [$\hat{X}$ is related to the usually employed Green's function $\hat{\Gamma}$ by $\hat{X}_{kk'}(z) = (\hat{\Gamma}_{kk'}(z^*))^+$.] Then, (2.7a) applied to $\hat{X}_{kk'}$ results in

$$\hat{G}_k^{0-1} X_{kk'} = (V/N) \sum_l \hat{\tau}_3 \hat{X}_{lk'} - (J/N) \sum_l \hat{T}_{lk'}, \quad (2.10)$$

containing the new Green's function

$$\hat{T}_{kk'} = \sum_{\mu\mu'\nu} \begin{pmatrix} \langle \mathbf{S} \cdot \mathbf{s}_{\mu\nu} c_{k\nu} | \mathbf{S} \cdot \mathbf{s}_{\mu'\mu} c^+_{k'\mu'} \rangle & -\mu \langle \mathbf{S} \cdot \mathbf{s}_{\mu\nu} c_{k\nu} | \mathbf{S} \cdot \mathbf{s}_{-\mu\mu'} c_{-k'\mu'} \rangle \\ -\mu \langle \mathbf{S} \cdot \mathbf{s}_{\nu-\mu} c^+_{-kv} | \mathbf{S} \cdot \mathbf{s}_{\mu'\mu} c^+_{k'\mu'} \rangle & \langle \mathbf{S} \cdot \mathbf{s}_{\nu-\mu} c^+_{-kv} | \mathbf{S} \cdot \mathbf{s}_{-\mu\mu'} c_{-k'\mu'} \rangle \end{pmatrix}. \quad (2.11)$$

Before solving Eqs. (2.8) and (2.10) for $\hat{G}_{kk'}$, we introduce the notations

$$\hat{F}_x(z) = (x/N) \sum_l \hat{G}_l^0(z), \quad (2.12)$$

where x may be V or J, and

$$\hat{F}_J{}^V(z) = \hat{F}_J(z)(1 - \hat{\tau}_3 \hat{F}_V(z))^{-1}. \quad (2.13)$$

Equation (2.10) then immediately yields

$$\sum_l \hat{X}_{lk'} = -\hat{F}_J{}^V \sum_l \hat{T}_{lk'} \quad (2.14)$$

and

$$\sum_{l'} \hat{X}_{kl'} = -(J/N) \hat{G}_k^0 (1 - \hat{\tau}_3 \hat{F}_V)^{-1} \sum_{ll'} \hat{T}_{ll'}, \quad (2.15)$$

and Eq. (2.8) similarly gives

$$\sum_{l'} \hat{G}_{kl'} = \hat{G}_k^0 (1 - \hat{\tau}_3 \hat{F}_V)^{-1} (1 + (J/2N) \sum_{ll'} \hat{T}_{ll'} \hat{F}_J{}^V). \quad (2.16)$$

Inserting (2.15) and (2.16) into Eq. (2.8), one finally obtains

$$\hat{G}_{kk'} = \hat{G}_k^0 \delta_{kk'} + \hat{G}_k^0 \hat{T} \hat{G}_{k'}^0, \quad (2.17)$$

where the T matrix

$$\hat{T} = \hat{T}_V + (1 - \hat{\tau}_3 \hat{F}_V)^{-1} (J/N) \hat{t} (1 - \hat{F}_V \hat{\tau}_3)^{-1} \tag{2.18}$$

describes the coherent potential and spin exchange scattering from one impurity. The two scattering mechanisms add up in the typical manner which was pointed out for the normal metal by Schotte [12].

$$\hat{T}_V = (V/N) \hat{\tau}_3 (1 - \hat{F}_V \hat{\tau}_3)^{-1} \tag{2.19}$$

corresponds to potential scattering only, and

$$\hat{t} = (J/2N) \sum_{u'} \hat{T}_{u'} \tag{2.20}$$

contains the spin exchange scattering amplitudes which, as will become clear in the next section, have to be calculated with electron states renormalized by the potential scattering.

III. Nagaoka Approximation

The Nagaoka decoupling [13, 14] is presented here in a form which differs from the traditional one. The present method, which was discussed for the normal metal case by Kawamura [27], has several advantages. First, the decoupling to be used will be manifestly correct to third order in the spin exchange energy J, where the first logarithmic divergence appears; the traditional procedure decouples in second order and only happens to be correct up to third order. Secondly, Hamann's integral equation [28] is obtained from the decoupled equation of motion right away without much algebra—we may say that the algebra has been performed in the last section already—and, third, the effect of potential scattering is very easily included. The present decoupling scheme is also accessible much more easily to a direct comparison with diagrammatic approaches to the problem.

In applying (2.7a) to the Green's function (2.11), one obtains a higher Green's function containing an additional electron–hole pair which is decoupled unambiguously in the usual fashion.[†]

[†] The components of the higher Green's function are of the type $\langle Sccc \mid Sc \rangle$, where c denotes an arbitrary electron creator or annihilator. Decoupling this in all possible ways and observing rotational invariance leads to terms of the form $\langle Scc \rangle \langle c \mid Sc \rangle$ and $\langle cc \rangle \langle Sc \mid Sc \rangle$. These terms correspond to $\hat{X}(t = -0) \cdot \hat{X}$ and $\hat{G}(t = -0) \cdot \hat{T}$, respectively, in Eq. (3.1).

12. MAGNETIC IMPURITIES IN SUPERCONDUCTORS

The resulting equation of motion is

$$\hat{G}_k^{0-1}\hat{T}_{kk'} = \frac{S(S+1)}{2}\delta_{kk'} - \hat{X}_{kk'}(t=-0) + \frac{V}{N}\sum_l \hat{\tau}_3 \hat{T}_{lk'}$$

$$+ \frac{J}{N}\sum_l \left\{\left[\frac{1}{2}\sum_{l'} \hat{X}_{kl'}(t=-0) - \frac{S(S+1)}{4}\right]\hat{X}_{lk'}\right.$$

$$\left. + \left[\frac{1}{2} - \sum_{l'} \hat{G}_{kl'}(t=-0)\right]\hat{T}_{lk'}\right\}. \quad (3.1)$$

From this, one obtains an equation for the quantity of interest (2.20) by summing over momenta and inserting Eqs. (2.14)–(2.16). After applying the identities

$$(z - \omega)\hat{G}_k^0(z)\hat{G}_k^0(\omega) = \hat{G}_k^0(\omega) - \hat{G}_k^0(z) \quad (3.2)$$

and

$$(1 - \hat{F}_V(z)\hat{\tau}_3)^{-1}(\hat{F}_J(\omega) - \hat{F}_J(z))(1 - \hat{\tau}_3\hat{F}_V(\omega))^{-1} = \hat{F}_J^V(\omega) - \hat{F}_J^V(z), \quad (3.3)$$

the familiar form of Hamann's integral equation [28]

$$\left[1 - \frac{S(S+1)}{4}\hat{F}_J^{V2}(z) + \mathcal{F}_\omega\left\{\frac{\hat{F}_J^V(\omega) - \hat{F}_J^V(z)}{z - \omega}(1 + \hat{t}(\omega)(\hat{F}_J^V(\omega) - \hat{F}_J^V(z)))\right\}\right]\hat{t}(z)$$

$$= \frac{S(S+1)}{4}\hat{F}_J^V(z) + \mathcal{F}_\omega\left\{\frac{\hat{F}_J^V(\omega) - \hat{F}_J^V(z)}{z - \omega}\hat{t}(\omega)\right\} \quad (3.4)$$

emerges, where the operation $\mathcal{F}$ denotes the principal value of a Fermi sum:

$$\mathcal{F}_\omega\{A(\omega)\} = (1/\beta)P\sum_n A(i\omega_n), \quad \omega_n = (\pi/\beta)(2n+1). \quad (3.5)$$

As one observes, the effect of potential scattering is simply to replace $\hat{F}_J$ by $\hat{F}_J^V$. Here $\hat{F}_J$ can be explicitly written as

$$\hat{F}_J(z) = \gamma \int_{-\infty}^\infty d\varepsilon \frac{\rho(\varepsilon)}{z^2 - \Delta^2 - \varepsilon^2}\begin{pmatrix} z+\varepsilon & -\Delta \\ -\Delta & z-\varepsilon \end{pmatrix}, \quad (3.6)$$

where

$$\gamma = JN_0 \quad (3.7)$$

is the dimensionless coupling constant of the spin exchange scattering, N_0 is the density of conduction-electron states per atomic cell at the Fermi energy $\varepsilon = 0$, and $\rho(\varepsilon)$ is the density of states at energy ε, normalized to unity at $\varepsilon = 0$. A short calculation shows that $\hat{F}_V^{\gamma}$ is obtained from Eq. (3.6) by replacing $\rho(\varepsilon)$ by

$$\rho_{\text{loc}}(\varepsilon) = \frac{\rho(\varepsilon)}{|1 - F_V^{(n)}(\varepsilon)|^2}, \quad (3.8)$$

where $F_V^{(n)}$ means the (1, 1) element of $\hat{F}_V$ for a normal metal ($\Delta = 0$). The function ρ_{loc} has the physical significance of a local density of states at the impurity which deviates from the bulk density of states due to the potential scattering from this impurity. The most remarkable feature of this renormalization is that $\rho_{\text{loc}}(0)$ differs from $\rho(0) = 1$ which results in a renormalization of the coupling constant (3.7). In the following we will no longer bother with potential scattering and assume that the renormalization has been accounted for by simply writing $\hat{F}$ for $\hat{F}_V^{\gamma}$. The factors associated with $(J/N)\hat{t}$ in Eq. (2.18) essentially do the job of renormalizing the extra exchange coupling J which has been extracted from the scattering amplitudes.

Before we turn to solving the integral equation (3.4), we should point out its relationship to earlier theories of magnetic impurities in superconductors which do not take the Kondo effect into account. The first of these theories is of course the second Born approximation of Abrikosov and Gor'kov [15] which one recognizes in Eq. (3.4) as

$$\hat{t}(z) = \frac{S(S+1)}{4} \hat{F}(z). \quad (3.9)$$

The physical features of this theory have to be recovered from the Nagaoka theory in the limit of weak spin exchange coupling γ (except for zero temperature).

An extension of the AG theory up to infinite order in γ was worked out by Shiba [16] for a classical impurity spin. Formally, the classical limit can be obtained from a quantum spin theory by taking the limits $\gamma \to 0$ and $S \to \infty$, while keeping γS constant, because commutators of spin components are then of relative order $1/S$. As the classical limit of Eq. (3.4), we immediately recover Shiba's result

$$\hat{t}(z) = \left[1 - \left(\frac{S}{2}\hat{F}(z)\right)^2\right]^{-1} \left(\frac{S}{2}\right)^2 \hat{F}(z). \quad (3.10)$$

There is one remarkable difference between the AG and Shiba scattering amplitudes. No matter how small the coupling γS, the scattering amplitudes in Shiba's theory exhibit two poles within the energy gap at energies

$$z_0 = \pm \Delta \frac{1 - (\pi \gamma S/2)^2}{1 + (\pi \gamma S/2)^2}. \tag{3.11}$$

Analogous bound states will show up in the solution of Eq. (3.4) and are of central importance for the appearance of the Kondo effect in superconductors. The presence of just two bound states is related to the pure s-wave character of the spin exchange interaction in Eq. (2.4); Rusinov [29] showed that an interaction of finite range leads to a pair of bound states for each partial wave.

IV. Solution of Hamann's Equation

The matrix integral equation (3.4) has not actually been solved in full generality, which would be quite desirable. The only restriction, however, under which an exact solution has been obtained was to assume particle–hole symmetry for the conduction electrons, that is, to assume that the density of states function $\rho(\varepsilon)$ [or rather, $\rho_{\text{loc}}(\varepsilon)$, if potential scattering is included] satisfies the symmetry relation

$$\rho(-\varepsilon) = \rho(\varepsilon). \tag{4.1}$$

It can then be inferred via a canonical transformation that the diagonal components as well as the off-diagonal components of the Nambu t-matrix are identical and we may write

$$\hat{t}(z) = \begin{pmatrix} t_1(z) & t_2(z) \\ t_2(z) & t_1(z) \end{pmatrix}. \tag{4.2}$$

Furthermore, the individual components satisfy the symmetry relations

$$t_1(-z) = -t_1(z), \qquad t_2(-z) = t_2(z). \tag{4.3}$$

Since now all matrices in Eq. (3.4) are of the type

$$\begin{pmatrix} a & b \\ b & a \end{pmatrix},$$

one can use the identity

$$\begin{pmatrix} a & b \\ b & a \end{pmatrix} \begin{pmatrix} 1 & -1 \\ -1 & 1 \end{pmatrix} = (a - b) \begin{pmatrix} 1 & -1 \\ -1 & 1 \end{pmatrix} \tag{4.4}$$

to derive a scalar integral equation for the difference

$$t(z) = t_1(z) - t_2(z). \tag{4.5}$$

Because of Eq. (4.3), t_1 and t_2 can be obtained individually from t as

$$t_1(z) = \tfrac{1}{2}[t(z) - t(-z)], \qquad t_2(z) = -\tfrac{1}{2}[t(z) + t(-z)]. \tag{4.6}$$

The equation for $t(z)$ is identical to Hamann's equation for the normal metal except that the quasiparticle excitation spectrum of the superconducting state is introduced via the function

$$F(z) = \gamma(z + \Delta) \int_{-\infty}^{\infty} \frac{\rho(\varepsilon)\, d\varepsilon}{z^2 - \Delta^2 - \varepsilon^2}. \tag{4.7}$$

If, for later convenience, we introduce the notations

$$R_\nu(z) = \mathscr{F}_\omega\left\{\frac{F^\nu(\omega)}{z-\omega}\right\}, \quad L_\nu(z) = \mathscr{F}_\omega\left\{\frac{F^\nu(\omega)}{z-\omega} t(\omega)\right\} - \frac{S(S+1)}{4}\delta_{\nu 0}, \tag{4.8}$$

the integral equation for $t(z)$ can be written as

$$\phi(z)t(z) = N(z) \tag{4.9}$$

with

$$N = L_1 - F \cdot L_0 \tag{4.10a}$$

and

$$\phi = 1 + R_1 - F \cdot R_0 + L_2 - 2F \cdot L_1 + F^2 \cdot L_0. \tag{4.10b}$$

As we shall see later, Eq. (4.9) [or Eq. (3.4)] without any further specifications does not have a unique solution. What makes the solution unambiguous is a well-known physical requirement which follows from the definition of t as a Green's function, namely, that $t(z)$ be analytic off the real axis (Im $z \neq 0$). This condition of analyticity is, in fact, not implied by the integral equation. Whereas the functions N and ϕ are analytic off the real axis by definition, Eq. (4.9) allows t to have poles wherever ϕ has zeros. Therefore, the analyticity condition for t has to be kept in mind explicitly in the following.

In order to understand the analytic behavior of t on the real axis, we next look at Eq. (4.7). For the sake of a simpler discussion we will assume that the density of states $\rho(\varepsilon)$ is given by one analytic function in some neighborhood of the entire real ε axis (take a Lorentzian, for instance). The function $F(z)$ then has cuts extending from $-\infty$ to $-\Delta$ and from Δ

12. MAGNETIC IMPURITIES IN SUPERCONDUCTORS

to ∞. If, by $(z^2 - \Delta^2)^{1/2}$, we mean a function with the same cuts and with a positive imaginary part (to specify the sign), we find

$$F(z) = \gamma \frac{z+\Delta}{(z^2-\Delta^2)^{1/2}} \int_{-\infty}^{\infty} d\varepsilon \frac{\rho(\varepsilon)}{(z^2-\Delta^2)^{1/2} - \varepsilon}. \qquad (4.11)$$

Equation (4.11) obviously implies that there exists an analytical continuation of F onto a second "unphysical" sheet of the complex z plane which is given by

$$\bar{F}(z) = F(z) + 2\pi i \gamma \frac{z+\Delta}{(z^2-\Delta^2)^{1/2}} \rho((z^2-\Delta^2)^{1/2}) \qquad (4.12)$$

and on which the singularities of the density of states function ρ will show up. The scattering amplitude $t(z)$ has a corresponding analytic continuation onto the unphysical sheet, and that is where the singularities that customarily appear in connection with the Kondo effect can be found. The analytic continuation of ϕ, for example,

$$\bar{\phi} = 1 + R_1 - \bar{F}R_0 + L_2 - 2\bar{F}L_1 + \bar{F}^2 L_0, \qquad (4.13)$$

exhibits poles at the poles of the Fermi function which are present in the functions R_ν and L_ν according to Eq. (4.8).

The clue to the solution of the integral equation (4.9) is contained in the function

$$X(z) = 1 + R_1 - FR_0 + \chi, \qquad (4.14)$$

where

$$\chi(z) = L_2 - (F+\bar{F})L_1 + F\bar{F}\cdot L_0. \qquad (4.15)$$

While this function is a functional of the unknown scattering amplitude t, just as $\phi(z)$ is, the dependence of $\chi(z)$ on t has a particularly simple form. The reason is that $\chi(z)$ combines in itself the virtues of not having cuts across the real axis, since $F + \bar{F}$ and $F\bar{F}$ obviously do not have cuts, and of canceling the poles of the individual $L_\nu(z)$, like $\phi(z)$. Accordingly, $\chi(z)$ has only one isolated singularity in the neighborhood of the real axis, namely, a simple pole at $z = \Delta$ due to the pole of $F \cdot \bar{F}$. We may write $\chi(z)$ as

$$\chi(z) = \gamma a \frac{\Delta}{z-\Delta} + \chi_\rho(z), \qquad a = 2\pi^2 \gamma L_0(\Delta), \qquad (4.16)$$

where $\chi_\rho(z)$ reflects the singularities of the density of states function ρ only and can be written as a nonsingular integral over t; it is therefore a small perturbative correction of order γ^2, just as the total function χ is for the case of a normal metal. In contrast to this, there is no reason why the pole term in Eq. (4.16) should be of the same order and, indeed, it will turn out to be very essential that it is not. The major shortcoming of various earlier attempts to solve the Nagaoka or Suhl equations for a superconductor was the failure to appreciate the importance of this term.

Since the preceding discussion shows that $X(z)$ is practically known up to the constant a, the integral equation (4.9) would be solved more or less by expressing $N(z)$ and $\phi(z)$ by $X(z)$. This is what will be done next.

First, we can express N in terms of X and ϕ via

$$(F - \bar{F})N = X - \phi. \tag{4.17}$$

In order to obtain ϕ we need the analytical continuation of X onto the second sheet which is simply

$$\bar{X} = 1 + R_1 - \bar{F}R_0 + \chi. \tag{4.18}$$

We then establish the identity

$$\phi\bar{\phi} = X\bar{X} + (F - \bar{F})^2 H, \tag{4.19}$$

where

$$H(z) = (1 + R_1 + L_2)L_0 - (R_0 + L_1)L_1 \tag{4.20}$$

is a quadratic form in the functions R_v and L_v which do not have any singularities except simple poles at the positions $z_n = i\omega_n$. Because of an obvious cancellation of the second-order pole terms, the corresponding poles of H are simple, too. We now arrive at the point where the integral equation is decisively employed: The residues of H at z_n turn out to be $\phi(z_n) \cdot t(z_n) - N(z_n) = 0$, which means that H does not have any singularities. From the easily verified behavior of H at infinity we then conclude

$$H(z) \equiv -\frac{S(S+1)}{4}. \tag{4.21}$$

Equation (4.21) is a crucial relation for finding the final solution. It gives us the equation

$$\phi(z)\bar{\phi}(z) = K(z), \tag{4.22}$$

where

$$K(z) = X(z) \cdot \bar{X}(z) + (\pi\gamma)^2 S(S+1) \frac{z + \varDelta}{z - \varDelta} \rho^2((z^2 - \varDelta^2)^{1/2}) \tag{4.23}$$

can be considered as essentially known, in the specific sense discussed above. In addition, Eq. (4.21) implies some information on the question

12. MAGNETIC IMPURITIES IN SUPERCONDUCTORS

of zeros of ϕ which is not contained in Eq. (4.22). Let us assume that at some energy z_0 both N and ϕ vanish. Equations (4.10) inserted into Eq. (4.20) then lead to $H(z_0) = 0$, which contradicts Eq. (4.21). Consequently, N and ϕ cannot vanish simultaneously. We infer two important facts from this result. First, for nonreal z_0 any zero of ϕ would imply that N vanishes simultaneously, since t should not have poles; thus $\phi(z)$ cannot have any zero off the real axis. Second, each zero of ϕ within the energy gap ($-\Delta < z_0 < \Delta$) necessarily does correspond to a pole of t. To complete the picture, zeros of ϕ on the real axis outside the gap are excluded by Eq. (4.22), which reads

$$\phi(\omega + i\delta)\phi(\omega - i\delta) = K(\omega) > 0 \tag{4.24}$$

for real ω with $|\omega| > \Delta$.

To proceed with finding the solutions for ϕ we realize that

$$\Psi(z) = \exp\left[\frac{(z^2 - \Delta^2)^{1/2}}{2\pi i} \left(\int_{-\infty}^{-\Delta} - \int_{\Delta}^{\infty}\right) \frac{d\omega}{z - \omega} \frac{\ln K(\omega)}{|\omega^2 - \Delta^2|^{1/2}}\right] \tag{4.25}$$

satisfies the relation we found for ϕ, Eq. (4.24). The function Ψ is very similar to ϕ in another respect, too; it is analytic except for the usual cuts along the real axis and it has the same behavior as ϕ at the points $z = \pm\Delta$, as may be verified by direct inspection. The features of Ψ can be stated more precisely by listing the porperties of the function

$$G(z) = \phi(z)/\Psi(z). \tag{4.26}$$

G is also analytic but for the real-axis cuts and may have zeros only within the energy gap. We know in addition:

$$G(\omega + i\delta)G(\omega - i\delta) = 1, \quad \omega \text{ real}, \ |\omega| \geqslant \Delta, \tag{4.27a}$$

$$G(z) = \text{sgn } X(-\Delta) + O((z^2 - \Delta^2)^{1/2}), \quad z \to -\Delta, \tag{4.27b}$$

$$G(z) = \text{sgn}(\gamma a) + O((z^2 - \Delta^2)^{1/2}), \quad z \to \Delta, \tag{4.27c}$$

$$G(z)\Psi(\infty) = 1 + O(1/z), \quad z \to \infty. \tag{4.27d}$$

All functions that have the stated properties of G can be classified by their zeros. To see this we look at the related function

$$f(z) = \frac{G'(z)}{G(z)} (z^2 - \Delta^2)^{1/2}. \tag{4.28}$$

One observes that, due to the stated properties of G, f is analytic everywhere except for a simple pole of residue $i \, | \, (\Delta^2 - \omega_j{}^2)^{1/2} \, |$ at each zero

$z = \omega_j$ of G and that f vanishes at infinity. This enables us to conclude that

$$f(z) = \sum_j \frac{i(\Delta^2 - \omega_j^2)^{1/2}}{z - \omega_j}, \tag{4.29}$$

where the sum includes all zeros of G, that is, all zeros of ϕ. Equation (4.28) is now easily integrated to give

$$G(z) = \text{sgn } X(-\Delta) \cdot \prod_j A_j(z), \tag{4.30a}$$

where

$$A_j(z) = \frac{l(\omega_j) - l(z)}{1 - l(\omega_j)l(z)}, \quad l(x) = \frac{x}{\Delta - i(x^2 - \Delta^2)^{1/2}}. \tag{4.30b}$$

At this point we have established the full analytic form of the solution of Eq. (4.9), the essential formulas being Eqs. (4.17), (4.25), (4.26) and (4.30). The only things we do not yet know are the number of poles and their positions ω_j and the function $\chi(z)$. There are, in fact, two more conditions which our most general solution does not satisfy identically, and these will yield the full solution. The first condition is obtained from Eq. (4.22). Since $\tilde{\phi}(z)$ is analytic within the gap by definition, Eq. (4.13), any zero of $\phi(z)$ has to be a zero of $K(z)$, too. The condition

$$K(\omega_j) = 0 \tag{4.31}$$

provides us with a relation between the values of $\chi(z)$ and the possible zeros of $\phi(z)$ and, furthermore, restricts the number of zeros to a maximum of four, because $K(z)$ never has more than four zeros within the gap. The second condition that has not been exploited yet is Eq. (4.27d), which gives explicitly

$$\text{sgn } X(-\Delta) \prod_j \exp\left[-i\left(\frac{\pi}{2} - \sin^{-1}\frac{\omega_j}{\Delta}\right)\right]$$

$$\cdot \exp\left[\frac{i}{2\pi} \int_\Delta^\infty \frac{d\omega}{(\omega^2 - \Delta^2)^{1/2}} \ln \frac{K(\omega)}{K(-\omega)}\right] = 1. \tag{4.32}$$

Equation (4.32) represents a relation between $\chi(z)$ and the actual zeros ω_j of $\phi(z)$. It turns out that Eqs. (4.31) and (4.32) are sufficient to settle the question of zeros completely in the weak coupling limit $|\gamma| \ll 1$, where $\chi_0(z)$ [see Eq. (4.16)] can be neglected.

That this limit is by no means a trivial one, at least for antiferromagnetic coupling ($\gamma < 0$), becomes obvious from a discussion of $X(z)$.

12. MAGNETIC IMPURITIES IN SUPERCONDUCTORS

One obtains

$$X(z) = 1 + \frac{\gamma}{g} + \gamma y(y+1) r(y^2) + \frac{\gamma a}{y-1} + i \frac{\pi \gamma}{2} \frac{y+1}{(y^2-1)^{1/2}} \tanh \frac{\beta \Delta y}{2}, \tag{4.33}$$

where

$$r(y^2) = \frac{1}{\beta \Delta} \sum_n \frac{-\pi}{[1+(\omega_n/\Delta)^2]^{1/2}} \left(y^2 + \frac{\omega_n^2}{\Delta^2} \right)^{-1}, \quad y = \frac{z}{\Delta}. \tag{4.34}$$

The most important term in Eq. (4.33) corresponding to the logarithmic term for the normal metal is γ/g with an effective superconducting coupling constant g given by

$$\frac{1}{g} = \int_0^\infty \frac{dx}{(x^2+\Delta^2)^{1/2}} \rho(x) \tanh \frac{\beta}{2} (x^2+\Delta^2)^{1/2} = \frac{1}{\lambda N_0} + \ln \frac{D}{\omega_D}, \tag{4.35}$$

where D denotes the conduction-electron bandwidth. Evidently, perturbation theory breaks down as soon as $|\gamma/g|$ becomes of order 1. Both coupling parameters γ and g are usually small compared to 1, but their ratio may have any value. It is useful to introduce the parameter

$$\tau = \frac{1}{\gamma} + \frac{1}{g}, \quad \tau = \ln \frac{T_K}{T_c} \quad (\gamma < 0) \tag{4.36}$$

measuring the relative magnitude of γ and g, because all results will depend on τ only.

Unfortunately, it is not possible to solve Eqs. (4.31) and (4.32) analytically in general, but one can handle them without numerical computation in a number of limiting cases. The solution has the simple feature that under all circumstances $\phi(z)$ has exactly one zero which we will call $\omega_0 = \Delta \cdot y_0$; it also turns out that always [see Eq. (4.30a)]

$$\operatorname{sgn} X(-\Delta) = +1. \tag{4.37}$$

When the spin exchange scattering is switched on ($|\gamma| \ll g$), the zero of ϕ emerges from the upper edge of the gap and is given by

$$y_0 = 1 - \frac{\pi^2}{2\tau^2} \left\{ \left[S(S+1) + \frac{1}{4} \tanh^2 \frac{\beta \Delta}{2} \right]^{1/2} - \frac{1}{2} (\operatorname{sgn} \gamma) \tanh \frac{\beta \Delta}{2} \right\}^2 + O\left(\frac{1}{\tau^3}\right), \tag{4.38}$$

while the residue of $\chi(z)$ is

$$a = -\frac{\pi^2 S(S+1)}{2\tau} + O\left(\frac{1}{\tau^2}\right). \tag{4.39}$$

For $\gamma > 0$, the parameter τ continues to be large compared to 1 regardless of the magnitude of γ ($\tau > 1/g \gg 1$). Thus, Eqs. (4.38) and (4.39) tell the whole story in the case of ferromagnetic spin exchange coupling. For $\gamma < 0$, τ ceases to be large as soon as $|\gamma|$ approaches g and Eqs. (4.38) and (4.39) are no longer meaningful. In the limit $\beta\Delta \to 0$ (i.e., $T \to T_c$), one now obtains

$$y_0 = \frac{-\tau}{[\tau^2 + \pi^2 S(S+1)]^{1/2}} \tag{4.40}$$

and

$$a = \tau + [\tau^2 + \pi^2 S(S+1)]^{1/2}, \tag{4.41}$$

which are valid for all values of τ. The zero traverses the gap as $|\gamma|$ increases, crossing the Fermi energy just when $\tau = 0$ or $T_K = T_c$ and a is no longer of order γ. For $-\gamma > g$, when τ becomes large again, one finds

$$y_0 = -1 + \frac{\pi^2}{2\tau^2}\left\{\left[S(S+1) + \frac{1}{4}\tanh^2\frac{\beta\Delta}{2}\right]^{1/2} - \frac{1}{2}\tanh\frac{\beta\Delta}{2}\right\}^2 + O\left(\frac{1}{\tau^3}\right) \tag{4.42}$$

and

$$a = 2\tau + \frac{\pi^2 S(S+1)}{2\tau} + O\left(\frac{1}{\tau^2}\right) \tag{4.43}$$

at all temperatures ($0 \leqslant \beta\Delta \leqslant \infty$).

Another limiting case, that can be treated analytically for all temperatures and all τ ($|\tau| \ll \pi S$), is the limit $S \to \infty$. One obtains

$$y_0 = \frac{1}{\pi(S+\frac{1}{2})}\left\{\tau_0 - \tau - \frac{\tau_1}{(S+\frac{1}{2})}\tanh\frac{\beta\Delta y_0}{2} + O\left(\frac{1}{S^2}\right)\right\} \tag{4.44}$$

and

$$a = \tau + \pi\left\{\left[S(S+1) + \frac{1}{4}\tanh^2\frac{\beta\Delta y_0}{2}\right]^{1/2} - \frac{1}{2}\tanh\frac{\beta\Delta y_0}{2}\right\} + O\left(\frac{1}{S}\right), \tag{4.45}$$

where

$$\tau_0 = \int_1^\infty \frac{dy}{y^2(y^2-1)^{1/2}}\tanh\frac{\beta\Delta y}{2} = \begin{cases}(\pi/4)\beta\Delta, & \beta\Delta \to 0, \\ 1, & \beta\Delta \to \infty,\end{cases} \tag{4.46a}$$

and

$$\tau_1 = \int_1^\infty \frac{dy}{y^4}(y^2-1)^{1/2}\tanh\frac{\beta\Delta y}{2} = \begin{cases}(\pi/8)\beta\Delta, & \beta\Delta \to 0, \\ \frac{1}{3}, & \beta\Delta \to \infty.\end{cases} \tag{4.46b}$$

Remarkably, the variation of y_0 with τ develops a wiggle around $y_0 = 0$ at lower temperatures due to the hyperbolic tangent term which at zero temperature results in the zero y_0 lying strictly at the Fermi energy over a whole range of τ values around $\tau = 1$. That this feature persists for all spin values can be seen in Fig. 1, which presents numerical

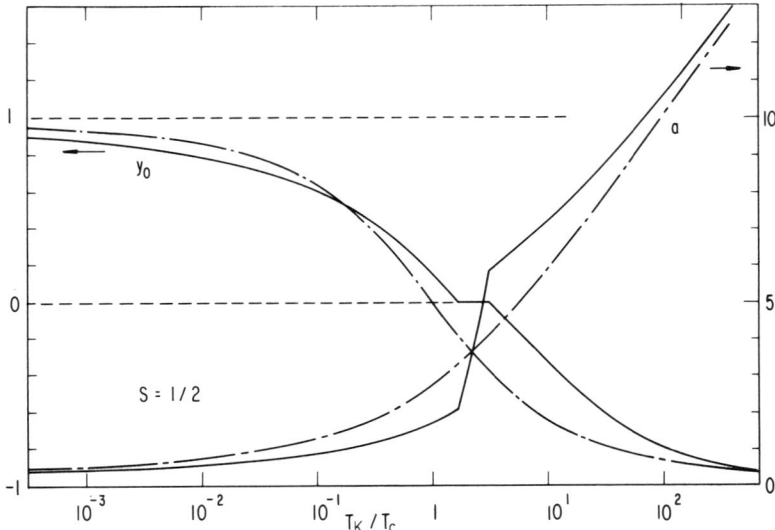

FIG. 1. Zero of ϕ within the gap and residue of χ for $T = 0$ (solid lines) and $T = T_c$ (dashed lines).

results for $S = \frac{1}{2}$ at zero temperature (solid lines) compared to the high-temperature results (dashed lines) of Eqs. (4.40) and (4.41). The zero lies at the Fermi energy for $1.65 \leqslant T_K/T_c \leqslant 3.09$.

The problem of poles having been settled by the preceding discussion, the integral equation (4.9) can be considered completely solved and we therefore turn to a discussion of the physical significance of our results.

V. Results at Low Impurity Concentration

In this section we shall assume an impurity concentration so low that we can treat it by simply multiplying the T matrix in Eq. (2.17) by the number of impurities N_i. Once we have solved for $t(z)$, the way back to physical reality leads through Eqs. (4.6), (4.2), (2.18), and (2.17). The first observation is that the pole of $t(z)$ at $z = \omega_0$ corresponds to two poles of the scattering amplitudes $t_1(z)$ and $t_2(z)$ at the energies $z = \pm\omega_0$. For ferromagnetic exchange coupling ($\gamma > 0$) these poles never separate appreciably from the gap edges. For antiferromagnetic coupling ($\gamma < 0$), however, they move well into the gap when the Kondo temperature T_K and the critical temperature approach each other, and they may cross over at the Fermi energy upon varying T_K/T_c or the temperature T (see Fig. 1); in this case they will add up in t_1 and cancel in t_2 for $\omega_0 = 0$. In order to get an insight into the physical significance of these poles it is useful to calculate the change in the electron density of states due to the presence of the impurities. The density of one-electron states for both spin directions per atomic cell is given by

$$N(\omega) = -\frac{2}{\pi N} \operatorname{Im} \sum_k G_{kk}^{(11)}(\omega + i\delta) \tag{5.1a}$$

$$= N_{\text{BCS}}(\omega) + \frac{N_i}{N} n(\omega), \tag{5.1b}$$

where

$$N_{\text{BCS}}(\omega) = 2N_0 \operatorname{Re} \frac{\omega}{(\omega^2 - \Delta^2)^{1/2}} \tag{5.2}$$

is the familiar density of states of a BCS superconductor. It follows from particle–hole symmetry that the impurity contribution $n(\omega)$ is an even function of energy and that

$$\int_{-\infty}^{0} n(\omega) \, d\omega = 0, \tag{5.3}$$

which means that the chemical potential is not changed by adding magnetic impurities. For the contribution of one impurity, one obtains

$$n(\omega) = -2\gamma\Delta \operatorname{Im} \left(\frac{\Delta t_1(z) - zt_2(z)}{(\Delta^2 - z^2)^{3/2}} \right)_{z=\omega+i\delta} \tag{5.4a}$$

$$= -\frac{1}{2}(\delta(\omega - \Delta) + \delta(\omega + \Delta)) + \frac{\alpha}{2}(\delta(\omega - \omega_0) + \delta(\omega + \omega_0)) + n_{\text{cont}}(\omega). \tag{5.4b}$$

Thus, for each impurity, half a state is removed from each of the edges of the gap, $\alpha/2$ states appear at the energies $\pm\omega_0$ within the gap, and a continuum contribution n_{cont} is, in general, added to the BCS continuum $|\omega| > \Delta$. In the limit $\beta\Delta \to 0$, there is no such continuum contribution and α equals 1 [18]. This result remains valid at all temperatures $T < T_c$ in Shiba's theory of a classical impurity spin. It indicates that the impurity spin lowers the binding energy 2Δ of one Cooper pair to become $2|\omega_0|$. The physical implications of this have been discussed in more detail by Sakurai [30]. Generally, however, one obtains

$$\alpha = \frac{2X(\omega_0)}{\Psi(\omega_0)}, \qquad (5.5)$$

which as we shall see may differ from 1. From Eqs. (4.31) and (4.37), one easily infers the exact relation

$$X(\omega_0) = \pi |\gamma| \left(\frac{1+y_0}{1-y_0}\right)^{1/2}$$

$$\times \left\{\left[S(S+1) + \frac{1}{4}\tanh^2\frac{\beta\Delta y_0}{2}\right]^{1/2} + \frac{1}{2}\operatorname{sgn}\gamma \cdot \tanh\frac{\beta\Delta y_0}{2}\right\}, \quad (5.6)$$

whereas $\Psi(\omega_0)$ has to be evaluated approximately in various limits. One finds for $\beta\Delta \to 0$ (all τ) or for $|\tau| \to \infty$ (all $T < T_c$)

$$\Psi(\omega_0) = 2\pi |\gamma| \left(\frac{1+y_0}{1-y_0}\right)^{1/2} \cdot \left[S(S+1) + \frac{1}{4}\tanh^2\frac{\beta\Delta y_0}{2}\right]^{1/2}, \qquad (5.7a)$$

and for $S \to \infty$ ($|\tau| \ll \pi S$)

$$\Psi(\omega_0) = 2\pi |\gamma| \left(\frac{1+y_0}{1-y_0}\right)^{1/2}$$

$$\times \left\{\left[S(S+1) + \frac{1}{4}\tanh^2\frac{\beta\Delta y_0}{2}\right]^{1/2} - \frac{1}{4}\tanh\frac{\beta\Delta y_0}{2} + O\left(\frac{1}{S}\right)\right\}. \quad (5.7b)$$

Consequently, for ferromagnetic coupling α increases from 1 to $(2S+2)/(2S+1)$ as the temperature drops from T_c to zero. For $\gamma < 0$, however, α is somewhere between $2S/(2S+1)$ and $(2S+2)/(2S+1)$ at absolute zero, depending on the value of $\tau = \ln T_K/T_c$. Actually α increases monotonically and continuously with τ. In particular, on the τ interval on which $y_0 = 0$, α increases from $S/(S+\frac{1}{4}) + O(1/S^2)$ to $(S+1)/(S+\frac{3}{4}) + O(1/S^2)$, as one finds from Eq. (5.7b). The variation of α with τ for $S = \frac{1}{2}$ (numerical computation) is depicted in Fig. 2.

The physical picture suggested by the preceding discussion is that not only one Cooper pair is involved in the effect of a quantum impurity spin on a superconductor. Whereas exactly one Cooper pair still appears to be removed from the gap edges, the new excitation of energy $2|\omega_0|$ which is created by the impurity has a spectral weight between $S/(S+\frac{1}{2})$ and $(S+1)/(S+\frac{1}{2})$. Moreover, the shape of the BCS continuum is modified, too.

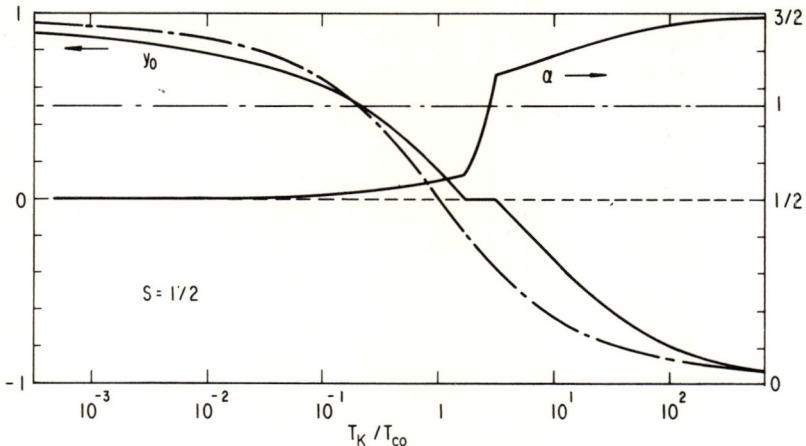

FIG. 2. Pole of the scattering amplitudes within the gap and its spectral weight for $T = 0$ (solid lines) and $T = T_c$ (dashed lines).

It is also to be noted that both the energy y_0 and the spectral weight of the excitation within the gap are temperature dependent. For a limited range of Kondo temperatures slightly above T_c, there are zero-energy excitations indicating a gapless superconductor for arbitrarily low impurity concentrations. There is no local accumulation of charge around the impurity [18], although the state of energy ω_0 is associated with a localized wave function (see, also, Rusinov [29]). Altogether, as in the normal metal, the physical picture of what happens is far from simple and seems to exclude a common quasiparticle description. It should be mentioned that the results of the present theory disagree with the findings of the ground-state approach [7], in particular for $T_K \gg T_c$. The same type of disagreement is found in the normal metal case and its origin is not well understood. The criticism of Nagaoka and Matsuura about the thermodynamic Green's function approach—they believe they found a self-contradiction—seems precipitate.

Another observable quantity which can be calculated is the change of the transition temperature from magnetic impurities [17]. Starting from Eq. (2.3) one obtains

$$\frac{T_c - T_{c0}}{T_{c0}} = -\bar{c} \cdot \sum_{\omega_n > 0} \left(\frac{2\pi T_c}{\omega_n}\right)^2 \cdot \alpha(\omega_n), \tag{5.8}$$

where

$$\bar{c} = \frac{N_i/N}{(2\pi)^2 N_0 T_{c0}} \tag{5.9}$$

12. MAGNETIC IMPURITIES IN SUPERCONDUCTORS

is the concentration of impurities measured in convenient units and

$$\alpha(\omega_n) = \lim_{\Delta \to 0} \operatorname{Im} \frac{2\pi\gamma}{\Delta} [i\omega_n t_2(i\omega_n) - \Delta t_1(i\omega_n)] \tag{5.10}$$

is the pair breaking parameter per unit concentration which, in contrast to the theories of Abrikosov and Gor'kov [15] and Shiba [16], is energy dependent. Evaluating Eq. (5.10) with the present solution for the scattering amplitudes yields

$$\alpha(\omega_n) = \frac{\phi_0(0)}{\phi_0(i\omega_n)} \left[1 - X_0(i\omega_n) \int_{-\infty}^{\infty} d\omega \, \frac{\omega_n/\pi}{\omega_n^2 + \omega^2} \frac{X_0(\omega)}{K_0(\omega)} \right]. \tag{5.11}$$

Here the subscript 0 denotes the $\lim_{\Delta \to 0}$ of the respective function which is identical with the corresponding function for a normal metal. The energy sum in Eq. (5.8) over the rather involved expression in Eq. (5.11) can be performed using some mathematical tricks, with the final result

$$\frac{T_c - T_{c0}}{T_{c0}}$$
$$= -\bar{c} \cdot \frac{P}{\pi} \int_{-\infty}^{\infty} \frac{dx}{x^2} \frac{\pi^2 S(S+1)(2g_+ - xg_+') - (\tau - g_-)(g_+ - xg_+')g_+}{(\tau - g_+)(\tau - g_-) + \pi^2 S(S+1)}, \tag{5.12}$$

where P denotes the principal part of the integral and

$$g_\pm = \psi(\tfrac{1}{2} \mp ix) - \psi(\tfrac{1}{2}) \tag{5.13}$$

is related to the digamma function. The integral in Eq. (5.12) has to be computed numerically (Fig. 3). A simple analytic approximation may, however, be obtained by expanding it for large spin:

$$\frac{T_c - T_{c0}}{T_{c0}} = -\bar{c} \cdot \frac{\pi^2}{2} \left[1 - \frac{1}{\pi^2 S(S+1)} \left(\tau^2 - \frac{48}{\pi^2} \tau \sum_{n=0}^{\infty} \frac{1}{(2n+1)^3} \right. \right.$$
$$\left. \left. + \frac{2\pi^2}{3} + \frac{16}{\pi^2} \sum_{k=1}^{\infty} \frac{1}{k^2} \sum_{n=k}^{\infty} \frac{1}{(2n+1)^2} \right) + O\left(\frac{1}{S^4} \right) \right]. \tag{5.14}$$

Extending this expansion into a geometric series, one finds

$$\frac{T_c - T_{c0}}{T_{c0}} = -\bar{c} \cdot \frac{\pi^2}{2} \frac{\pi^2 S(S+1)}{(\tau - 2.558)^2 + 0.498 + \pi^2 S(S+1)}, \tag{5.15}$$

which is in excellent numerical agreement with Eq. (5.12) for all spins.

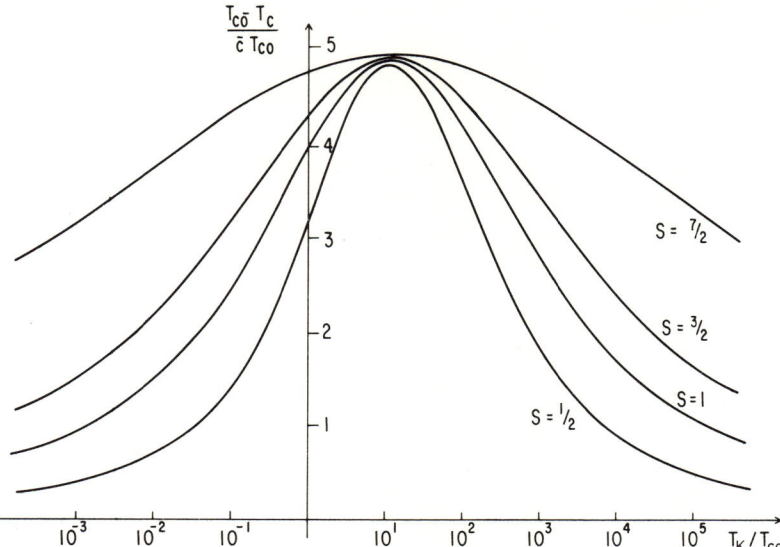

FIG. 3. Depression of the transition temperature for small impurity concentration [Reprinted, by permission, from E. Müller-Hartmann and J. Zittartz, Theory of magnetic impurities in super conductors, II. Z. Phys. 234, 58–69 (1970). Springer-Verlag, Berlin, Heidelberg, and New York.]

Notably, the transition temperature is always depressed by magnetic impurities. In the limit $T_K/T_c \to 0$, one recovers the second Born approximation result of AG. There is a maximum depression, independent of the strength of the exchange coupling, when $T_K \approx 12.9 T_c$, and for higher Kondo temperatures the depression is again smaller. This shows that the pair breaking activity of the impurities is most pronounced at temperatures somewhat lower than T_K. It is instructive to compare the present theory for ΔT_c to the second Born approximation result, which is shown in Fig. 4, with $g = 0.1$ and $S = \frac{1}{2}$. The AG result is altered strikingly for both antiferromagnetic and ferromagnetic exchange coupling.

Since the calculation of ΔT_c does not require the full knowledge of a Green's function for finite-order parameter, there are a number of different approaches to it. The first one is to calculate the Green's function (2.5) [or the scattering amplitudes (4.2)] just up to first order in Δ, which was attempted by Zuckermann [8]. Expanding Eq. (4.9) with $t = t^{(0)} + \Delta t^{(1)}$, $t^{(0)}$ understandably turns out to be the normal metal amplitude, whereas one gets a new integral equation for $t^{(1)}$. The reason why Zuckermann obtained instead an algebraic equation is

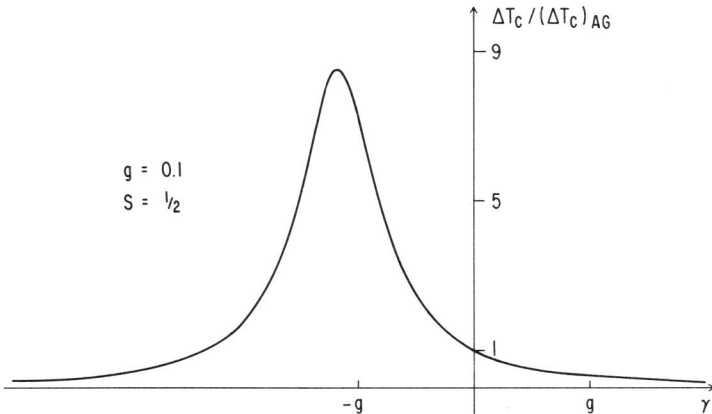

FIG. 4. Comparison of the depression of T_c obtained here with that of AG [Reprinted, by permission, from E. Müller-Hartmann and J. Zittartz, Theory of magnetic impurities in superconductors, II. Z. Phys. 234, 58–69 (1970). Springer-Verlag, Berlin, Heidelberg, and New York.]

that for the self-consistent thermal averages, he used a Green's function averaged over impurity positions, thereby deleting the off-diagonal part (in momentum space) of the Green's function (2.5) [cf. his Eqs. (3.13) and (3.19)]. Another way of obtaining ΔT_c is to look at the electron–electron interaction vertex which was first done in this context by Sólyom and Zawadowski [31]. They calculated the expression of T_c up to third order in $1/\tau$. Their result agrees with the expansion of Eq. (5.12),

$$\frac{T_c - T_{c0}}{T_{c0}} = -\bar{c} \cdot \frac{\pi^2}{2} \frac{\pi^2 S(S+1)}{\tau^2} \left[1 + \frac{48}{\pi^2 \tau} \sum_{n=0}^{\infty} \frac{1}{(2n+1)^3} + O\left(\frac{1}{\tau^2}\right)\right], \quad (5.16)$$

except for a factor of 3 which they dropped in their third-order term.

There remains some work to be done in the limit of low impurity concentration. The change in free energy due to impurities has not been calculated so far. In particular, it would be highly desirable to compute the specific heat jump at $T = T_c$ in the presence of magnetic impurities.

VI. Treatment of Finite Impurity Concentrations

The results of the previous section apply only as long as the impurity concentration $\bar{c}$, defined by Eq. (5.9), is much smaller than unity. If $\bar{c}$ does not satisfy this condition, which, due to the small factor $N_0 T_{c0}$

corresponds to pretty low absolute impurity concentrations N_i/N, the impurities have to be represented by a proper self-energy rather than by scattering amplitudes. After averaging over impurity positions the Green's function $\hat{G}_{kk'}$ in Eq. (2.5) is diagonal in momentum, the diagonal elements being given by the Dyson equation

$$\hat{G}_k^{-1}(z) = \hat{G}_k^{0-1}(z) - \hat{\Sigma}(z). \tag{6.1}$$

To first order in the impurity concentration the self-energy is simply $\hat{\Sigma}(z) = N_i \cdot \hat{T}(z)$. But since $\hat{T}(z)$ varies strongly over energies of order Δ, it is absolutely crucial to replace all intermediate electron propagators in $\hat{T}(z)$ by exact ones, when $\bar{c}$ is of order 1, thereby obtaining self-consistent scattering amplitudes such that

$$\hat{\Sigma}(z) = N_i \hat{T}_{\text{sc}}(z). \tag{6.2}$$

In the theories of AG and of Shiba, self-consistency is easily achieved by replacing $\hat{F}(z)$ in Eq. (3.9) or (3.10) — see also Eq. (2.12) — by

$$\hat{F}_{\text{sc}}(z) = (J/N) \sum_l \hat{G}_l(z). \tag{6.3}$$

This led to the well-known AG variation of T_c with impurity concentration, to gapless superconductivity, and to Shiba's impurity bands within the gap.

In the theory of magnetic impurity scattering discussed in this chapter, the self-consistent treatment of intermediate electron propagators results in a new equation, replacing Eq. (3.4) or Eq. (4.9). If we employ the convenient notation

$$\hat{G}_k^{-1}(z) = \begin{pmatrix} \tilde{z}(z) - \varepsilon_k & \tilde{\Delta}(z) \\ \tilde{\Delta}(z) & \tilde{z}(z) + \varepsilon_k \end{pmatrix}, \tag{6.4}$$

which was already introduced by AG, we obtain

$$(\tilde{z}(z) - \tilde{z}(\omega))\, \hat{G}_k(z)\, \hat{G}_k(\omega) = \hat{G}_k(\omega) - \hat{G}_k(z) \tag{6.5}$$

instead of Eq. (3.2). The equation for the self-consistent scattering amplitude corresponding to Eq. (4.9) is then explicitly given by

$$\left[1 - \frac{S(S+1)}{4} F_{\text{sc}}^2(z) + \mathscr{F}_\omega \left\{ \frac{F_{\text{sc}}(\omega) - F_{\text{sc}}(z)}{\tilde{z}(z) - \tilde{z}(\omega)} (1 + t_{\text{sc}}(\omega)(F_{\text{sc}}(\omega) - F_{\text{sc}}(z))) \right\} \right] t_{\text{sc}}(z)$$
$$= \frac{S(S+1)}{4} F_{\text{sc}}(z) + \mathscr{F}_\omega \left\{ \frac{F_{\text{sc}}(\omega) - F_{\text{sc}}(z)}{\tilde{z}(z) - \tilde{z}(\omega)} t_{\text{sc}}(\omega) \right\}. \tag{6.6}$$

Equations (6.1)–(6.4) and (6.6) determine $\tilde{z}$ and $\tilde{\Delta}$ as functions of energy z and order parameter Δ. These solutions have to be inserted into the self-consistency condition (2.3) for the order parameter which can be written in terms of the present notation as

$$\ln \frac{T}{T_{c0}} = \frac{2\pi}{\beta} \sum_{\omega_n > 0} \left(\frac{\tilde{\Delta}(i\omega_n)/\Delta}{(\tilde{\Delta}^2(i\omega_n) + \tilde{z}^2(i\omega_n))^{1/2}} - \frac{1}{\omega_n} \right). \quad (6.7)$$

A solution of the whole self-consistent program described above would be highly desirable, but has not yet been obtained. In particular, the variation of the transition temperature T_c with impurity concentration is of interest, because it will differ from the universal variation obtained by AG and Shiba. For experimental work possibly related to the present theory, we refer to Maple and Kim [32] and Kim [33].

The density of states (5.1a) of the superconducting alloy is now given by

$$N(\omega) = 2N_0 \operatorname{Re} \frac{\tilde{\omega}(\omega)}{[\tilde{\omega}^2(\omega) - \tilde{\Delta}^2(\omega)]^{1/2}}. \quad (6.8)$$

It should exhibit a conspicuous structure within the conduction band gap due to impurity bands, if T_K and T_{c0} are of the same order of magnitude. Gapless superconductivity is obtained then at impurity concentrations much lower than those predicted by the AG theory. This has been observed experimentally by Edelstein [34], Tsuda [35], Culbert and Edelstein [36], and Smith [37].

Some approximate treatments of the self-consistent theory discussed above have been put forward. The author has tried [38] to neglect the renormalization of intermediate energies ω_n in the $\mathcal{F}_\omega$ sums in Eq. (6.6), in the following sense: If we introduce a function t by

$$t_{sc}(z, \Delta) = t(\tilde{z}(z), \tilde{\Delta}(z)), \quad (6.9)$$

then $t(z, \Delta)$ satisfies the integral equation for scattering from just one impurity (4.9), except that the $\mathcal{F}_\omega$ sums (3.5) extend over $i\tilde{\omega}_n = \tilde{z}(i\omega_n)$ instead of over $i\omega_n$. Ignoring this difference amounts to approximating $t(z, \Delta)$ in Eq. (6.9) by the one-impurity scattering amplitude. From Eq. (6.7) and the Dyson equation (6.1), one then obtains the implicit equation

$$\ln \frac{T_c}{T_{c0}} = \sum_{n=0}^{\infty} \left(\frac{1}{n + \tfrac{1}{2} + [\bar{c}/(T_c/T_{c0})] \alpha(\tilde{\omega}_n)} - \frac{1}{n + \tfrac{1}{2}} \right) \quad (6.10)$$

for the transition temperature T_c. Here, $\alpha(\omega)$ is the pair breaking function from Eq. (5.11) taken at the temperature T_c and the renormalized energy $\tilde{\omega}_n$ determined from

$$i\tilde{\omega}_n = i\omega_n - J(N_i/N) \cdot t(i\tilde{\omega}_n, \tilde{\Delta} = 0); \qquad (6.11)$$

ω_n and t also have to be taken at the temperature T_c. The results of a numerical evaluation of Eqs. (6.10) and (6.11) are shown in Figs. 5

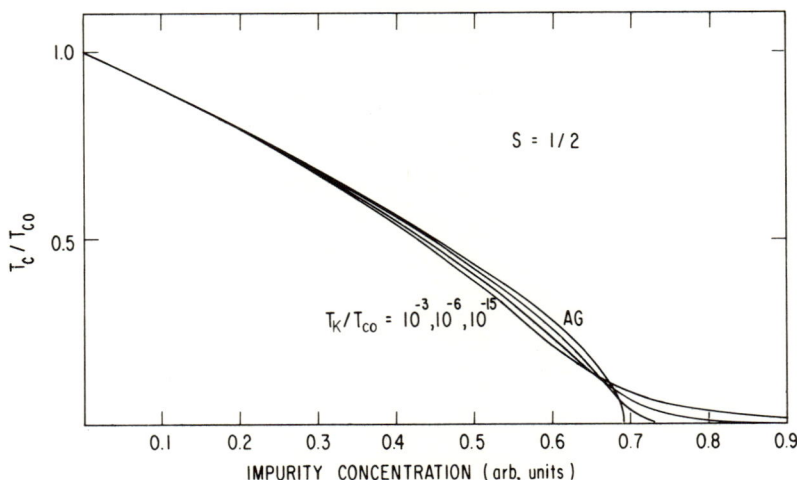

FIG. 5. Variation of T_c with impurity concentration according to Eq. (6.10) for small T_K.

through 7. Since the depression of T_c in absolute values of impurity concentration was already discussed in the last section, the initial slopes $|dT_c/d\bar{c}|_{\bar{c}=0}$ of all curves were normalized to 1 to facilitate comparison with the AG concentration dependence. It does not seem possible to give a simple physical explanation to the results obtained here. They do not look convincing and one should not have too much confidence in the approximate treatment of self-consistency discussed above.

Ludwig and Zuckermann [39] obtain an equation similar to (6.10) from an electron–electron interaction vertex approach. They do not discuss a fully self-consistent treatment of electron propagators. Therefore, in addition to the approximation we made in deriving Eq. (6.10), they obtain a pair breaking parameter taken at the unrenormalized energy ω_n instead of $\tilde{\omega}_n$; they also finally use the incorrect expression of Zuckermann [8] for $\alpha(\omega)$.

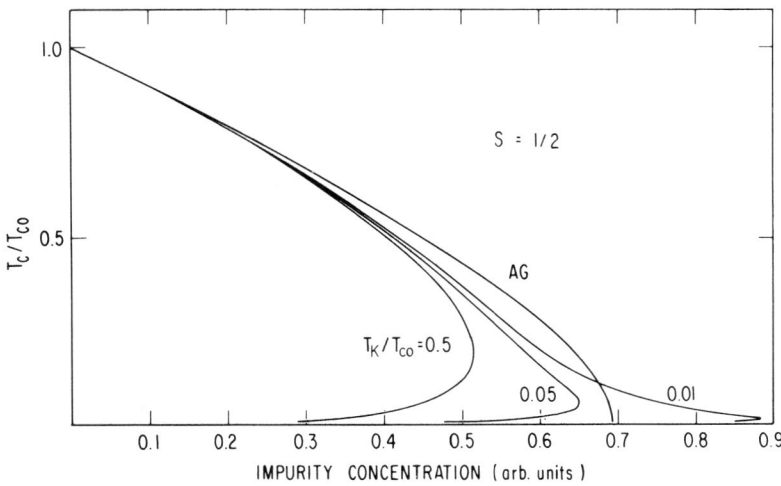

FIG. 6. Variation of T_c with impurity concentration according to Eq. (6.10) for medium T_K.

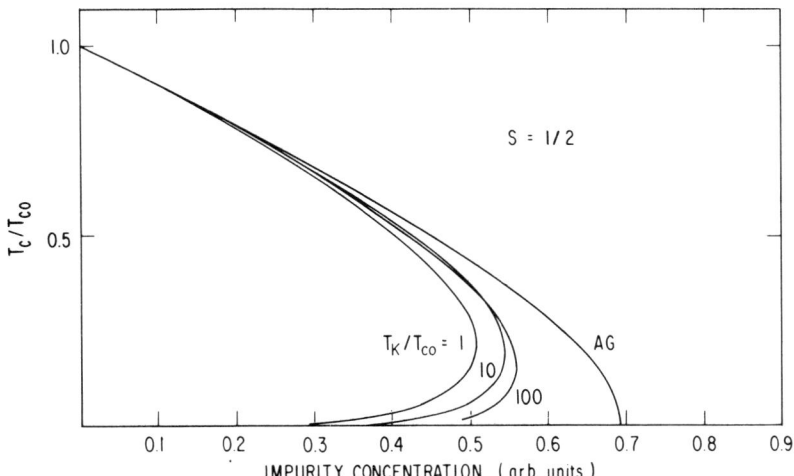

FIG. 7. Variation of T_c with impurity concentration according to Eq. (6.10) for large T_K.

Another approximation was proposed by Müller-Hartmann and Zittartz [40]. They used the scattering amplitudes (3.10) of Shiba, but whith a pole position depending on T_c in the manner of Eq. (4.40).

This leads to Eq. (6.10) with the energy $\tilde{\omega}_n$ replaced by zero, yielding

$$\alpha(0) = \frac{\pi^2 S(S+1)}{\ln^2(T_c/T_K) + \pi^2 S(S+1)}. \tag{6.12}$$

In this approximation, the AG theory is modified by the Kondo effect in a particularly simple fashion and the numerical evaluation of Eq. (6.10) with Eq. (6.12) is as trivial as in the AG theory. Although this approach can be viewed as an additional approximation to Eqs. (6.10) and (6.11), it seems more meaningful to describe it as an exact treatment of self-consistency making use of the slightly less accurate scattering amplitudes of Zittartz [18]. It is dangerous to make any compromise regarding self-consistency. We therefore think that the approach leading to Eq. (6.12) is more reliable than the other approximations discussed above. Actually, some experiments can be fitted quite convincingly (see Müller-Hartmann and Zittartz [40]). One striking result of this theory which one might truly call a "Kondo effect in superconductors" should be mentioned. For appropriate values of the Kondo temperature a superconducting alloy may have more than one transition temperature. This effect should be easily observable, if T_K is between one and two orders of magnitude below T_{c0}.

VII. Conclusion

We would like to conclude with a brief summary of some directions in which the work described in this chapter might be continued in the future. As pointed out earlier, it is a mathematical challenge to solve Hamann's equation for a superconductor, Eq. (3.4), without invoking particle–hole symmetry. There are, however, some other problems left which are of considerable physical interest. One of them is the calculation of the specific heat of a superconducting alloy in the present context. In particular, it should be possible to work out the specific heat jump at $T = T_c$ analytically, at low impurity concentrations. A second calculation which remains to be done is the solution of the full self-consistent many-impurity problem discussed in Section VI. A treatment of the Kondo problem for superconducting alloys in the presence of an external magnetic field would be desirable. In the most general case, one has to include into such a treatment the coupling of the magnetic field to both the orbital motion of the Cooper pairs and to the spins. The first of these requirements has not been attacked yet, to our knowledge. The second one leads to difficulties even in the normal state. It seems that a successful treatment of the Kondo effect in type II superconductors would require much further work.

ACKNOWLEDGMENT

The author wishes to thank Dr. M. B. Maple for carefully reading the manuscript and Professor J. Zittartz for valuable correspondence on the effect of many impurities.

References

1. P. W. Anderson, *Phys. Rev.* **124**, 41 (1961).
2. P. A. Wolff, *Phys. Rev.* **124**, 1030 (1961).
3. H. Suhl, *Phys. Rev. Lett.* **19**, 442 (1967).
4. A. B. Kaiser, *J. Phys. C* **3**, 409 (1970).
5. A. Griffin, *in* "Superconductivity" (P. R. Wallace, ed.), Vol. II, pp. 577–621. Gordon & Breach, New York, 1969.
6. M. Fowler and K. Maki, *Phys. Rev. B* **1**, 181 (1970).
7. Y. Nagaoka and T. Matsuura, *Progr. Theor. Phys.* **46**, 364 (1971).
8. M. Zuckermann, *Phys. Rev.* **168**, 390 (1968).
9. F. Takano and S. Matayoshi, *Progr. Theor. Phys.* **41**, 45 (1969).
10. J. Zittartz and E. Müller-Hartmann, *Z. Phys.* **232**, 11 (1970).
11. P. W. Anderson, *J. Phys. Chem. Solids* **11**, 26 (1959).
12. K.-D. Schotte, *Z. Phys.* **212**, 467 (1968).
13. Y. Nagaoka, *Phys. Rev. A* **138**, 1112, (1965).
14. Y. Nagaoka, *Progr. Theor. Phys.* **37**, 13 (1967).
15. A. A. Abrikosov and L. P. Gor'kov, *Zh. Eksp. Teor. Fiz.* **39**, 1781 (1960); *Sov. Phys. JETP* **12**, 1243 (1961).
16. H. Shiba, *Progr. Theor. Phys.* **40**, 435 (1968).
17. E. Müller-Hartmann and J. Zittartz *Z. Phys.* **234**, 58 (1970).
18. J. Zittartz Z., *Phys.* **237**, 419 (1970).
19. L. P. Gor'kov, *Zh. Eksp. Teor. Fiz.* **34**, 735 (1958); *Sov. Phys. JETP* **7**, 505 (1958).
20. T. Tsuzuki and T. Tsuneto, *Progr. Theor. Phys.* **37**, 1 (1967).
21. J. Heinrichs, *Phys. Rev.* **168**, 451 (1968).
22. T. Kitamura, *Progr. Theor. Phys.* **43**, 271 (1970).
23. T. Soda, *Progr. Theor. Phys.* **44**, 580 (1970).
24. Y. Kuroda, *Progr. Theor. Phys.* **46**, 688 (1971).
25. H. Keiter, *Z. Phys.* **223**, 289 (1969).
26. E. Müller-Hartmann, *Z. Phys.* **223**, 277 (1969).
27. K. Kawamura, *Progr. Theor. Phys.* **42**, 1058 (1969).
28. D. R. Hamann, *Phys. Rev.* **158**, 570 (1967).
29. A. J. Rusinov, *Pis'ma Zh. Eksp. Teor. Fiz.* **9**, 146 (1969); *Sov. Phys. JETP Lett.* **9**, 85 (1969).
30. A. Sakurai, *Progr. Theor. Phys.* **44**, 1472 (1970).
31. J. Sólyom and A. Zawadowski, *Z. Phys.* **226**, 116 (1969).
32. M. B. Maple and K. S. Kim, *Phys. Rev. Lett.* **23**, 118 (1969).
33. K. S. Kim, Ph.D. Thesis, University of California, San Diego, 1971, unpublished.

34. A. S. Edelstein, *Phys. Rev.* **180**, 505 (1969).
35. N. Tsuda, *J. Phys. Soc. Jap.* **27**, 1075 (1969).
36. H. V. Culbert and A. S. Edelstein, *Solid State Commun.* **8**, 445 (1970).
37. F. W. Smith, *J. Low Temp. Phys.* **5**, 683 (1971).
38. E. Müller-Hartmann, unpublished, 1971.
39. A. Ludwig and M. J. Zuckermann, *J. Phys. F* **1**, 516 (1971).
40. E. Müller-Hartmann and J. Zittartz, *Phys. Rev. Lett.* **26**, 428 (1971).

Author Index

Numbers in parentheses are reference numbers and indicate that an author's work is referred to although his name is not cited in the text. Numbers in italics show the page on which the complete reference is listed.

A

Abrikosov, A. A., 188(6), 212, 213, *214*, *215*, 235(22, 23), *236*, 238(11), *252*, 261, *285*, 289, 291, *323*, 328, 329, *351*, 354, 360, 373, *381*
Aio, T., 293(14), *307*, *308*, *323*
Alderson, J. E. A., 136(73), *146*
Alekseevskii, N. E., 132(60), *146*
Alexander, S., 78(21), *88*
Allali, V., 30(75), *53*
Alloul, H., 17, 21, *52*, 127(29), 144(29), *145*, 168(73, 74), 173(74), 175(73), 177(106, 107), 179(106, 107), *182*, 314(93, 94), *325*,
Ambegaokar, V., 294(23), *323*
Anderson, P. W., 5, 34(3), 47, *51*, 58, 64, 73(4, 14), 76(15), 78(21), 83(15), 85(34), 86(34, 36, 37), 87(15), *87*, *88*, 91(4b), 99(4b), 100(4b), 102(4b), 106, 108(4b), *116*, 123(5), 125(10), 131(5), *145*, 150(1), 161(42), 165(1), *180*, *181*, 218(2), 221(6), 222(7), 223(7), 224(7, 10), 227(13), 230, 231, 232(13, 17), 233, 234, *235*, *236*, 238(10), 240(10), 242(10), 249(16), 250(16, 17), *252*, 254(2, 3), 256, 265, 278, 281, *285*, 300(42), 301, 308, *324*, 327(4), 343, *351*, *352*, 354(1, 11), *381*
Andres, K., 297(31), 298(31), *323*, 333, *351*
Aoki, R., 123(2), 130(2), 131(2), 137(2), *144*, 175(93), *182*, 309, 311(77, 78), 314(78), *324*
Appelbaum, J. A., 126(21), *145*, 257(18, 19, 20), *285*
Armbrüsker, H., *307*
Asayama, K., 159(36), 160(36), 165(36), 175(92), *181*, *182*

Aslamazov, L. G., 345(51), *352*
Avenhaus, R., 346, *352*

B

Baber, W. G., 113(42), *117*
Babić, E., *311*
Bachmann, R., 39(118), 40(118), 44(118), *54*
Bader, S., 48(138), *55*
Baltensperger, W., 343, 346, 350, *352*
Barham, D. C., 165(65), *181*
Basters, F. B., *52*
Bates, L. F., 313(88), *325*
Bauer, G., 14(23a), *52*
Baym, G., 211(41), *215*
Béal-Monod, M. T., 91(3b), 96(3b), 100(14), 101, 103(19), 104, 105, 107, 108, 109, 110, 114(14), *116*, 136(72), 140(86a), 142(93, 94), 143(93), *146*, *147*, 237(2), *252*
Beaudry, B. J., 294(19), *323*
Beck, P. A., 339(36), *351*
Behringer, R. E., 168, *182*
Benda, R., *293*, 335, *351*
Bennemann, K. H., 135(65), *146*, 293(15), 317,*323*,*325*,332,334(24, 25, 26, 27), *351*
Benoit, H., 157, *181*
Bensel, J., 130(54), *146*
Berk, N. F., 30(71), *53*,91(3a), 96(3a), *116*
Bernier, P., 168(74), 173(74), 177(106, 107), 179(106, 107), *182*
Betbeder-Matibet, O., 291(8), 294(8), 312(8), *323*
Bijvoet, J., 44(129), *55*
Birgeneau, R. G., 45(137), *55*

Bjerkaas, A. W., *308*
Blandin, A., 58(3a,b), 61(3a,b), 64, 69(9, 10), 70(9), 71(3a,b), 73(3a,b, 10), 75(10), 76(10), 77(9, 10, 17b,d), 78(17b,d), *87*, *88*, 111(37), *117*, 123(4), 127(24), 131(4), *145*, 155(21), 156(25), 159(37), *180*, *181*, 309, 310(81), 311, 313, *324*, *325*,
Blandin, B., 5, 31, 48(6), *52*, *53*
Blaugher, R. D., 35(87), *54*
Blokhim, S. M., 44(128), 46(128), *55*
Bloch, C., 211(41), *215*
Blood, P., 36(93), *54*
Bloomfield, P. E., 123(8), 126(8), 128(38), 130(8), 132(8), 136(38), 137(8), *145*, 192(17), 206(27), *214*
Blum, N. A., 128(39), *145*, 161(47, 48, 49), 172(48), *181*
Boato, G., 127(22, 25), 131(22, 58), 132 (22), 139(58), 144(25), *145*, *146*, 308, 311(75, 76), 314(95), 322(122), *324*, *325*
Boerstoel, B. M., 137(76), 144(76), *146*, 340(41), *351*
Boes, J., 44, *55*
Bogoliubov, N. N., 105(21), *116*
Bongi, G., *308*, 348(60), *349*, *352*
Booth, J. G., 127(31), *145*
Borelius, G., 138(77), 139(77), 144(77), *146*
Brenig, W., 190(13), 192(13), 200, 208(29), 209(29), 210(35, 36), 211(13), 212, *214*
Brewig, E., 139(78), *146*
Brinkman, W., 91(3a), 96(3a), 110, *116*, *117*
Brock, J. C. F., 144(101), *147*, 170(89), *182*
Brog, K. C., 22, 26(52), *53*, 125(13), 127 (13, 31), 132(13), *145*, 158(34), 159(38), 165(38), 169(38), 176(38), 177(38), *181*
Brout, R., 25(58), *53*, 142(95), *147*, 340, *351*
Brown, M. E., 168(79), *182*
Bucher, E., 39(113, 115, 121), 44(115, 127), 45(137), *54*, *55*, 297, 298(31), *323*, 333, *351*
Buckel, W., 289(1), *323*
Buehler, E., 336, *351*
Bugo, M., 127(22), 131(22), 132(22), *145*
Bunch, M. D., 139(85), *146*
Buschow, K. H. J., 38, 44(110), *54*

C

Cadeville, M. C., 84(33), *88*

Cameron, J. A., 163(53), *181*
Campbell, I. A., 16, 19, *52*, 83(30), 84(33), *88*, 152(13), 163(53, 54, 59), *180*, *181*
Cape, J. A., 35(88), *54*
Capellmann, H., 298, *323*
Caplin, A. D., 11, 21, 32, *52*, 87(40), *88*, 127(22), 131(22), 132(22), *145*, 313, *325*
Cappelletti, R. L., 294(22), *323*
Caroli, B., 77(17a, b, c), 78(17a, b, c), *88*, 99, 111, *116*, 169, 174(90), *182*,
Caskey, G. R., 125(14), *145*
Caywood, L. P., Jr., 139(85), *146*
Chaikin, P. M., 304(63), 305, *324*
Chandrasekhar, B. S., 35(87), *54*, 329, *351*
Chapman, A. C., 168, *182*
Chatterjee, A., 39(117), *54*
Chelkowski, A., *294*
Chen, C. W., 84(32), *88*
Cheng, C. H., 339(36), *351*
Cheung, C. Y., 213(45), *215*
Chikazumi, S., 165(62), *181*
Chock, E. P., 12(22), *52*, 150(5), *180*
Chouteau, G., 35(91), *54*, 111(33, 36), *117*
Christenson, E. L., 139(80), *146*
Chui, R., 27(68), *53*
Clad, R., 127(29), 144(29), *145*
Claus, H., 163, *181*
Clogston, A. M., 9, 30(12), 34(85), 35(12), *52*, *53*, 64(7), 71(7), 73, 80(23), *87*, *88*, 97(8), 112, *116*, 254(3), *285*, 329, *351*
Cohen, R. L., 14(26), 44(26), 45, *52*
Coles, B. R., 11(16), 12(30), 27(20, 67), 34(86), 35(89), 36(95, 96), 37(100, 104), *52*, *53*, *54*, 83(27), *88*, 127(37), 130(52), *145*, *146*, 307(72), 314(72), 315(72), 316(72), 317(72), *324*,
Collins, M. F., 84(31), *88*
Compton, J. P., 16(28), 19(28), *52*, 163(53, 54, 56), *181*
Cooper A. S., 39(121), *54*
Cooper, J. R., 297(32. 33), 317(32), *323*
Coqblin, B., 5, 48(6), *52*, 69(9), 70(9), 71(13), 75(13), 77(9, 19), *87*, *88*, 130(56), *146*, 150(4), 180(4), *180*, 313, 318, 321 (118, 119), *325*
Corenzwit, E., 9(12), 30(12), 34(85), 35(12, 87), 38(107), *52*, *53*, *54*, 64(7), 71(7), *87*, 289(2), 292(2, 10), 293(11), 301(2), 302(2), *323*, 327(1, 2, 3), 328(3, 7), 329(2, 3), 330(2, 3), 332(3), 333(7), *351*

AUTHOR INDEX

Corradi, G., 314, *325*
Costa-Ribeiro, P., 127(36), *145*
Cotignola, J. M., 307(73), *315*(73), 317(73), *324*
Cox, D. E., 45, *55*
Craig, R. S., 44(134), *55*
Creveling, L., Jr., 22(49), 34, *53*, 159(35), 177(35), *181*
Crow, J. E., 293(12), 296(28), *323*, 332, 334(17, 22), *351*
Culbert, H. V., 305, *324*, 377, *382*
Cyrot, M., 334, *351*

D

Daniel, E., 33, 37, *53*
Davidov, D., 12(22), *52*, 150(5), *180*, 294, *295*
Daybell, M. D., 6, 25, 26, *52*, 122(1), 127(1), 128(40), 129(1), 130(49), 131(40, 59), 133(40, 59, 63), 135(63), 136(40), 139(84), 140(1), 143(63, 98, 99), 144(1, 40, 49, 59, 98), *144, 145, 146, 147*
Decker, W. R., 294(20, 21), 295, 311(21), 312, *323*
de Dominicis, C., 86(35),*88*, 220, 221, 224(4), *236*, 238(4), 246(4), *252*
de Fajet de Calteljau, P., 58(2) 60(2), 61(2), 62(2), 63(2), 69(2), 83(2), *87*
deGennes, P. G., 25(57), *53*, 77, *88*, 157, *181*, 291, 296, *323*, 334, *351*
DeLong, L. E., 303(62), 304(62), *324*
Deltour, R., 127(29), 144(29), *145*
Dempesey, C. W., 311(86), 312(86), 313(86), *325*
De Nobel, J., 19(37), *52*
De Vroede, E., *52*
Dick, G. J., 294, 306(24), *323*
Dietrich, M., *322*
Dimnock, J. O., 111(34), *117*
Domenicali, C. A., 139(86), *147*
Doniach, S., 9, 30(72), 36, 37, *52, 53*, 91 (3a), 96(3a), 110(31), 114, *116, 117*, 210(39), *214*
Donzé, P., 30(75), *53*, 328(6), 337(6), 338(6), 339(6), 342(6), *351*
Dreyfus, B., 22(44), *53*
Duchatenier, F. J., 19(37), *52*
Dunlap, B. D., 45(136), *55*

Durand, J., 84(33), *88*
Dworin, L., 99, 112, *116*, 165(63), 174(63), 176(95, 96), 177, *181, 182*
Dyson, F. J., 27(65), *53*, 105(23), *117*, 232, *236*
Dzyaloshinski, I. E., 188(6), *214*, 238(11), *252*

E

Edelstein, A. S., 130(57), 135(65), *146*, 301, 305(66, 67, 69), 306, *324*, 377, *381*
Eguchi, H., 38(109), 44, *54*, 127(30), *145*, 300(43, 44), 301(44, 54), 305(54), *324*
Ehrenreich, H., 23(55), 48(55), *53*
Eibschutz, M., 14(26), 44(26), 45(26), *52*
Ekström, H. E., 170(88), *182*
Engelsberg, S., 30(72), *53*, 91(3a), 96(3a), 110(31), *116, 117*
Ericsson, T., 165(66, 67), *181, 182*
Evenson, W. E., 26(64), *53*, 87(42), *88*, 103(18), 108(18, 26), *116, 117*, 238(6, 7), 238(6), 243(7), *252*
Everett, G. E., 40(123), 44(123), *55*
Everts, H. U., 155(20), *180*, 209(33), *214*

F

Falge, R. L., Jr. 293(11), *323*, 328(7), 333(7), *351*
Falk, D. S., 192(16), *214*
Farrell, T., 84(33), *88*
Ferrell, R. A., 344, *352*
Fert, A., 84(33), *88*
Fertig, W. A., 303(62), 304(62), *307, 324*
Feynman, R. P., 220, 225, *236*
Finnemore, D. K., 294(19, 20, 21, 22), 295, 301(45), 311(21, 87), 312, *317, 323, 324, 325*, 333, *351*
Fischer, K., 22, 37, *53*, 54, 126(19), 129(47), 132(19, 47), 139(47), 144(47), *145, 146*, 187(5), 198(5), 205(5), *214*
Fischer, Ø. *308*, 328(6), 337(6), 338(6), 339(6), 342(6), 346(56, 47, 58) 347(43), 348(43, 60), *349*(43), *351, 352*
Fisk, Z., 301(56), 307(71), *324*
Fitzgerald, R., 303(62), 304(62), *324*

Flouquet, J. 16, *52*, 151(7), 163(7, 55, 60, 61), 164(7), *180*, *181*, 301(48), *324*
Flükiger, R., *349*
Flynn, C. P., 155(23), 167, 168, *180*, *182*
Ford, P. J., 31(77), 32, *53*, 127(35), 132 (61, 62), 133(62), *145*, *146*, *311*
Fourneaux, R., 35(91), *54*, 111(33, 36), *117*
Fowler, M., 192(16), *214*, 233, 235, *236*, 354(6), *381*
Fradin, F. Y., 177(108), *182*
Franceschi, E., 43, 44(125), *55*
Frankel, R. B., 26(62), *53*
Frankel, R. B., 128(39), *145*, 161(48, 49), 172(48), *181*
Fredkin, D., 91(3b), 96(3b), 110(3b), *116*
Freeman, A. J., 111(34), *117*, 152(12), 153(12, 14), 155(15), 161(47), *180*, *181*
Frei, C., 348(60), *352*
Friedel, J., 5, 31, 35, *51*, *53*, 58(2, 3a), 60(2), 61(2, 3a) 62(2), 63(2), 64, 69(2), 71(3a), 73(3a), 81(25), 82(25), 83, 84(25), 85(25), *87*, *88*, 91(4a), 99(4a), 100(4a), 102(4a), 108(4a), *116*, 123(4), 131(4), *145*, 155, *180*, 300(41), 308, *324*
Friederich, A., 140(39), *147*
Fulde, P., 295, 296(25), 297, 298(30, 34, 35), 305, *323*, 331, 344, 347, *351*, *352*
Furdyna, A. M., 111(34), *117*

G

Gaidukov, Iu. P., 132(60), *146*
Gainon, D., 135(67), 139(81), *146*
Galleani d'Agliano, E., 314, *325*
Gallinaro, G., 308(75, 76), 311(75, 76), *324*
Gambino, R. J., 45(136), *55*
Ganguly, B. N., 155(20), *180*, 209(*33*), *214*
Gardner, J. A., 130(54), *146*, 155(23), 168(23), *180*
Gardner, W. E., 39(120), 40(120), *54*
Garland, J. W., 135(65), *146*, 334(25, 27), *351*
Gautier, F., 81(24), 82(24), 84(33), *88*
Gavoret, J., 238 (3), *252*
Geballe, T. H., 39(118), 40(118), 44(118), *54*
Gehman, B. L., 27(68), *53*
Gerstenberg, D., 9, 30(11), *52*

Gey, W., 301(47), 320(47), *322*, *324*
Gillespie, D. J., 111(35), 113(35), *117*
Ginsberg, D. M., *295*, *308*
Giovannini, B., 129(45), 135(64), *146*, 156, 157(26, 31), 179(26), *180*, *181*, 346(56) *352*
Gladstone, G., 156(26), 157(26), 169(86), 177(103), 179(26), *180*, *182*
Gobrecht, K., 111(33), *117*
Götze, W., 190(13), 192(13), 200, 208(29), 209(29), 210(35, 36, 38), 211(13), 212, *214*
Golibersuch, D. C., 128(44), *145*, 169(87), *182*
Gomès, A. A., 83(30), *88*
Gonzalez, J. A., 206(28), 208(29), 209(29), *214*
Gor'kov, L. P., 188(6), *214*, 238(11), *252*, 289, 291, *323*, 328, 329, *351*, 354, 355, 360, 373, *381*
Gorter, F. W., 139(83), *146*
Gossard, A. C., 21, 34, 39(121), *53*, 128(43), *145*, 155(24), 159(24), 169(24), 172(24), 173(24), 176(24), *180*
Grassic, A. D. C., 32(78), 33(83), 35(83), 37(83), *53*, 128(41), 130(41), *145*
Greedan, J. E., 43(126), 44(126, 134), *55*
Greig, D., 36(93), *54*, 84(33), *88*
Griffin, A., 294(23), *323*, 354, *381*
Griffiths, D., 12(20), 27(20, 67), *52*, *53*, 130(52), *146*
Grodzins, L., 161(47), *181*
Groff, R. P., 296(27), 297(27), *323*
Gruenberg, L. W.. 345(50), 346, *352*
Grüner, G., 32, *53*, 156(27), *181*
Gschneidner, K. A., 39(111), *54*
Gubbens, P. C. M., 22(50), 33, *53*, 130(50), *146*, 165(70), 177(70), *182*
Guénault, A. M., 139(79), *146*
Guertin, R. P., 293(17), 296(27, 28), 297, *323*, 332(17), 333, 334(17), *351*
Gunther, L., 345(50), *352*

H

Hake, R. R., 332(16), *351*
Hamann, D. R., 86(36), 87(41, 42), *88*, 102, 103, 107, 108, 109, *116*, *117*, 123(8),

125(10), 126(8, 21), 130(8), 132(8), 137(8), *145*, 161(42), *181*, 192(15, 17), 193(15), *214*, 230(17), 231(17), 232(17), *236*, 237(2), 238(5), 240(5), 241(5), 246 (5), 250(17), *252*, 265(27), 274(32), 278 (27, 32), 282, *285*, *286*, 358, 359, *381*
Hanabusa, M., 177(100, 101), *182*
Hargitai, C., 314, *325*
Harris, I. R., 39(112, 120), 40(120), *54*, 313(89), *325*
Harrison, R. J., 136(71), *146*
Hasegawa, H., 27(66), *53*
Hauser, J. J., 125(17), *145*, *311*
Hechler, K., 345(53), *352*
Hecht, R., 128(38), 136(38), *145*, 206(27), *214*
Hedgcock, F. T., 21, *53*, 131(58), 139(58), *146*, 314, *325*
Heeger, A. J., 6, 30, *52*, *53*, 128(44), 135(67), 140(88), 144(88), *145*, *146*, *147*, 156(26), 157(26, 31), 161(41), 169(41, 86, 87), 176(41), 177(99), 179(26), *180*, *181*, *182*, 257, *285*, 301(53), 308, *324*
Hein, R. A., 293(11), *323*, 328(7), 333, *351*
Heinrichs, J., 355, *381*
Helfand, E., 332(15), 345(15), 346(55), 348(15), *351*, *352*
Herring, C., 289, *323*
Hillenbrand, B., 327(5), 328(5, 6), 332(5), 336, 337(6, 31), 338(6, 35), 339(6), 341(42), 342(6), *349*, *351*
Hilsch, R., 289(1), *323*
Hirschkoff, E. C., 11, 18(17), 19(17), 20(17), 28, 32, *52*, 127(23), 130(23), 143(23, 97), 144(23), *145*, *147*
Hirst, L. L., 6, 12(20, 21), 15(27), 27(20), 49, 50(7), 51(7), *52*, 130(52), *146*, 151(8), 153, 180(111), *180*, *183*, 297(30), 298(30), *323*
Hirvonen, M. T., 165(66), *181*
Ho, J. C., 144(101), *147*, 170(89), *182*
Hoare, F. E., 340, *351*
Hoenig, H. E., 297, 298(34), *323*
Hohenberg, P. C., 332(15), 345(15), 348 (15), *351*
Holden, T. M., 25(61), *53*
Holliday, R. J., 165(64), *181*
Hopkins, D. C., 333(19), *351*
Hubbard, J., 93, *116*, 238, *252*
Huber, J. G., 11(16), 39(116), 40(116), 42(116), 44(116), *54*, 127(37), *145*, 303, 307(72, 73), 311(84, 85), 312, 313(85), 314(72, 99), 315(72, 73, 99), 316(72), 317(72, 73, 99), 322(123), *324*, *325*
Huiskamp, W. J., 163(57), *181*
Hull, G. W., 39(118), 40(118), 44(118), *54*, 297(31), 298(31), *323*
Hulm, J. K., 35(87), *54*
Huntley, D. J., 140(87), *147*
Hurault, J. P., 103(19), 104(19), 105(19), 107(19), 108(19), 109(19), *116*
Hurd, C. M., 130(53), 136(73), *146*
Hutchens, R. D., 43(126), 44(126), 134), *55*

I

Iandelli, A., *42*, 43, *55*
Ishii, H., 274(33, 35), 278(39), 281(39), 282, *286*
Iwahashi, K., 293(14), *323*
Izuyama, T., 344, *352*

J

Jaccarino, V., 22, 26, *53*, 158, 159(32), 177(97), *181*, *182*, 318(111), *325*, 346, *352*
Janossy, A., 32, *53*
Jayaraman, A., 39(113, 115, 117), 44(115, 127), *54*, *55*
Jensen, M. A., 30(70), *53*
Jensen, M. A., 169(86), 117(99), *182*, 257, *285*, 291(9), *323*, 328(10), *351*
Jerome, D., 44(133), *55*
Johansson, C. H., 138(77), 139(77), 144(77), *146*
Jones, H., *308*, 348(60), *349*, *352*
Jones, W. H., Jr., 22, 26(52), *53*, 125(13), 127(13, 31), 132(13), *145*, 158(34), 159(38), 165(38), 169(38), 176(38), 177(38), *181*
Johnson, D. L., 294(19), *323*
Junod, A., 328(6), 337(6), 338(6), 339(6), 342(6), *351*

K

Kaiser, A. B., 9, 36, 37, *52*, 114, *117*, 310, 311, 314, 315, *324*, *325*, 354, *381*

Kalvius, G. M., 45(136), *55*
Kanamori, J., 83(29), *88*
Karlsson, A., 24(56), *53*
Kasuya, T., 58(1), *87*, 154, *180*
Katila, T. E., 165(66, 67), *181*, *182*
Kato, A., 277(36), *286*
Kawamura, K., 358, *381*
Kazama, S., 176(94), *182*
Keesom, W. H., 138(77), 139(77), 144(77), *146*
Keiter, H., 104, *116*, 209(34), 210(40), 212(40, 43), *214*, *215*, 240(12), 241(12), *252*, 355(25), *381*
Keller, J., *293*, 298, *323*, 335, *351*
Kierspe, W., 139(78), *146*
Kim, D. J., 128(39), *145*, 161(48, 49), 172(48), *181*
Kim, K. S., 30(74), 38(74), 45(74), *53*, 71(13), 75(13), *87*, 301(46), 305(65), 311(84), 318(65, 116), 319(116), 320(46, 65, 116), *324*, *325*, 377, *381*
King, E., 39(112), *54*
Kirvonen, M. T., 165(67), *182*
Kitamura, T., 355, *381*
Kitchens, T. A., 19, 25(36), 35(36), *52*, 128(42), *145*, 161(45, 46), 172(46), *181*
Kittel, C., 58(1), *87*, 90(1), *116*, 154(16, 17), 168(79), *180*, *182*, 280(40), *286*
Kjekshus, A., 19, *52*
Kjøllerstrøm, B., 242(13), *252*
Klein, A. P., 30, *53*
Klein, M. W., 136(71), 142(95), *146*, *147*, 340, *351*
Klose, W., 344, *352*
Knapp, G. S., 36, *54*, 125(13, 15), 127(13), 132(13), *145*
Knauer, R. C., 160(39), *181*
Knight, W. D., 168(79), *182*
Kogure, O., 127(33), 139(33), *145*
Kohlstedt, D. L., 139(84), *146*
Koide, S., 153(14), 155(14), *180*
Kondo, J., 17, 37, *52*, *54*, 59(6), 76, *87*, 123, 126(20, 21), 131(3), 143(96), 144(104), *145*, *147*, 150(3), *180*, 185, 190(14), 198(1, 21), *214*, 217, 219(1), *235*, 237(1), *252*, 253, 254(4), 256(11), 257, 260(22), 264(25), *285*, 299, *323*
Korn, D., 18(34), *52*
Korringa, J., 157(29), *181*
Koster, G. F., 79, *88*

Kovàcs-Csetényi, E., 156(27), *181*
Kreuzer, H., 305(64), *324*
Kroo, N., 14(23, 23b), *52*
Kubo, R., 152(11), *180*
Kume, K., 32, *53*, 127(32, 33), 130(32), 132(32), 139(33), *145*, 176(94), 177(98), *182*
Kurata, Y., 282, *286*
Kuroda, Y., 237(2), *252*, 355, *381*
Kushida, T., 177(101), *182*
Kuwasawa, Y., 293(13), *323*
Kveselava, D. A., 224(9), *236*

L

Laborde, O., 37, *54*
Lagendijk, E., 163(57), *181*
Lang, N. D., 23(55), 48(55), *53*
Launois, H., 17, 21, *52*, 110(30), *117*, 168(73, 74), 173(74), 175(73), 177(107), 179(107), *182*, 314(93, 94), *325*
Lawson, A. C., 11(16), *52*, 127(37), *145*, 307(72), 314(72), 315(72), 316(72), 317 (72), *324*
Lawson, A. W., 38, *54*
Lecoanet, B., 22(48), *53*
Lederer, P., 9, 30(73), 36, *52*, *53*, 76(16a), 83, 87(16a), *88*, 92(5, 6), 95(5), 97(5, 6, 9), 99(5, 6, 9, 10, 12), 109(5, 28), 110, 111(12, 36), 112, 113(6, 41), 115(10, 41), *116*, *117*, 125(12), *145*, 174(90), *182*, 237(2), *252*, 317(105), *325*
Lee, J. A., 39(112), *54*
Lee, K. N., 39(118), 40(118), 44(118), *54*
Leonard, P., 84(33), *88*
Levine, M., 26(63), *53*, 87(41), *88*, 102, 103(16b, c), 106(16b, c), 107(16b, c), *116*, 237(2), *252*, 316(102), *325*
Levine, R., 177(105), *182*
Li, P. L., 21, *53*, 314, *325*
Linde, J. O., 138(77), 139(77), 144(77), *146*
Lines, R. A. G., 163(53), *181*
Liu, S. H., 340, *351*
Lock, J. M., 38(106), *54*
Loram, J. W., 32, 33, 35(83, 92), 37(83, 99), *53*, *54*, 86(39), *88*, 127(35), 128(41), 130(41), 132(61, 62), 133(62), *145*, *146*
Low, C. G. E., 84(31), *88*

Low, G. G., 14, 25(61), *52*, *53*
Lowe, I. J., 156(28), 178(28), *181*
Ludwig, A., 302(59), 307, *324*, 378, *382*
Luengo, C. A., *294*, *307*(73), *315*(73), 317 (73), *324*
Lumpkin, O. J., 177(104), *182*
Luo, H. L., 22(49), 34, *53*, 159(35), 177(35), *181*
Luther, A., 297(30), 298(30), *323*

M

Ma, S. K., 91(3b), 96(3b), 110(3b), *116*
MacDonald, D. K. C., 127(34), 139(34), *145*
McHenry, M. R., 178, 179, *183*
Mackliet, C. A., 111(32, 35), 113(35), *117*
MacLaughlin, D. E., 317, *325*
McMillan, W. L., 316, *325*
MacPherson, M. R., 40(123), 44(123), *55*
McWhan, D., 44(127), 45(135), 48(138), *55*
Mahan, G. D., 219(3), *235*, 238(3), *252*
Maines, R. G., 39(113, 115), 44(115), *54*
Maita, J. P., 39(121), *54*, 297(31), 298(31), *323*
Maki, K., 103(19), 104(19), 105(19, 22a, b), 107(19), 108(19), 109(19), *116*, *117*, 136(68), *146*, 295, 296(25), 305, *323*, 331, 347, *351*, 354(6), *381*
Maleev, S. V., 189, 190(10), *214*
Maletta, H., 9, *52*, 163, *181*
Maley, M. P., 127(27), *145*, 161, 162, 163, *181*
Malm, H. L., 139(82), *146*
Mamiya, T., 293, *323*
Maple, M. B., 4(1), 11, 21(1), 30(74), 38(1, 74), 39(114, 116, 119), 40(114, 116, 119, 123), 42(116), 44(116, 123, 131), *51*, *52*, *53*, *54*, *55*, 71(13), 75(13), *87*, 127(37), *145*, 293, *294*(16), 301(46, 56, 57), 302(57), 303(61, 62), 304(62), 305(65), *307*(72, 73), 311(84, 85), 312, 313(85), 314(72, 99), *315*(72, 73, 99), 316(72), 317(72, 73, 99), 318(65, 114, 116), 319, 320(46, 65, 116), 321(118), *322*(121, 123), *323*, *324*, *325*, 377, *381*
Maranzana, F. E., 44(132), *55*
Marsh, J. D., 163(58, 61), *181*
Marshall, W., 142(95), *147*, 340, *351*
Martin, D. L., 317(108), *325*

Masker, W. E., 296(27), 297(27), *323*
Masuda, Y., 293(14), *307*, *308*, *323*
Matayoshi, S., 354, *381*
Matho, K., 142(94), *147*
Matsuura, T., 354, 372(7), *381*
Matthias, B. T., 9(12), 30(12), 34(85), 35(12, 87), 38(107), *52*, *53*, *54*, 64(7), 71(7), *87*, 289(2, 4), 292(2, 10), 293(11), 301(2), 302(2), 307(71), *323*, *324*, 327(1, 2, 3), 328(3, 7), 329, 330(2, 3), 332(3), 333(7), *351*
Matthias, E., 16(29), *52*
Mattis, D. C., 70(11), *87*, 100(15), *116*, 202, *214*, 230, 234, *236*, 310(83), *325*
Mattuck, R. D., 213, *215*
Meier, R., 305(64), *324*
Migdal, A. A., 213(46), *215*, 235(23), *236*
Mihalisin, T. W., 296(27), 297(27), 304(63), 305, *323*, *324*
Millet, W. E., 161(45), *181*
Mills, D. L., 9, 30(73), 36, *52*, *53*, 76(16a), 83, 87(16a), *88*, 92(5, 6), 95(5), 97(5, 6, 9), 99(5, 6, 9, 10), 100(14), 101, 104, 108, 109, 110, 112, 113(6, 41, 44), 114(14, 44), 115(10, 41, 44), *116*, *117*, 125(12), *145*, 237(2), *252*, 317(105), *325*
Miwa, H., 255(7), 264(26), *285*
Mizuno, K., 168, 176(94), *182*
Mössbauer, R. L., 9, *52*, 151(8), *180*
Monod, P., 136(70), 140(39), *146*, *147*
Moore, R., 206(26), *214*
Morandi, G., 317, *325*
Moriya, T., 35, *54*, 69(8), 78(8), *87*, 112, *117*, 151(6), *180*
Mota, A. C., 303(62), 304(62), *324*
Mott, N. F., 84(33), *88*
Mozumder, S., 37(100), *54*
Mrstik, B. J., *295*, *308*
Mueller, F. M., 111(34), *117*, 334(26, 27), *351*
Mühlschlegel, B., 238, *252*
Müller-Hartmann, E., 137(75), *146*, 192(18), 195, 198, 201(18), 202(18), 203(18, 22, 24), 204(24), 208(31), 209(31), *214*, 299(39), 302(58), *307*(58, 74), 320, *323*, *324*, 354, 355(36), 372(17), *374*, *375*, 377, 379, 380, *381*, *382*
Muir, W. B., 131(58), 139(58), *146*
Murani, A., 13(130), 44, *55*
Murata, K. K., 135(66), *146*

Muskhelishvili, N. I., 192(19), 197(19), 214, 224, 236, 247, 252
Myers, H. P., 24(56), 53, 170(88), 182

N

Nagaoka, Y., 126, 132(18), 145, 186, 189, 192, 193(3), 214, 274(31), 286, 354, 358 (13, 14), 372(7), 381
Nagasawa, H., 127(26, 28), 132(28), 139 (26), 145, 158(33), 181, 301(50), 324
Nagle, D. E., 161(45), 181
Nakajima, S., 261, 285, 334(24), 351
Nakamura, Y., 176(94), 182
Nam, S. B., 274(34), 278, 286
Nap, G. M., 52
Narath, A., 17, 21, 32, 34, 52, 53, 99, 111 (38), 112, 116, 117, 128(43), 144(90, 105), 145, 147, 152(9, 10), 155(24), 159(24, 38), 165(38, 63, 65, 68, 69), 166, 167(68, 72), 169(10, 24, 38), 172(24), 173(24, 69), 174(63), 175(91), 176(10, 24, 38), 177, 180, 181, 182
Narayanamuriti, V., 39(113, 115), 44(115), 54
Newmann, M. M., 313(88), 325
Newrock, R. S., 127(25), 144(25), 145
Nickerson, J. C., 39(118), 40(118), 44, 54
Nicolas-Francillon, M., 44(133), 55
Niesen, L., 163(57), 181
Nieuwenhuys, G. J., 19, 21(41), 33(41), 52
Norman, M., 54
Nowik, I., 163(52), 181
Nozières, P., 86(35), 88, 220, 221, 224(4), 235, 236, 238(3, 4), 246(4), 252

O

Oda, Y., 175(92), 182
Ogawa, T., 278(38), 286
Ohtsuka, T., 123(2), 130(2), 131(2), 137(2), 144, 175(93), 182, 309, 311(77, 78), 314(78), 324
Okiji, A., 160(40), 165(62), 181, 255(6), 257(14), 277, 285, 286
Olcesse, G. L., 43, 44(125), 55
Orbach, R., 12(22), 27(68), 52, 53, 150(5), 180, 294, 295
Ortega, S., 322
Ortelli, J., 328(6), 337(6, 34), 338(6), 339(6), 342(6), 351
Ostenson, J. E., 294(19), 323
Owen, J., 168(79), 182

P

Paderno, Yu. B., 44(128), 46(128), 55
Page, D. F., 311(87), 325
Palmer, P. E., 333(19), 351
Parks, R. D., 293(12), 296(26, 27, 28), 297(27), 323, 332(17), 333, 334(17, 22, 23), 347(23), 351
Passell, L., 45(137), 55
Paulson, R., 129(45), 146
Pearson, W. B., 19, 52, 127(34), 139(34), 145
Penfold, J., 39(120), 40(120), 54
Perrier, J., 212, 215
Petalas, P., 350, 352
Peter, M., 9(12), 30(12), 34(85), 35(12), 52, 53, 64(7), 71(7), 87, 328, 337(6, 34), 338, 339, 342(6), 344, 346(54, 56, 57, 58), 351, 352
Peters, J. J., 167(71), 182
Peterson, D. T., 294(20), 311(87), 317, 323, 325
Phillips, J. C., 291, 323
Phillips, N. E., 48(138), 55, 137(74), 138(74), 144(101), 146, 147, 161(44), 170(89), 181, 182
Pincus, P., 156(26), 157(26), 179(26), 180
Polya, G., 223(8), 236
Pomeroy, A. R., 84(33), 88
Potts, J. E., 169(84, 85), 170, 171, 177 (85, 102), 182
Pouget, J. P., 168(74), 173(74), 182
Powell, R. L., 139(85), 146
Pratt, W. P., Jr., 143(98, 99), 144(98), 147

R

Radhakrishna, P., 37, 54
Ramakrishnan, T. V., 102(16c), 103(16c), 106(16c), 107(16c), 116, 237(2), 249, 252
Rao, G. N., 16, 52,
Rao, K. R. P., 163(52), 181
Rao, V. U. S., 43, 44(126, 134), 55

Ratto, C. F., 71(13), 75(13), *87*, 309, 310(81), 311, 314(97, 98), 318, *324*, *325*
Raynor, G. J., 313(89), *325*
Read, M., 139(79), *146*
Reif, F., 294, 305, 306(18, 24), *323*
Reivari, P., 165(66), *181*
Rettori, C., *294*, *295*
Riblet, G., 303, *307*, *308*, 316, *324*, *325*
Rice, M. J., 113(43, 45), *117*
Ricks, B., *295*
Rigney, D. A., 155(23), 168(23), *180*
Rivier, N., 9, 29(13), *52*, 76(16b), 87(16b), *88*, 99(11), 110, 115, *116*, 125(9), 130, *145*, 237(2), *252*, 317, *325*
Rizzuto, C., 11, 21, 31(77), 32(18, 77, 139), *52*, *53*, *55*, 87(40), *88*, 127(22), 131(22), 132(22), *145*, 308(75, 76), *311*(75, 76), 313, *322*(122), *324*, *325*
Ron, A., 135(65), *146*
Ross, M. H., 189(11), *214*
Roth, M., *322*
Roulet, B., 238(3), *252*
Rowlands, J. A., 36, *54*
Ruderman, M. A., 58(1), *87*, 154, *180*, 280(40), *286*
Rump, R. B., 311(87), *325*
Rusby, R., 37(100, 104), *54*
Rusinov, A. I., 328, 329, *351*, 361, 372, *381*

S

Saint-James, D., 99(12), 111(12), *116*, 174(90), *182*
Sakai, N., 158(33), *181*
Sakurai, A., 371, *381*
Salamoni, E., 31(77), 32(77), *53*, *311*
Sales, B., 39(119), 40(119), *54*
Sanctuary, C. J., 163(59), *181*
Sarachik, M. P., 36(97), *54*, 125(15), *145*
Sarma, G., 291, 296, *323*, 334, 344, *351*, *352*
Saur, E., 345(53), *352*
Savage, W. R., 125(16), *145*
Scalapino, D. J., 129(46), *146*
Schindler, A. I., 34(86), 36, *54*, 111(32, 35), 113(35, 43, 44), 114(44), 115(44), *117*
Schneider, J., 305(64), *324*
Schotte, K. D., 125(11), 130(48), *145*, *146*, 161(43), *181*, 187(4), 189(4), 190(4), 193(4), *214*, 225(11), 228, *236*, 354(12), 358, *381*
Schotte, U., 130(48), 139(78), *146*, 161(43), *181*, 225(11), 228(14), *236*
Schrieffer, J. R., 25(59, 60), 26(64), 30(71), *53*, 59(5), 70(11, 12), 72(13), 74, 75(12), 76(12), 77, 87, *87*, *88*, 91(3a), 96(3a), 100(15), 103(18), 107, 108(18, 26, 27), 109(27), *116*, *117*, 123(6), 129(45), 130(56), 131(6), 136, *145*, *146*, 150(2, 4), *180*(4), 238(6, 7), 238(6), 243(7), 245(14), 250(14), *252*, 254(5), *285*, 300(40), 310(83), *324*, *325*
Schriempf, T. J., 113(44), 114(44), 115(44), *117*
Schuster, K., *349*
Schwartz, B. B., 26(62), *53*, 128(39), *145*, 161(48, 49), 172(48), *181*, 346, *352*
Schwartz, G. P., 144(101), *147*, 170(89), *182*
Schweitzer, J. W., 125(16), *145*
Seidel, E. R., 15(27), *52*, 151(8), *180*
Seitz, E., 14(23a), *52*
Sehkizawa, K., 293(13), *323*
Sellmyer, D. J., 125(14), *145*
Sereni, J., 307(73), *315*(73), 317(73), *324*
Serin, B., 127(25), 144(25), *145*
Sessler, A. M., 211(41), *214*
Seymour, E. F. W., 168, *182*
Shaltiel, D., 110, *117*, *295*
Shenoy, G. K., 45, *55*
Sherwood, R. C., 9(12), 30(12), 34(85), 35(12), *52*, *53*, 64(7), 71(7), *87*, 301(55), *324*
Shiba, H., 265, 269, 277(36), *286*, 354, 360, 373, *381*
Shirkov, D. V., 105(21), *116*
Shirley, D. A., 16(29), *52*
Shull, C. G., 14, *52*
Sierro, J., 139(81), *146*
Sievert, P. R., 128(38), 136(38), *145*, 206(27), *214*
Silbernagel, B. G., 178(109), 179(109, 110), *183*
Silhouette, D., 157, 168(75), *181*, *182*
Singh, A. K., 39(117), *54*
Skalski, S., 291, 294(8), 312, *323*
Slater, J. C., 79, *88*
Smith, F. W., 306(70), 317(107), *324*, *325*, 377, *382*

Smith, T. F., 38(108), 39(112, 120), 40(120), *54*, 71(13), 75(13), *87*, 318, *325*
Soda, T., 355, *381*
Sólyom, J., 375, *381*
Souletie, J., 19, 22(44, 45, 46), 33(40, 45), 37, *52*, *53*, 78(20), 86(39), *88*, 127(36), 141(92), *145*, *147*
Spedding, F. H., 294(19), *323*
Spencer, H. J., 210(39), *214*
Star, W. M., 12, 19, 21(41), 33(19, 41), *52*, 137(76), 144(76), *146*
Stassis, C., 14, *52*
Steglich, F., *307*
Steyert, W. A., 6, 19(36), 25(8, 36), 26, 35(36), *52*, 122(1), 127(1), 128(40, 42), 129(1), 130(49), 131(40, 59), 133(40, 59), 136(40), 139(84), 140(1), 143(98, 99), *144*(1, 40, 49, 59, 98), *145*, *146*, *147*, 161(45, 46), 172(46), *181*
Stoner, E. C., 91, *116*
Strässler, S., 343, 346, *352*
Stratonovich, R. L., 238, 241(8), *252*
Sugawara, T., 38(109), 44, *54*, 127(26, 30), 130(55), 132(55), 139(36), *145*, *146*, 168(77), *182*, 300(43, 44), 301(44, 49, 50, 54), 305(54), 320(49), *324*
Suhl, H., 5, 25(58), 26(63), 30(70), 38(107), *52*, *53*, *54*, 76(16b), 86, 87(16b, 41), *88*, 102, 103(16a, b), 106, 107, *116*, 123(7), 126(7), 132(7), 133(7), 139(7), 140(91), 143(100), *145*, *147*, 186, 189, 190(7, 8, 9), 192(7, 8), 206(26), *214*, 237(2), *252*, 274(30), *286*, 289(2, 4), 291(9), 292(2, 10), 301(2), 302(2), 316(101, 102), 321, 322(117), *323*, *325*, 327(1, 2, 3, 4), 328(3, 10), 329(2, 3), 330(2, 3), 332, *351*, 354(3), *381*
Sullivan, S., 110(30), *117*
Sunjie, M., 99(11), *116*
Susz, C., 337(34), *351*
Swallow, G. A., 32(78), 35(92), *53*, *54*, 128(41), 130(41), *145*
Swartzendruber, L. J., 158, *181*
Sweedler, A. R., 307(73), *315*(73), 317(73), *324*
Symko, O. G., 11(17), 18(17), 19(17), 20(17), 28(17), 32(17), *52*, 127(23), 130(23), 143(23), 97), 144(23), *145*, *147*

Szegö, G., 223(8), *236*
Szentirmay, Z., 14(23b), *52*

T

Tagayama, H., 105(22b), *117*
Takada, S., 344, 345, *352*
Takanaka, K., 310(80), *324*
Takano, F., 278(38), *286*, 310(80), *324*, 354, *381*
Tang, T. Y., 38, *54*
Tao, L. J., 12(22), *52*, 150(5), *180*, *295*
Tari, A., 36(95), *54*
Taylor, R. D., 19(36), 25(36), 35(36), *52*, 127(27), 128(42), *145*, 161(45, 46, 50), 162, 163, 172(46), *181*
Templeton, I. M., 127(34), 139(34), *145*
Theumann, A., 322(120), *325*
Tholence, J. L., 22(44, 47), *53*, 130(51), *146*, 169(83), 170(83), *182*
Thorpe, A., 110(30), *117*
Thouless, D., 232, *236*
Thoulouze, D., 127(36), *145*
Ting, C. S., 209(32), *214*, 229, *236*
Tomonaga, S., 225, *236*
Tompa, K., 32, *53*, 156(27), *181*
Toulouse, G., 71(13), 75(13), *87*, 321 (118, 119), *325*
Tournier, R., 22(44, 45, 46, 47, 48), 33(45), 35(91), *53*, *54*, 78(20), *88*, 111(33, 36, 37), *117*, 127(24), 130(51), 141(92), *145*, *146*, *147*, 159(37), 169(83), 170(83), *181*, *182*
Toxen, A. M., 45(136), *55*
Treyvaud, A., 30(75), *53*, *308*, 328(6), 337(6), 338(6), 339(6), 342(6), 348(60), *349*, *351*
Triplett, B. B., 137(74), 138(74), *146*, 161(44), *181*
Tse, D., 156(28), 178(28), *181*
Tsuchida, T., 39(122), *54*
Tsuda, N., 305(68), *324*, 377, *382*
Tsuneto, T., 355, *381*
Tsuzuki, T., 355, *381*
Typpi, V. K., 165(67), *182*

U

Umlauf, E., 301(47), 305(64), 320(47), *322*, *324*
Usui, N., 293(13), *323*

AUTHOR INDEX

V

Vainshtein, E. E., 44(128), 46, *55*
Van Baarle, C., *52*, 139(83), *146*
Van Daal, H. J., 38, 44(110), *54*
Van Dam, A. J., 44(129), *55*
Van Dam, J. E., 6, 22(50), 33, *52*, *53*, 130 (50), *146*, 165(70), 177(70), *182*
van den Berg, G. J., 6, 33, *52*, *53*, 144, *147*
Van Rongen, A., 144(102), *147*
Van Vleck, J. H., 23(54), 29(54), 48(54), *53*
Vassel, C. R., 156(27), *181*
Vig, J., 127(25), 131(58), 139(58), 144(25), *145*, *146*, 314(95), *325*
von Minnigerode, G., *307*

W

Wada, S., 159(36), 160(36), 165(36), *181*
Wagner, D., 139(78), *146*
Walker, C. W. E., 140(87), *147*
Walker, E., 328(6), 337(6, 34), 338(6), 339(6), 342(6), *351*
Walker, L. R., 22, 26, *53*, 158, 159(32), *181*, 318(111), *325*
Wallace, W. E., 39(122), 44(134), *54*, *55*
Walldén, L., 24(56), *53*
Walstedt, R. E., 167(72), 168(76), 177(97), *182*
Wang, S. Q., 26(64), *53*, 87(42), *88*, 103, 108 (18, 26), *116*, *117*, 238(6, 7), 238(6), 243(7), *252*
Watson, G. N., 267(29), *286*
Watson, H. L., *317*
Watson, R. E., 152(12), 153(12, 14), 155(14), *180*
Weaver, H. T., 175(91), *182*
Wei, C. T., 339, *351*
Weiner, R. A., 102(16c), 103(16c), 106(16c), 107(16c), *116*, *117*, 136(72), 140(86a), *146*, *147*, 237(2), *252*
Weiss, P. R., 291(8), 294(8), 312(8), *323*
Welsh, L. B., 169(84, 85, 86), 170, 171, 177(85, 99, 102), *182*
Wernick, J. H., 110(29), *117*, 168(76), 177(97), 178(109), 179(109), *182*, *183*, 301(55), *324*

Wert, C. A., 167(71), *182*
Werthamer, N. R., 332(15), 345(15), 346(55), 348, *351*, *352*
West, K. W., 14(26), 44(26), 45(26), *52*
Weyhmann, W., 165(64), *181*
Whall, T. E., 127(35), 132(61, 62), 133(62), *145*, *146*
Wheatley, J. C., 11(17), 18(17), 19(17), 20(17), 28(17), 32(17), *52*, 127(23), 130(23), 143(23, 97), 144(23), *145*, *147*
White, J. A., 301(55), *324*
White, R. J., 33(83), 35(83), 37(83), *53*
White, R. M., 39(118), 40(118), 44(118), *54*
Whittaker, E. T., 267(29), *286*
Wilhelm, M., 327(5), 328(5, 6), 332(5), 336, 337(6, 31), 338(6, 35), 339(6), 342(6), *349*, *351*
Willens, R. H., 336, *351*
Williams, G., 12(20, 21), 27(20), 35(92), *52*, *54*, 130(52), *146*
Williams, H. J., 9(12), 30(12), 34(85), 35(12), *52*, *53*, 64(7), 71(7), *87*, 110(29), *117*, 301(55), *324*
Williams, I. R., 16(28), 19(28), *52*, 163(54, 56, 59), *181*
Wilson, G. V. H., 16(28), 19(28), *52*, 163(53, 54, 56, 59), *181*
Winzer, K., 303, *307*, *308*, *324*
Wittig, J., 30(74), 38(74), 44, 45(74), *53*, *55*, 71(13), 75(13), *87*, 318(116), 319(116), 320(116), 322, *325*
Wölfe, P., 208(29), 209(29), 210(35, 36, 37, 38), *214*
Wohlleben, D., 18(35), 19(38), 30(35), 39(114, 116, 119), 40(114, 116, 119, 123), 42(116), 44(116, 123), *52*, *54*, *55*, 303(62), 304(62), *324*
Wolff, P. A., 5, 25, *52*, *53*, 59(5), 74, 82(26), *87*, *88*, 91(4c), 93, 99(4c), 100(4c), 102(4c), 108(4c), *116*, 150(2), *180*, 245(14), 250(14), *252*, 254(5), *285*, 300(40), *324*, 354, *381*
Wollan, J. J., 301(45), *324*
Wong, D., 123(7), 126(7), 132(7), 133(7), 139(7), *145*, 189(8), 190(8), 192(8), *214*, 274(30), *286*, 321, 322(117), *325*
Woo, J. W. F., 274(34), 278, *286*
Woods, S. B., 139(82), *146*
Woolf, M. A., 294, 305, 306(18), *323*

Y

Yamagata, H., 175(92), *182*
Yasukochi, K., 293(13), *323*
Yeo, Y. K., 133(63), 135(63), 143(63), *146*
Yoshida, S., 127(26), 130(55), 132(55), 139(26), *145*, *146*, 301(49, 50), 320(49), *324*
Yoshimori, A., 126(21), *145*, 204, *214*, 257(15, 16, 17), *285*
Yosida, K., 58(1), *87*, 126(21), 136(69), *145*, *146*, 154, 160(40), 165(62), *180*, *181*, 204, *214*, 255(6), 257(12, 13, 16, 17), 264(26), 274(33), 280(41), *285*, *286*
Yuval, G., 86, *88*, 125(10), *145*, 161(42), *181*, 221(6), 222(7), 223(7), 224(7), 227(13), 230(17), 231(17), 232(13, 17), *236*, 249(16), 250(16, 17), *252*, 265(27), 278(27), *285*

Z

Zawadowski, A., 233, 235, *236*, 375, *381*
Zener, C., 58(1), *87*
Zimmerman, J. E., 340, *351*
Zittartz, J., 137(75), *146*, 190(12), 191, 192(18), 193(12), 194(20), 195, 201(18, 20), 202(18), 203(18, 24), 204(24), 208(30), 213(30), *214*, 278(37), *286*, 299(39), 302(58), *307*, 320, *323*, *324*, 354, 355, 370(18), 372(17, 18), *374*, *375*, 379, 380, *381*, *382*
Zrudsky, D. R. 125(16), *145*
Zuckermann, M. J., 9, 29(13), *52*, 76(16b), 87(16b), *88*, 99(11), 110, 115, *116*, 125(9), 130, *145*, 237(2), *252*, 299(38), 302(59), *307*, 309, 311, 318, 320, *323*, *324*, *325*, 354, 374, 378, *381*, *382*

Subject Index

A

Abriksov–Gor'kov curve, 334
Abriksov–Gor'kov theory, 354, 360, 380
  experimental verification of, 291–295
Alloys
  dilute, see Dilute alloys
  specific heat changes in, 110–111
  static susceptibility of, 109–110
Anderson Hamiltonian, 64–68, 254
Anderson model, 68–69
Anisotropic exchange interaction, bound state for, 264–269
Annihilation operator, 225
Antiferromagnetic coupling, 73–78
  Kondo anomalies in, 201
Antiferromagnetic exchange, effective, 73–75

B

BCS theory, 327–328
Bethe–Salpeter equation, 102
Born series, 217
Bose gas, Fermi gas transformation to, 225
Bound state
  for anisotropic exchange interaction, 264–269
  in magnetic field, 274–278

C

Charge density, 270–274
Clogson limit, 329
Coexistence, in superconductors, 336–343
Conduction-electron-moment interactions, 10
Conduction-electron polarization, 28
Conduction-electron spin density, 141
Conduction-electron spins, Zeeman energy of, 27
Coulomb potential, 95
Coulomb repulsion, 23
Crystal field split energy levels, 297–298
Curie constant, 208
Curle law susceptibility, 160
Curie temperature, defined, 339
Curie–Weiss law, 7–8, 15, 17, 63 n

D

Decoupling procedure, Kondo problem and, 192–194
Dilute alloys
  hyperfine interactions in, 150–157
  local spin fluctuations and, 109–116
  nuclear magnetic resonance data for, 175–176
  properties of, 121
Dispersion equations, solution of, 190–192
Dispersion theory, Green's function and, 189–192

E

Effective antiferromagnetic exchange, 73–75
Effective exchange Hamiltonian, 74, 76
Electrical conductivity, scattering cross section and, 205
Electronic magnetic moment, hyperfine coupling and, 149, see also Local magnetic moment
Electronic susceptibility, 213
Electron paramagnetic resonance (EPR), 12
Electrons, localized vs. conduction, 73–78
Electron scattering, by spin fluctuations, 113–114, see also Scattering

Entropy, Kondo anomalies and, 201–203
Environment, characteristic temperature and, 22
Equation of motion, Kondo problem and, 192–197
External field, exchange field and, 343–350

F

Fermi energy, 150
  transitional impurities and, 62–63
Fermi gas
  elimination of, 221–224
  transformation of to Bose gas, 225
Fermi level
  conduction band density states and, 74
  state densities at, 90
  transition impurities and, 76
Ferromagnetic matrices, impurities in, 83–85
Ferromagnetic superconductors, recent work on, 327–350, see also Superconductors
Feynman diagrams, 211–212
Feynman's path-integral methods, 219–220
Flipping, defined, 221 n
$4d$ metals, moment formation of Fe in, 25–26
Fractional moment effects, 29
Free energy, specific heat from, 340–342
Friedel–Anderson Hamiltonian, 99
Friedel–Anderson parameters, 24
Friedel model, 69–71
Friedel's sum rule, 59–61, 155
Friedel virtual bound state, 34
Fugacity prefactor, 250
Functional integral method, 103–104
  in magnetic impurity problem, 237–251

G

Goldstone diagram, 212
Green's functions, 67
  adiabatic part of, 247
  Hamiltonian and, 186–189
  many-electron, 221
  one-electron, 222–223
  superconductors and, 354–358
Ground-state energy, 203–204

H

Hall coefficient, 140
Hamann's equation, 359, 361–369, 380
Hamiltonian
  Anderson, see Anderson Hamiltonian
  electron–electron interaction in, 95
  Green's functions and, 186–189
  Kondo problem and, 219
  one-electron, 85
  s–d exchange, 254
Hartree–Fock hyperfine fields, 153
Hartree–Fock method, 66–68, 70–73, 155
  for ferromagnetic matrices, 83–85
  impurity in, 59
  local magnetic moments and, 57–87
  in narrow bands, 82–83
  phase diagrams for, 62, 87
  single impurity and, 91
Hirst model, local magnetic moments and, 49–51
"Hopping path," 248–249
Host hyperfine effects, 154–157
Host nuclear spin-lattice relaxation, 177–179
Host polarization, 167–174
Hund's rule, 8–9, 15, 23–25, 29, 37, 48, 290
Hyperfine interactions
  in dilute alloys, 150–157
  dynamic response studies in, 174–179
  host, 154–157
  impurity and, 151–153
  in local magnetic moments, 14–17
  nuclear magnetic resonance and, 16–17
Hyperfine linewidth studies, 168–170

I

Impurity
  finite concentrations of, 375–380
  Hartree–Fock method and, 59–60
  interactions in, 18–31
  longitudinal and transverse, 157
  low concentration of, 369–375
  magnetic, 63 n
  magnetic–nonmagnetic transitions in, 318–322
  mean field description and, 92–100
  nonmagnetic, 21
  nuclear magnetic resonance at site of, 111–113

SUBJECT INDEX

with one electron or hole in local shell, 30–31
paramagnetic, 289–322
spin fluctuations around, 89–116
spin polarization around, 58, 73
in superconductors, 289–322, 353–381
transitional, 58, 61–64
Impurity hyperfine effects, 151–153
Impurity-impurity interactions, Kondo anomalies and, 123–124
Impurity mangetization studies, 157–174
Impurity nuclear spin-lattice relaxation, 174–177
Impurity problem, functional integral methods in, 237–251
Impurity spin relaxation time, 237
Integral valence phase, mixing in, 48
Interaction energy, Kondo problem and, 226
Intermediate valence phase, 39, 43
 resistivity and resistance anomalies in, 44
 temporal vs. spatial mixture in, 45
Intermetallic compounds, abnormal susceptibility of, 39–42
Intra-atomic Coulomb interaction, 93–94
"Intrahop" term, 249
Intraionic Coulomb repulsion, 23
Invariant coupling, 235
Iron, moment formation of in $4d$ metals, 25–26
Iron atoms, magnetic moments in, 64

K

Knight shift, in nuclear magnetic resonance 15
Kondo alloys, 123
Kondo anomaly
 parameters of, 132–133
 specific heat and, 137–138
 strong and weak interactions in, 142–144
 susceptibility and, 128–131
 thermoelectric power and, 138–140
Kondo condensed state, 72
Kondo effect, 77, 140, 150, 167
 Abriksov–Gor'kov theory and, 380
 divergences in, 237
 s–d model and, 121–144, 149–180
 s-matrix theory and, 189
 in superconductors, 299–207, 363

Kondo limit, 76
Kondo model, thermal and transport properties in, 125–140
Kondo problem, 58, 72
 asymptotically exact methods in, 217–235
 discrete path-integral approach to, 219–221
 equation of motion method in, 192–197
 entropy and specific heat in, 201–203
 finite temperatures in, 227–228
 heart of, 218, 353
 integral equation in, 195–197
 local magnetic moments and, 75
 physical implications of, 233
 "renormalization group" methods in, 233–235
 scaling method and, 230–232
 Schotte–Schotte method in, 225
 summing up over the paths in, 226–227
 thermal properties in, 200–204
 thermodynamic properties in, 198–199
Kondo resonance, magnetoresistance and, 206
Kondo singular function, 211
Kondo-Suhl expression, 25
Kondo system
 vs. classical system, 228
 numerical results in, 228–230
Kondo temperature, 122, 126–128, 150, 163, 217, 232, 241
 alloy behavior above and below, 251
 inverse of, 250
 for $3d$ elements in simple metal hosts, 32–33

L

Larmor period, 151, 164
Local electron distribution, 278–284
Local exchange enhancement factor, 97
Local magnetic moment
 defined, 7–10
 d-electron concentration and, 64–65
 experimental observation in, 7–31
 experimental techniques in, 10–17
 formation of, 3–51
 Hartree–Fock theory of, 57–87
 in impurity magnetization studies, 157–160
 magnetic susceptibility and, 10–12

measurements of via hyperfine interaction, 14–17
neutron scatter and, 13–14
nonmagnetic rare earth shells and, 47–49
in rare earth metals, 37–51
qualitative aspects in, 22–31
single-impurity effects in, 17–22
in superconductors, 291–307
Local spin fluctuations
dilute alloys and, 109–116
random phase approximation and, 99
renormalized theories of, 100–109
superconductivity and, 313–318
theory of, 92–100

M

Magnetic environment, characteristic temperature and, 22
Magnetic field
bound state in presence of, 274–278
s–d model and, 205–210
Magnetic hyperfine interaction, 149–180
in dilute alloys, 150–157
Magnetic imputiry, see also Impurity
defined, 63 n
one-electron (hole) type, 30–31
Magnetic moment
local, see Local magnetic moment
susceptibility and, 71–73
Magnetic response, measurement of, 149
Magnetic solution, phase shifts and, 74
Magnetic state, effective width of, 26, 48
Magnetic susceptibility, 10–12, see also Susceptibility
Magnetoresistance
local electron distributions and, 278–284
variation in, 136
Zeeman splitting and, 206
Many-body theory, random phase approximation and ,99
Many-field theory, breakdown of, 101
Metals
local magnetic moments and, 3–51
one-electron theory of, 89–90, 93
transitional impurities and, 88
Molecular field, superconductor and, 343
Moment formation, critical energy parmaeters for, 24, see also Local magnetic moments
Môssbauer effects, 14–16
absorption spectra for, 158
magnetic response and, 148
Motion equation, Kondo problem and, 192–197

N

Neutron scattering, local magnetic moments and, 13–14, see also Scattering
Nonmagnetic environment, characteristic temperature and, 22
Nonmagnetic impurity, 21, see also Impurity
ionic scheme and, 29
Nonmagnetic rare earth shells, local moment formation and, 47–49
Nuclear Larmor period, 151
Nuclear magnetic moments, hyperfine coupling and, 149
Nuclear magnetic resonance (NMR), 16–17, 32–33, 170
for dilute aluminum alloys, 175
disadvantage of, 151–152
at impurity site, 111–113, 158
linewidth in, 168
magnetic response and, 149
shifts in, 172–173
Nuclear orientation. magnetic response and, 149
Nuclear spin-lattice reaction, 174–177

O

One-electron (hole) Green's function, 222
One-electron (hole) impurities, 30
Orbital degeneracy, 69–70
Orbital hyperfine interactions, 165

P

Paramagnetic impurities, local spin fluctuations and, 92–100, see also Impurity
Paramagnons, 104–105
Partition function, 238–240, 243
Pauli paramagnetism, 122, 242
Pauli polarization term, 296

SUBJECT INDEX

Perturbation theory
  breakdown in, 353
  Kondo problem and, 234–235
  singlet wave function and, 257–264
Phase transition, Hartree–Fock solution in, 70–71
Phase diagrams, Kondo problem and, 75
Pseudo-Fermions, 210, 212, 235

Q

Quadrupole effect "wipe-out" numbers, 32
Quantum field theory, 105

R

Random phase approximation, of many-body theory, 99–100
Rare earth impurities, effective exchange for, 76–77
Rare earth ions, "soft" moments of, 38–39
Rare earth metals, local moment formation in, 37–51
Rare earth monochalcogenides, lattice constants for, 43
Rare earth shells, nonmagnetic, 47–49
Rare earth systems, resistivity in, 44–45
Renormalization, phenomenological, 108
Renormalized local spin fluctuation, 107
Renormalized random phase approximation, 102–103
Resistivity
  Kondo anomaly and, 131–136
  in s–d models, 142
  single-impurity effects and, 19–21
  temperature dependence of, 33
  transitional impurities and, 58
Resonance scattering, virtual bound states of, 59–64
RKKY (Roderman–Kittel–Kasuya–Yosida) interaction, 18–21, 27, 38, 58, 154, 157, 179, 328

S

Scattering, electrical conductivity and, 205
Scattering matrix, 211
Scattering problem, in narrow bands, 79–82
Scattering theory, 59–60
Schrieffer–Wolff transformation, 25
s–d exchange Hamiltonian, 140, 254
s–d hybridization, 152
s–d model
  Green's function theory and, 185–213
  ground state and, 253–285
  interactions in, 140–144
  Kondo effect and, 121–144, 149–180
  perturbative theory and, 185–213
  scattering theory and, 185–213
Simple hosts, solid solutions of $3d$ elements in, 31–34
Single impurity effects, see also Impurity
  Hartree–Fock theory of, 91
  RKKY interaction and, 18–21
  standard systems and, 17–22
Single-particle free electron states, 28
Singlet ground state, perturbation theoretic approach for, 258–264
Slater–Pauling curve, 83
Soft rare earth moments, 38–39
Specific heat
  free energy and, 340–342
  Kondo anomaly and, 137–138, 201–203
  s–d model and, 142
Spin correlation densities, 270–274
Spin-dependent potential, 61
Spin–echo spectra, 159
Spin–flip amplitude, for scattering of electrons, 186
Spin, fluctuations, dynamics of, 242
Spin–orbit scattering time, 331
Spin polarization, around impurity, 73
Spin polarization density, 270–274
Spin–spin correlation, magnetic susceptibility and, 107–208
Spin susceptibility, 160–164
Static impurity susceptibility, 160–164
Superconductivity
  localized spin fluctuations and, 313–318
  nonmagnetic resonant states in, 308–313
Superconductors
  BCS theory and, 327–238
  coexistence in, 336–343
  crystal-field split energy levels in, 297–298
  ferromagnetic, see Ferromagnetic superconductors
  finite impurity concentrations in, 375–380

Green's function and, 355–358
Hamann's equation for, 359, 361–369, 380
impurities in, 353–381
Kondo effect and, 299–307, 380
long-lived local moments and, 291–307
low-concentration limit in, 328–335
low impurity concentration in, 369–375
magnetic moment effects in, 287–375
magnetic-nonmagnetic transitions of impurities in, 318–322
magnetic ordering in, 328–335
in molecular field, 343–350
multiple pair breaking effects in, 295–297
Nagoka approximation in, 358–361
paramagnetic impurities in, 289–322
Susceptibility
Curie–Weiss law of, 17
electronic, 213
of intermetallic compounds, 39–42
Kondo model and, 128–131
magnetic moments in, 71–73
orbital, 164–167
pressure-dependence of, 42
s–d interactions and, 142
single-impurity effects and, 18–21
spin–spin correlations and, 207
static impurity and, 160–167
temperature dependence of, 33
zero-frequency local, 245–246

T

Temperature, magnetic environment and, 22
Temporal mixture, in intermediate valence phase, 45–47

Thermal properties, Kondo anomalies and, 125–140
Thermodynamic Green's function, 188
Thermoelectric power, Kondo anomaly and, 138–140
Thermopower, Kondo anomaly and, 142–143
$3d$ elements
solid solutions of in simple hosts, 31–34
transition metals and, 34–37
Transition impurities
effective exchange for, 76–77
spin-dependent potential and, 61–64
in transition metals, 78–87
Transition metals
with $3d$ solutes, 34–37
transition impurities in, 78–87
Transition region, mean field theory and, 101
Transport properties, 113–116
Kondo anomalies and, 125–140

V

Vertex corrections, 104–107
Virtual bound states, resonance scattering and, 60

Z

Zeeman energy, 11, 15, 94, 187
of conduction-electron spins, 27
Zeeman level, 28–29
Zeeman period, 12
Zeeman split iron spectra, 158
Zero-temperature saturation moment, 26

QC
753
R3
v.5

OCT 26 1973